T0182201

Introduction to Digital Systems Design

Giuliano Donzellini · Luca Oneto
Domenico Ponta · Davide Anguita

Introduction to Digital Systems Design

 Springer

Giuliano Donzellini
Dipartimento di Ingegneria Navale, Elettrica,
    Elettronica e delle Telecomunicazioni
    (DITEN)
Università degli Studi di Genova
Genoa
Italy
e-mail: giuliano.donzellini@unige.it

Luca Oneto
Department of Informatics, Bioengineering,
    Robotics and Systems Engineering
    (DIBRIS)
Università degli Studi di Genova
Genoa
Italy
e-mail: luca.oneto@unige.it

Domenico Ponta
Università degli Studi di Genova
Genoa
Italy
e-mail: ponta@unige.it

Davide Anguita
Department of Informatics, Bioengineering,
    Robotics and Systems Engineering
    (DIBRIS)
Università degli Studi di Genova
Genoa
Italy
e-mail: davide.anguita@unige.it

ISBN 978-3-030-06520-1   ISBN 978-3-319-92804-3   (eBook)
https://doi.org/10.1007/978-3-319-92804-3

Originally published in Italian as "Introduzione al Progetto di Sistemi Digitali", in 2018 by Springer Milano, Italy, Print ISBN 978-88-470-3962-9, Online ISBN 978-88-470-3963-6. The rights for the Italian language version of the text are owned by Springer-Verlag Italia S.r.l. 2018.

Printed on acid-free paper

This Springer imprint is published by the registered company Springer International Publishing AG part of Springer Nature
The registered company address is: Gewerbestrasse 11, 6330 Cham, Switzerland

Foreword of Prof. Filippo Sorbello

It is a great pleasure to present the book *Introduction to Digital Systems Design* by my friends and colleagues Donzellini, Oneto, Ponta, and Anguita. This textbook is suited for first year students of Engineering and Computer Science. It starts from the theoretical bases of digital systems, chosen and treated at the right level of depth, proceeds toward the analysis and synthesis of combinational and sequential logic to reach its target of designing and simulating controller–datapath systems. A very high number of examples and exercises with related solutions are provided.

The evolution of electronic technologies has brought a wide diffusion of digital systems in every field of everyday life. Speed, density, and complexity of current digital circuits have been made possible by automatic design methodologies and technological progress.

The knowledge of the theoretical bases of logic networks is necessary to achieve a complete mastery of digital system architectures of different complexities and also for the correct use of automatic design tools based on hardware description languages (HDLs).

First year students possess neither the adequate programming and abstraction abilities nor the physics and electronics knowledge necessary to use HDLs properly. In this textbook, this difficulty is overcome by employing a simulation tool (*Deeds*), developed by one of the authors, which uses an user-friendly interface. *Deeds* is employed to simulate the behavior both of the circuits proposed in the textbook and of those that the learner will autonomously design and then verify. *Deeds* projects can be exported in HDL and tested on FPGA circuits.

The use of languages for hardware description, together with the knowledge of the theoretical bases of logic circuits, represent the keys to understanding the digital world.

I think that this book is a good tool to face this challenge, given the ability that the authors have shown in transferring the necessary theoretical and professional know-how in a text with clear contents, smooth layout, and pleasant aspect.

Palermo, Italy   Filippo Sorbello
March 2018

Foreword of Prof. Mauro Olivieri

The textbook written by Giuliano Donzellini, Luca Oneto, Domenico Ponta, and Davide Anguita is characterized by two features that distinguish it in the wide field of university textbooks introducing digital design.

The first feature is the focus on a well-defined group of notions and tools representing the basis for digital systems: combinational and sequential logic synthesis and the topics strictly connected with them. The book covers neither electronic circuits nor microprocessor systems, and by remaining within these limits, it allows great clarity, precision, completeness, and consistency for the learners, as it appears immediately to the reader by inspecting the high number of schematics and timing diagrams provided with the textbook. Besides, the availability of solved exercises, together with the simulation software *Deeds*, represents a key element that is always appreciated and requested by students.

The second feature is the balance between the theoretical structure and the practical implementation, which allows solid learning. Even though in the textbook the word "voltage" is never cited, a student using the book always has the impression he/she is studying an electronic system formalization. At the same time, the textbook avoids presenting the topics only as a series of practical design examples without a theoretical basis.

This peculiarity characterizes the approach of the "school" of digital systems at the University of Genoa, in respect of which I consider myself an outsider.

For these reasons, the textbook of Donzellini, Oneto, Ponta, and Anguita is a precious tool for a student willing to deeply understand the concepts belonging to the big world of digital electronics design.

Roma, Italy   Prof. Mauro Olivieri
March 2018

Preface

The large and ever-growing complexity of today's digital systems places heavy demands on educational systems that are in charge of training the new generations of designers or just providing a solid understanding of the digital world. Academic institutions struggle to keep the pace of technological advancements, and people, like the authors of this book, who are in charge of introductory- or intermediate-level education, have the responsibility to face the problem and make choices.

It is certainly obvious that a digital designer must be trained in the use of hardware description languages (HDLs) and it is nowadays a common practice to introduce them very early in the courses, substituting the traditional approach based on components and schematics. The choice to describe digital systems by HDL matches very well with the adoption of Field-Programmable Gate Arrays (FPGA) for the practical implementation of projects, using prototype boards provided by chip producers.

Nevertheless, it is our opinion that the adoption of HDL in a beginner course of logic networks with limited resources in terms of credits (as in our case) may present problems. We believe that it is not easy to build a solid understanding of the foundations by completely replacing logic components and schematics with HDL, which requires a level of abstraction and a familiarity with programming that beginner students generally do not possess.

What is more, employing a simulation and synthesis software developed by FPGA chip producers presents other problems. Tools developed for digital systems designers may not necessarily satisfy learning needs: their use is not immediate for students, who may end up using them partially and mechanically, with the risk of missing important basic concepts, hidden under the technicalities of HDL and tools.

It is therefore necessary that students acquire a solid foundation on which to build design abilities and, at the same time, adapt to the fast rate of technological innovation, while gaining familiarity with languages and design tools.

For these reasons, the textbook maintains a traditional approach to logic networks, described and designed through symbols and schematics, while taking into account today's state of the art when choosing topics and, especially, exercises and projects. This feature allows an optimal use of the book in university curricula that contain only one course on digital hardware, while providing a solid foundation for higher-level studies.

The book takes an original approach in introducing FPGA devices and VHDLs. The last chapter shows how projects similar to the ones presented earlier and tested by simulation only can be practically and quickly implemented on FPGA boards, using Deeds tools. The procedure offers the opportunity of an "hands-on" introduction to FPGA devices and VHDL.

The book is self-sufficient, since it supports the theoretical part with a huge number of examples and exercises, complete with their solutions. In courses that have room for a laboratory session, the symbiosis with the *Digital Electronics Education and Design Suite* (*Deeds*) simulation tool can be exploited, with important advantages. *Deeds* was developed recently by one of the authors (Giuliano Donzellini), with the precise target of supporting learning and laboratory activities for Information Engineering students. The strong connection with *Deeds* represents an important strength and the originality of our work, since all the schematics, examples, and design exercises included, from the easiest to the most complex, were created with *Deeds* and are available online for an immediate simulation.

The *Deeds* environment covers all the principal aspects of digital systems design, from combinational and sequential logic to finite state machines and embedded systems, thus allowing for the design and simulation of complex networks containing standard logic, finite state machines, user-defined components, and microprocessors, including their programming in *assembly* language.

Deeds has been developed with the idea of matching ease of use and almost professional features. The main differences between *Deeds* and a professional tool are represented by the friendliness of the user interface and the availability of a wide collection of teaching material and projects. Furthermore, *Deeds* is an "alive," continuously evolving system: updates are periodically available to improve existent tools and add new ones. The same is true for teaching materials.

The transition toward FPGA devices is supported by *Deeds* that allows to export any of its projects to a professional tool, in order to test it in FPGA hardware. *Deeds* bypasses the complexity of the process that is normally required by a specific professional software and does not require writing HDL code, which is automatically generated by *Deeds*. The rich teaching material of *Deeds* is hence redirected toward FPGA implementation, without substantial modifications.

However, after practicing with the automatic HDL code generation by *Deeds*, students can directly interact with FPGA tools, thus having the possibility to observe, modify, and reuse HDL code (VHDL in our case), making a gradual transition toward current design techniques.

Teaching Objectives

According to authors' experience, the whole content of the book, together with design exercises and simulations based on *Deeds*, may be developed in an introductory course to digital systems of at least nine credits.

In the following, we briefly report the content of the chapters, indicating in italics the topics that can be avoided without loss of continuity with the teaching project, for courses with a smaller number of credits:

1. Boolean Algebra and Combinational Logic

   - Classic approach that does not require preliminary knowledge.
     It is possible to skip Shannon's theorems.

2. Combinational Network Design

   - Synthesis and minimization with Karnaugh maps.
   - Standard combinational logic.
   - Propagation delays.
     Variable-Entered Maps and hazards may be omitted.

3. Numeral Systems and Binary Arithmetic

   - Classic approach.
   - Arithmetic networks.
     Binary negative numbers may be omitted, as well as BCD arithmetic.

4. Complements in Combinational Network Design

   - Minimization of expressions with Quine–McCluskey method.
     The entire chapter may be omitted.

5. Introduction to Sequential Networks

   - Intuitive transition from combinational to sequential logic.
   - Structure and operation of principal flip-flop types.
   - Dynamic flip-flop characteristics.
     It is possible to consider just "D" and "E" logic types and to skip their circuital details.

6. Flip-Flop-BasedSynchronous Networks

   - Introduction to synchronous flip-flop networks.
   - Sequential networks: registers and counters.
   - Techniques for timing analysis of synchronous networks.
     Counters and registers section may be reduced, as well as timing analysis of sequential networks.

7. Sequential Networks as Finite State Machines

   - FSM project, realized through ASM diagrams.
   - Solved exercises of ASM diagrams.
   - FSM synthesis with state tables and maps.
     FSM synthesis may be reduced, by omitting variable-entered maps, or completely left aside.

8. The Finite State Machine as System Controller

   - Design of Controller–Datapath systems.
   - Solved exercises on controller–datapath systems.
     This chapter applies all the material presented in the book to develop controller–datapath systems. The projects may be chosen according to the needs and level of the class.

9. Introduction to FPGA and HDL Design

   - Introduction to FPGA.
   - System prototyping on FPGA with *Deeds* tools.
   - Introduction to VHDL.
   - Examples of FPGA prototyping projects.
     This chapter requires the use of Deeds, and it is fully exploited when accompanied by laboratory activities.

How to Use the Book

The strict connection between this book and the *Deeds* tool suggests using it together with the simulation tools, both to verify and test concepts and procedures in an active way and to have a support for the solution of the exercises and the design of systems.

This "learning by doing" practice allows students to progressively build the analytic and design capabilities that represent the target to reach.

Giuliano Donzellini
Luca Oneto
Domenico Ponta
Davide Anguita
Genova, Italy

Digital Contents for the Book

This textbook contains theoretical parts, examples, exercises, and solutions. All the examples have also been implemented with the *Deeds* simulator that can be downloaded from the link:

https://www.digitalelectronicsdeeds.com

The Web site contains a description of the *Deeds*'s features, tutorials, and learning materials. The simulator does not require an Internet connection.

On the same Web site, as additional material, it is possible to find almost all the schematics and charts included in the book:

https://www.digitalelectronicsdeeds.com/books

Thanks to this material, it is possible to simulate with *Deeds* the proposed circuits and the exercises. The material has been organized by following the same structure of the book in order to make it easier to access. On the same Web site, future updates, corrections, and additions will be made available.

Contents

Chapter 1
Boolean Algebra and Combinational Logic

Abstract This chapter introduces to the idea of digitally representing analog quantities and goes step by step through the main concepts of the Boolean algebra: variables, functions, truth tables, operations, and properties. The chapter is quite detailed and accompanied by many examples and exercises in order to provide a precise framework of the fundamentals of digital design. It includes the theorems which constitute the foundation for the application of the Boolean algebra to logic networks, with a precise focus on their application for combinational network design.

1.1 Analog and Discrete Variables

In every field of human knowledge, information's observation, memorization, elaboration, and communication is something everyone has to deal with. The definition of the word *"information"* may sound obvious, since this term is commonly used in everyday language, but for our aim we need a definition that leaves no space to any ambiguous interpretation. We hence refer to R. V. L. Hartley (1888–1970), one of the fathers of *Information Theory*, who helps us with the following definition:

> *Information is a reduction of uncertainty.*

From this sentence, it is clear that information is associated with "before" and "after," in relation to an event having a probability to happen; thanks to this probability, an observer reduces his uncertainty related to the event itself. From this definition, it can be easily derived that information can be conveyed through the employment of physical quantities variables, changing in time or space. For example, we can refer to information transmitted by a computer screen through images, defined as variations of luminosity and color in time and space, or the information transmitted by an earphone, through a sound, defined as variations of air pressure over time.

To the aim of studying information processing, we will not refer directly to a physical quantity variable, but rather to its numeric representation, indicated with *"G,"* and its variation over time, indicated with *"T."*

This approach allows us to divide the representation into two families: if G can vary continuously between a value and another one, assuming all the infinite intermediate

© Springer International Publishing AG, part of Springer Nature 2019

G. Donzellini et al., *Introduction to Digital Systems Design*,

https://doi.org/10.1007/978-3-319-92804-3_1

values, we are employing an "analog" representation. If, instead, G can assume only a limited number of values, we are employing a "discrete" (or "digital") representation, becoming "binary" (or "Boolean") if the numbering system uses only two symbols.

Sometimes, the distinction between a physical quantity variable and its numeric representation may create confusion, but we are interested only in the latter. Light, for example, may be described both through a discrete representation (*photons*) and a continuous one (*electromagnetic waves*), but identifying the "true" representation, if it exists, is beyond our scope: we gladly leave this objective to philosophers. For an engineer, what matters is using the most appropriate tool to solve the problem under analysis.

In the picture, four possible representations obtained using discrete or continuous values for G and T are provided [red lines are given for reference]:

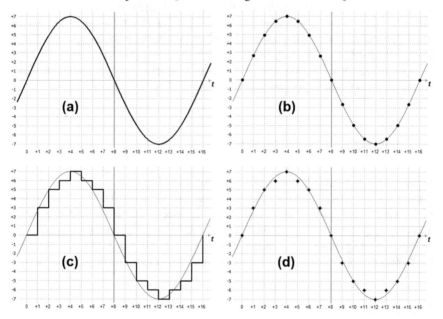

(a) continuous quantity variable changing over time;
(b) continuous quantity variable sampled over time;
(c) quantity variable quantized in amplitude and continuous over time;
(d) quantity variable quantized in amplitude and sampled over time.

In the analog case, the main tool at our disposal is math, involving real numbers and functions defined over them: i.e., algebra and infinitesimal calculus. The analog representation is effective at dealing with natural physical quantities variables at the macroscopic level.

This does not mean that natural quantities' variables are analog, but only that in this case the analog representation is the most suitable and effective. In fact, if we go deeper toward the microscopic level, matter reveals its discrete nature (atoms and particles) and the analog representation is not necessarily the most convenient. Generally, a device able to convert continuous quantities variables into digital ones, and vice versa, is necessary. These devices are called *Analog to Digital Converters* (ADC) and *Digital to Analog Converters* (DAC) respectively, but they are beyond the scope of this book.

The digital case may be faced with discrete math, generally more complex than real numbers math. If we limit to the case of *G* assuming only two values, i.e., the binary case, we can refer to the *Boolean algebra* that takes its name from its creator, the Irish logician and mathematician George Boole (1815–1864). *Boolean algebra* will be sufficient for our scope, that is, putting the basis of the combinational logic and digital systems.

Binary variables are usually indicated with $\{0, 1\}$ or also with other symbols, like $\{-1, +1\}$, $\{L, H\}$ (*Low* and *High*) or $\{T, F\}$ (*True* or *False*), depending on the context.

In the digital field, we will make use of both the representation over continuous time, called *"asynchronous,"* and the one over discrete time, called *"synchronous."* A logic network is called *synchronous* if its parts operate simultaneously, according to a common synchronization signal; it is instead defined as *asynchronous* if its parts operate in an autonomous mode among each other.

The binary (digital) representation possesses advantages and disadvantages with respect to the analog one:

- an analog value is a pure mathematical abstraction, since it requires infinite precision to be expressed.
- a discrete value (binary) is easily storable, since it requires a finite number (two) of the physical variable values to be memorized.
- the management and processing of binary variables are less sensitive to a possible damaging of the signals that occur during its processing and transmission. In fact, if the damage is not large enough to alter the distinction between the two signal levels *(high/low)*, there is no damage in the information carried by signals.
- the precision of the system can be easily controlled by choosing the number of bits that code the information.
- devices processing digital information, namely digital systems, are simpler to design, though the practical realization requires a higher number of circuital components.

1.2 Boolean Variables

Let X be a certain discreet variable. We will call *Boolean variable* any discreet variable that can assume only two values. These values are denoted as follows:

$$X = 0 \quad \text{false}$$
$$X = 1 \quad \text{true}$$

In the following, the values 0, 1 will be used.

1.3 Boolean Functions

If we have the Boolean variables X_1, X_2, ..., X_n, the following:

$$f(X_1, X_2, \ldots, X_n)$$

is called a Boolean function, and it can assume only the values 0 and 1. This function associates a Boolean value to every element in its domain.

The domain of a function of n-variables is composed of all the 2^n combinations of their values. Therefore, domain's elements are countable. Two functions are equivalent if they assume the same value for any combination of their variables' values.

1.4 Truth Tables

The Principle of Perfect Induction (that is, carrying out all the calculations) makes it possible to prove the value of f for all the 2^n points of the domain. The function is represented in the truth table.

Let's assume a three-variable function X_1, X_2, X_3. We can construct a table with all the values assumed by f:

X_1	X_2	X_3	f
0	0	0	
0	0	1	values
0	1	0	
0	1	1	assumed
1	0	0	
1	0	1	by
1	1	0	
1	1	1	f

Observation: To write the $2^3 = 8$ elements of the domain, we begin at the farthest right column (X_3), from the top and alternate between one 0 and one 1. In the next column, we alternate between two 0s and two 1s, while in the column after that, four 0s and four 1s and so on, doubling the number of 0s and 1s with each new column.

Examples:

Derive the **truth tables** from the *verbal definitions*:

1. U is true if C is true or if B and A are both true.

2. Z is true if the number of ones in the inputs M, G, D is equal to two.

	C	B	A	U			M	G	D	Z
	0	0	0	0			0	0	0	0
	0	0	1	0			0	0	1	0
	0	1	0	0			0	1	0	0
1.	0	1	1	1	2.		0	1	1	1
	1	0	0	1			1	0	0	0
	1	0	1	1			1	0	1	1
	1	1	0	1			1	1	0	1
	1	1	1	1			1	1	1	0

1.5 Definition of Boolean Algebra

Boolean algebra provides the necessary tools to calculate and interpret information presented in binary form. Boolean algebra is an *algebraic system* (a set of elements to which a set of operations is associated), defined by:

- The set of values $\{0,1\}$;
- The operations *OR*, *AND*, and *NOT*;
- The equivalence operator "=", along with the properties reflexive, symmetric, and transitive.

The three operations are defined as follows:

Operation:	OR	AND	NOT
	(logical sum)	(logical product)	(negation)

Algebraic symbols:			
	$X + Y$	$X \cdot Y = XY$	$\overline{X}$
	$X \vee Y$	$X \wedge Y$	$!X$
	$X \cup Y$	$X \cap Y$	$-X$
	X or Y	X and Y	$\text{not}(X)$

Truth table:

X	Y	$X + Y$
0	0	0
0	1	1
1	0	1
1	1	1

X	Y	$X \cdot Y$
0	0	0
0	1	0
1	0	0
1	1	1

X	$\overline{X}$
0	1
1	0

Circuit diagram symbols:

1.6   The Fundamental Properties of Boolean Algebra

Conventions

- $X, Y, Z, X_1, X_2, X_3, \ldots, X_n$, are considered Boolean variables.
- The parentheses establish the calculation priorities as in regular algebra.
- AND is prioritized over OR (e.g., $X + YZ = X + (YZ)$).

This is also analogous to regular algebra. All the properties can be demonstrated through Perfect Induction, that is, by verifying the validity of each combination of values assumed by the variables that make up the expression.

Example: $X \cdot 0 = 0$ is verified through the truth table:

X	0	$X \cdot 0$
0	0	0
1	0	0

Duality Principle

If a given expression is valid, its dual expression is also valid. The dual expression is obtained by switching the OR with the AND and the 0 constants with the 1 constants from the original expression. For example:

$$X + 1 = 1$$
$$\text{(dual:)} \quad X \cdot 0 = 0$$

$$X + 0 = X$$
$$\text{(dual:)} \quad X \cdot 1 = X$$

Idempotent Law

$$X + X = X$$
$$\text{(dual:)} \quad X \cdot X = X$$

Commutative Law

$$X + Y = Y + X$$
$$\text{(dual:)} \quad X \cdot Y = Y \cdot X$$

Associative Law

$$(X + Y) + Z = X + (Y + Z) = X + Y + Z$$
$$\text{(dual:)} \quad (X \cdot Y) \cdot Z = X \cdot (Y \cdot Z) = X \cdot Y \cdot Z.$$

The associative law makes it possible to extend fundamental operations to more than two variables. The circuit symbols for the first expression are:

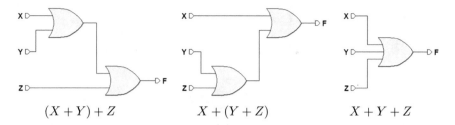

$$(X + Y) + Z \qquad\qquad X + (Y + Z) \qquad\qquad X + Y + Z$$

As there is no distinction between the first and second circuits, it makes sense to generally define an OR of three or more inputs. The same holds true for the AND, so it is really the property of Associativity that allows us to make sense of OR and AND gates with more than two inputs.

We can redefine the OR and AND operations with n inputs:

- An OR with n inputs gives a 0 as output only if all the n inputs are 0, otherwise it gives a 1 as output.
- An AND with n inputs gives a 1 as output only if all the n inputs are 1, otherwise it gives a 0 as output.

Distributivity

$$\text{Factoring law} \qquad (X + Y) \cdot (X + Z) = X + (Y \cdot Z)$$
$$\text{Distributive law (dual:)} \quad (X \cdot Y) + (X \cdot Z) = X \cdot (Y + Z)$$

Proof of the factoring law:

$$(X + Y) \cdot (X + Z) = X \cdot X + X \cdot Z + X \cdot Y + Y \cdot Z =$$
$$= X + X \cdot Z + X \cdot Y + Y \cdot Z$$
$$= X \cdot (1 + Y) + X \cdot Z + Y \cdot Z$$
$$= X + X \cdot Z + Y \cdot Z$$
$$= X \cdot (1 + Z) + Y \cdot Z$$
$$= X + (Y \cdot Z)$$

It would also be possible to demonstrate this law through Perfect Induction (i.e., verifying all the possible combinations for X, Y, Z):

X Y Z	$Y \cdot Z$	$X + Y \cdot Z$	$X + Y$	$X + Z$	$(X + Y)(X + Z)$
0 0 0	0	0	0	0	0
0 0 1	0	0	0	1	0
0 1 0	0	0	1	0	0
0 1 1	1	1	1	1	1
1 0 0	0	1	1	1	1
1 0 1	0	1	1	1	1
1 1 0	0	1	1	1	1
1 1 1	1	1	1	1	1

It is clear that columns $X + Y \cdot Z$ and $(X + Y)(X + Z)$ are equal.

Complementation

$$X + \overline{X} = 1$$
$$\text{(dual:)} \quad X \cdot \overline{X} = 0$$

Absorption

First form:

$$X + X \cdot Y = X$$
$$\text{(dual:)} \quad X \cdot (X + Y) = X$$

Second form:

$$X + (\overline{X} \cdot Y) = X + Y$$
$$\text{(dual:)} \quad X \cdot (\overline{X} + Y) = X \cdot Y$$

Proof:

$$X + X \cdot Y = X \cdot (1 + Y) = X \cdot 1 = X$$
$$X \cdot (X + Y) = X \cdot X + X \cdot Y = X + X \cdot Y = X$$
$$X + \overline{X} \cdot Y = X + X \cdot Y + \overline{X} \cdot Y = X + Y(X + \overline{X}) = X + Y$$
$$X \cdot (\overline{X} + Y) = X \cdot \overline{X} + X \cdot Y = X \cdot Y$$

Logic Adjacency

$$Y X + Y \overline{X} = Y$$
$$\text{(dual:)} \quad (Y + X) \cdot (Y + \overline{X}) = Y$$

Proof:

$$Y X + Y \overline{X} = Y \cdot (X + \overline{X}) = Y \cdot 1 = Y$$
$$(Y + X) \cdot (Y + \overline{X}) = Y + (X \cdot \overline{X}) = Y + 0 = Y$$

Consensus

$$X \cdot Y + Y \cdot Z + Z \cdot \overline{X} = X \cdot Y + Z \cdot \overline{X}$$
$$\text{(dual:)} \quad (X + Y)(Y + Z)(Z + \overline{X}) = (X + Y)(Z + \overline{X})$$

Proof:

$$X \cdot Y + Y \cdot Z + Z \cdot \overline{X} =$$
$$= X \cdot Y + Y \cdot (X + \overline{X}) \cdot Z + Z \cdot \overline{X} =$$
$$= (X \cdot Y + X \cdot Y \cdot Z) + (Z \cdot \overline{X} \cdot Y + Z \cdot \overline{X}) =$$
$$= X \cdot Y + Z \cdot \overline{X}$$

$$(X + Y)(Y + Z)(Z + \overline{X}) =$$
$$= (X + Y)[(X + Y + Z)(\overline{X} + Y + Z)](Z + \overline{X}) =$$
$$= [(X + Y)(X + Y + Z)][(Z + \overline{X} + Y)(Z + \overline{X})] =$$
$$= (X + Y)(Z + \overline{X})$$

Involution

Also known as *Double Complement law*: $\overline{\overline{X}} = X$.

Duality or De Morgan's Theorem

A logical product of two variables can be substituted by the negation of their logical sum. Dual: a logical sum of two variables can be substituted by the negation of their logical product:

$$X \cdot Y = \overline{\overline{X} + \overline{Y}}$$
$$\text{(dual:)} \quad X + Y = \overline{\overline{X} \cdot \overline{Y}}$$

This theorem is important: it allows us to obtain an AND through an OR gate and vice versa. The theorem tells us that either one of the two functions is superfluous according to the definition of Boolean algebra.

Generalized De Morgan's Theorem

The theorem applies to any number of variables:

$$X_1 \cdot X_2 \cdot \ldots \cdot X_n \quad = \quad \overline{\overline{X_1} + \overline{X_2} + \ldots + \overline{X_n}}$$
$$\text{(dual :)} \quad X_1 + X_2 + \ldots + X_n \quad = \quad \overline{\overline{X_1} \cdot \overline{X_2} \cdot \ldots \cdot \overline{X_n}}.$$

1.7   Other Operations

In this paragraph, we define other operations in Boolean algebra: NAND, NOR, and EXOR.

NAND

The NAND operation is equivalent to an AND whose output is negated:

$$X \text{ nand } Y = \overline{(X \cdot Y)}$$

X Y	(X nand Y)
0 0	1
0 1	1
1 0	1
1 1	0

Circuital symbols:

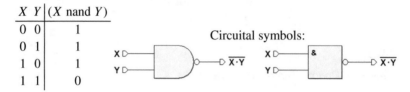

NOR

The NOR operation is equivalent to an OR whose output is negated:

$$X \text{ nor } Y = \overline{(X + Y)}$$

X Y	(X nor Y)
0 0	1
0 1	0
1 0	0
1 1	0

Circuital symbols:

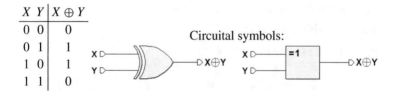

NAND and NOR are *commutative* but <u>not</u> *associative*.

XOR (Exclusive OR)

The XOR operation is said "anticoincidence" (it provides 1 when the inputs are different):

$$X \oplus Y = X \text{ xor } Y = X \overline{Y} + \overline{X} Y$$

X Y	X ⊕ Y
0 0	0
0 1	1
1 0	1
1 1	0

Circuital symbols:

The XOR is *commutative* and *associative*. If we negate its output, we obtain the "coincidence" function (equivalence of inputs):

$$\overline{X \oplus Y} = X\,Y + \overline{X}\,\overline{Y}$$

Generalized XOR

Is a multiple inputs XOR, written thus:

$$X_1 \oplus X_2 \oplus \ldots \oplus X_n = \begin{cases} 1 \text{ if there is an odd number of inputs} = 1 \\ 0 \text{ if there is an even number of inputs} = 1 \end{cases}$$

They are made with the typical structure of the XOR *tree*:

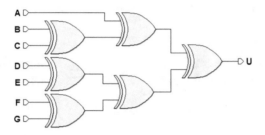

1.8 Functionally Complete Operation Sets

We have seen that Boolean algebra is based on a set of two elements {0, 1} and a set of operations: OR, AND, NOT. Also, De Morgan's Theorem shows that one of the two AND or OR operations can be considered superfluous and the sets of {OR, NOT} or {AND, NOT} are a sufficient basis to construct all of Boolean algebra. Let's broaden the subject by discussing other sets of operations that allows to construct Boolean algebra (named, for this reason, *Functionally Complete Operation Sets*):

1. {AND, OR, NOT}
2. {NOR}
3. {NAND}
4. {OR, NOT}
5. {AND, NOT}
6. {EXOR, AND}
6. {EXOR, OR}

Note: in practice, only {NOR} and {NAND} sets are used.

{NOR} Set

We can obtain OR and NOT from NOR gates. If we connect a NOR as in the figure below, we obtain a NOT. Given that the X and Y inputs are connected together, we obtain the following from the NOR table:

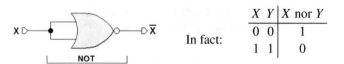

X	Y	X nor Y
0	0	1
1	1	0

In fact:

However, we obtain the OR gate by negating the NOR output with a NOT:

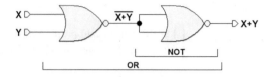

To obtain the AND, we apply De Morgan: $X \cdot Y = \overline{\overline{X} + \overline{Y}}$. We have:

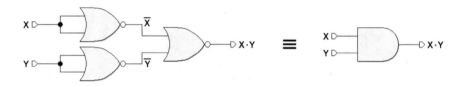

{NAND} Set

Similar to the above, the NOT is obtained as follows, taking into account the two lines of the NAND table where the two X and Y inputs are equal:

X	Y	X nand Y
0	0	1
1	1	0

In fact:

Therefore, to obtain the AND, it is sufficient to connect the NAND to a NOT made with a NAND.

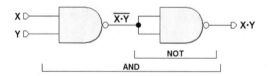

Finally, by De Morgan, we obtain the OR:

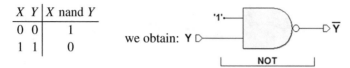

Take note: there is another way to obtain the NOT by the NAND. By connecting one of the inputs to the constant $X = 1$:

X Y	X nand Y
0 0	1
1 1	0

we obtain:

Similarly, if we posit $X = 0$ for the NOR we get:

X Y	X nor Y
0 0	1
0 1	0

we obtain: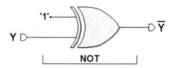

{OR, NOT} Set

The AND is obtained by De Morgan's Theorem.

{AND, NOT} Set

The OR is obtained by De Morgan's Theorem.

{XOR, AND} Set

The NOT is obtained by the XOR as follows:

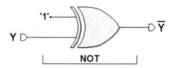

From the XOR truth table, we get:

X Y	$X \oplus Y$
0 0	0
0 1	1
1 0	1
1 1	0

positing $X = 1$:

X Y	$X \oplus Y$
1 0	1
1 1	0

Note: If we change the constant $X = 1$ in 0, we get the identity. Therefore, we obtain an inverting/identity function, "programmable" by the input X.

{XOR, OR} Set

The NOT is obtained through the XOR and the AND by using De Morgan's Theorem.

Identity

Identity can be obtained in the following ways:

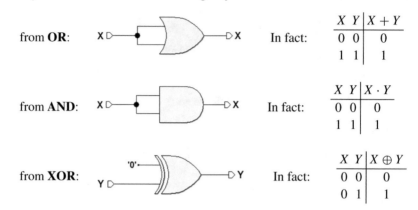

from **OR**:   In fact:

X Y	X + Y
0 0	0
1 1	1

from **AND**:   In fact:

X Y	X · Y
0 0	0
1 1	1

from **XOR**:   In fact:

X Y	X ⊕ Y
0 0	0
0 1	1

1.9 Shannon's Expansion Theorem

First Form

A Boolean function can be broken down this way:

$$f(X_1, X_2, X_3, \ldots, X_n) = \overline{X_1} \cdot f(0, X_2, X_3, \ldots, X_n) + X_1 \cdot f(1, X_2, X_3, \ldots, X_n)$$

The first of the two terms obtained is equivalent to the starting function but only when $X_1 = 0$, so it is conditioned by $\overline{X_1}$. Likewise, the second term, which applies to $X_1 = 1$, is conditioned by X_1.

Now that we have seen the process, we can extract all the variables of the function:

$$f(X_1, X_2, X_3, \ldots, X_n) = \overline{X_1} \cdot \overline{X_2} \cdot f(0, 0, X_3, \ldots, X_n) + \\ \overline{X_1} \cdot X_2 \cdot f(0, 1, X_3, \ldots, X_n) + \\ X_1 \cdot \overline{X_2} \cdot f(1, 0, X_3, \ldots, X_n) + \\ X_1 \cdot X_2 \cdot f(1, 1, X_3, \ldots, X_n) = \\ = \ldots$$

In the end, every $f(0, 1, \ldots)$-type entry will turn out to be a (0 or 1) constant. From an n-variable function, we obtain 2^n product terms in OR, where each term consists of all the direct and negated variables.

As an example, let's break down a $f(C, B, A)$ into three variables:

$$
\begin{aligned}
f(C, B, A) = \ & \overline{C}\,\overline{B}\,\overline{A} \cdot f(0, 0, 0)+ \\
& \overline{C}\,\overline{B}\,A \cdot f(0, 0, 1)+ \\
& \overline{C}\,B\,\overline{A} \cdot f(0, 1, 0)+ \\
& \overline{C}\,B\,A \cdot f(0, 1, 1)+ \\
& C\,\overline{B}\,\overline{A} \cdot f(1, 0, 0)+ \\
& C\,\overline{B}\,A \cdot f(1, 0, 1)+ \\
& C\,B\,\overline{A} \cdot f(1, 1, 0)+ \\
& C\,B\,A \cdot f(1, 1, 1)
\end{aligned}
$$

The expanded form of the function is called *sum of products*, or *first canonical form*, or *AND–OR form*.

Example

We want to derive the analytical expression from a Boolean function $f(C, B, A)$ defined through the truth table using the first form of the theorem.

C B A	f
0 0 0	0
0 0 1	1
0 1 0	0
0 1 1	0
1 0 0	0
1 0 1	1
1 1 0	1
1 1 1	0

$\Rightarrow$

$$
\begin{aligned}
f = \ & \overline{C}\,\overline{B}\,\overline{A} \cdot f(0, 0, 0)+ \\
& \overline{C}\,\overline{B}\,A \cdot f(0, 0, 1)+ \\
& \overline{C}\,B\,\overline{A} \cdot f(0, 1, 0)+ \\
& \overline{C}\,B\,A \cdot f(0, 1, 1)+ \\
& C\,\overline{B}\,\overline{A} \cdot f(1, 0, 0)+ \\
& C\,\overline{B}\,A \cdot f(1, 0, 1)+ \\
& C\,B\,\overline{A} \cdot f(1, 1, 0)+ \\
& C\,B\,A \cdot f(1, 1, 1)
\end{aligned}
=
\begin{aligned}
& \overline{C}\,\overline{B}\,\overline{A} \cdot 0 + \\
& \overline{C}\,\overline{B}\,A \cdot 1 + \\
& \overline{C}\,B\,\overline{A} \cdot 0 + \\
& \overline{C}\,B\,A \cdot 0 + \\
& C\,\overline{B}\,\overline{A} \cdot 0 + \\
& C\,\overline{B}\,A \cdot 1 + \\
& C\,B\,\overline{A} \cdot 1 + \\
& C\,B\,A \cdot 0
\end{aligned}
=
\begin{aligned}
& \overline{C}\,\overline{B}\,A + \\
& C\,\overline{B}\,A + \\
& C\,B\,\overline{A}
\end{aligned}
$$

Beginning by the definition, we substitute all the $f(\ldots)$-type constants with the real value of the function, taken directly from the truth table. We observe that in the logical sum, the terms with 0 in AND can be omitted since they are always 0. We then simplify the remaining terms corresponding to the lines with output 1 and eliminate the product for the constant. The remaining expression is what we're looking for. It analytically expresses the behavior of the function in the *first canonical form*.

Second Form

The second form allows us to break down a function f into a product of sums:

$$
\begin{aligned}
f(X_1, X_2, X_3, \ldots, X_n) = \ & (X_1 + f(0, X_2, X_3, \ldots, X_n)) \cdot \\
& (\overline{X_1} + f(1, X_2, X_3, \ldots, X_n))
\end{aligned}
$$

The first sum term is equivalent to the starting function after substituting $X_1 = 0$ and applies if $X_1 = 0$ (otherwise the whole term is 1). The second sum term is equivalent to the starting function after substituting $X_1 = 1$ and applies if $X_1 = 1$ (otherwise the whole term is 1). In other words, for a certain value of X_1, one of the two terms is always 1, while the other assumes the value of the function.

If we go through the break-down procedure until we exhaust all the f argument variables, we obtain 2^n terms in AND. Each term is composed of the logical sum of all the direct or negated variables and the value of the function for that specific combination:

$$
\begin{aligned}
f(X_1, X_2, X_3, \ldots, X_n) = (X_1 + X_2 + f(0, 0, X_3, \ldots, X_n)) \cdot \\
(X_1 + \overline{X_2} + f(0, 1, X_3, \ldots, X_n)) \cdot \\
(\overline{X_1} + X_2 + f(1, 0, X_3, \ldots, X_n)) \cdot \\
(\overline{X_1} + \overline{X_2} + f(1, 1, X_3, \ldots, X_n)) = \\
= \ldots
\end{aligned}
$$

Consider a three-variable function $f(C, B, A)$; we obtain 2^3 OR terms in AND:

$$
\begin{aligned}
f(C, B, A) = (C + B + A + f(0, 0, 0)) \cdot \\
(C + B + \overline{A} + f(0, 0, 1)) \cdot \\
(C + \overline{B} + A + f(0, 1, 0)) \cdot \\
(C + \overline{B} + \overline{A} + f(0, 1, 1)) \cdot \\
(\overline{C} + B + A + f(1, 0, 0)) \cdot \\
(\overline{C} + B + \overline{A} + f(1, 0, 1)) \cdot \\
(\overline{C} + \overline{B} + A + f(1, 1, 0)) \cdot \\
(\overline{C} + \overline{B} + \overline{A} + f(1, 1, 1)
\end{aligned}
$$

A function expanded this way takes on the *second canonical form*, or *product of sums*, or *OR–AND form*. Any Boolean function can be expressed this way. For a function with n-variables, we obtain 2^n factors to multiply.

Example

Through the second form of Shannon's Expansion Theorem, we derive the analytical expression of a Boolean function $f(C, B, A)$, given the truth table that describes it.

C	B	A	f
0	0	0	1
0	0	1	1
0	1	0	0
0	1	1	0
1	0	0	0
1	0	1	1
1	1	0	1
1	1	1	1

$\Rightarrow$

$$
\begin{aligned}
f = (C + B + A + f(0, 0, 0)) \cdot \\
(C + B + \overline{A} + f(0, 0, 1)) \cdot \\
(C + \overline{B} + A + f(0, 1, 0)) \cdot \\
(C + \overline{B} + \overline{A} + f(0, 1, 1)) \cdot \\
(\overline{C} + B + A + f(1, 0, 0)) \cdot \\
(\overline{C} + B + \overline{A} + f(1, 0, 1)) \cdot \\
(\overline{C} + \overline{B} + A + f(1, 1, 0)) \cdot \\
(\overline{C} + \overline{B} + \overline{A} + f(1, 1, 1)
\end{aligned}
$$

$=$

$$
\begin{aligned}
f = (C + B + A + 1) \cdot \\
(C + B + \overline{A} + 1) \cdot \\
(C + \overline{B} + A + 0) \cdot \\
(C + \overline{B} + \overline{A} + 0) \cdot \\
(\overline{C} + B + A + 0) \cdot \\
(\overline{C} + B + \overline{A} + 1) \cdot \\
(\overline{C} + \overline{B} + A + 1) \cdot \\
(\overline{C} + \overline{B} + \overline{A} + 1)
\end{aligned}
$$

We have applied the definition of the second form of the theorem by substituting all the $f(\ldots)$-type constants with the values of the truth table. Given that the constant term 1 in a logical sum absorbs the other variables and that adding 0 is redundant, the function's final expression is:

$$f = (C + \overline{B} + A) \cdot (C + \overline{B} + \overline{A}) \cdot (\overline{C} + B + A)$$

The expression expresses how the function behaves in the *second canonical form*.

1.10 Level of Boolean Expressions

The level is the maximum number of cascading operations made on the input variables. For example:

$$\begin{aligned}
f &= a + b &&\text{is a one-level expression}\\
f &= ab + c &&\text{is a two-level expression}\\
f &= ab + cd &&\text{is a two-level expression}
\end{aligned}$$

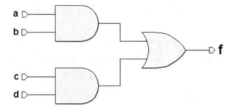

Take note: the levels are important for technical reasons. The more levels there are, the longer the delays; we will focus mainly on syntheses of two-level networks. When the number of levels has to be computed, we suppose all the input variables and their complemented forms to be available. Thus, the expression $f = \overline{a}b + \overline{c}d$ is a two-level Boolean expression.

1.11 Literals

Literals are the number of input variables that make up a Boolean expression (not to be confused with the number of variables).
For example: if $f(a, b)$ is a logical function with two binary variables a and b:

$$\begin{aligned}
f &= a + b &&\text{has 2 literals}\\
f &= ab + \overline{a}b &&\text{has 4 literals.}
\end{aligned}$$

1.12   Minterms

If an AND term in a Boolean expression contains all the direct or negated variables in the entire expression, it is called a *fundamental product*, or *minterm*. For example:

$$f(X_1, X_2, X_3) = X_1 \cdot X_2 \cdot \overline{X_3} \qquad \text{is a minterm.}$$

An n-variable function has 2^n minterms since every variable in the function must be part of a minterm, in its direct or negated form. Note that among all the possible combinations of variables, there is only one for which a certain minterm equals 1 (e.g., $X_1 \cdot X_2 \cdot \overline{X_3} = 1$ if and only if $X_1 = 1$, $X_2 = 1$, $X_3 = 0$).

1.13   Maxterms

If an OR term in a Boolean expression contains all the direct or negated variables in the entire expression, it is called a *fundamental sum*, or *maxterm*. As above, if there are n-variables, there are 2^n maxterms. For example:

$$f(X_1, X_2, X_3) = X_1 + \overline{X_2} + X_3 \qquad \text{is a maxterm.}$$

Remember that there is only one combination of variables for which a certain maxterm equals zero (e.g., $X_1 + \overline{X_2} + X_3 = 0$ if and only if $X_1 = 0$, $X_2 = 1$, $X_3 = 0$).

1.14   Implicants

Given the Boolean expressions f and g, g is an implicant of f: g implies f ($g \Rightarrow f$) or f covers g ($f \supset g$) if f always $= 1$ when $g = 1$.

In this example: $f(X, Y, Z) = XY + Z$ we have $XY \Rightarrow f$
$$Z \Rightarrow f$$

Every time Z and/or XY equal 1, f also equals 1. XY and Z are therefore implicants of f. X does not imply f: in fact if X equals 1 f does not necessarily equal 1.

1.15   Prime Implicants

g is a *prime* implicant of f if:

- $g \Rightarrow f$ ($f \supset g$);

- g is not covered by another implicant with fewer literals.

In other words, an implicant is prime if it equals 1 when no other implicant equals 1. An implicant, which *is not* prime, can be removed from the expression since it is superfluous. In the example:

$$f = XY + X + Z$$

X and Z are prime implicants while XY is a non-prime implicant of f. In fact, to have $XY = 1$ it is necessary that $X = 1$, but if $X = 1$ then $f = 1$ anyway, since X already implies f $(X \supset XY)$.

1.16 Combinational Networks

A *combinational network* is defined as a logical circuit whose output depends only on the combination of its inputs. In Chap. 5, we will discuss *sequential networks* whose output does not only depend on the values of the inputs in that time but also on the "inputs history." In other words, we will see that these networks have memory capacity.

A combinational network can be described in terms of a Boolean function. Note some examples of combinational networks.

1.16.1 Example: Logical Network Analysis

We want to analyze the following circuit, obtaining its truth table:

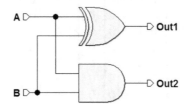

It is composed of two known gates, so we can complete the truth table directly:

A	B	Out1	Out2
0	0	0	0
0	1	1	0
1	0	1	0
1	1	0	1

We will find this circuit again in Chap. 2, being an arithmetic circuit.

1.16.2   Example: Two-Level Logical Network Analysis

As above, we will analyze the circuit, compiling its truth table:

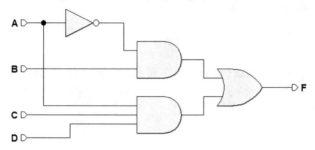

By analyzing all the circuit paths, we must determine the output value F for all the combination of A, B, C, and D. It is useful to include the intermediate outputs in the truth table. The result of the analysis is:

A	B	C	D	$\overline{A}$	$\overline{A}\,B$	$A\,C\,D$	F
0	0	0	0	1	0	0	0
0	0	0	1	1	0	0	0
0	0	1	0	1	0	0	0
0	0	1	1	1	0	0	0
0	1	0	0	1	1	0	1
0	1	0	1	1	1	0	1
0	1	1	0	1	1	0	1
0	1	1	1	1	1	0	1
1	0	0	0	0	0	0	0
1	0	0	1	0	0	0	0
1	0	1	0	0	0	0	0
1	0	1	1	0	0	1	1
1	1	0	0	0	0	0	0
1	1	0	1	0	0	0	0
1	1	1	0	0	0	0	0
1	1	1	1	0	0	1	1

1.16.3   Example: Circuit Schematic of a Logical Network (1)

We want to draw the logical schematic of a circuit, given its Boolean expression:

$$F = A\,D + \overline{C}\,B$$

The result is:

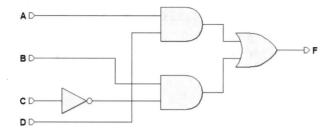

1.16.4   Example: Circuit Schematic of a Logical Network (2)

If we draw the logical schematic of the circuit, given this Boolean expression:

$$F = A\,D\,C + \overline{C}\,(\overline{A}\,B + A\,(B + D))$$

The result is a five-level network:

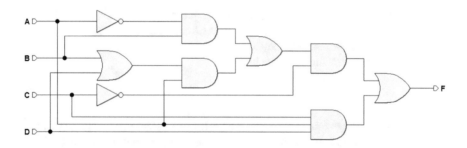

For practice, if we derive the truth table of this circuit and the one of the previous example, we see that the truth tables are identical! Perfect Induction shows that the two networks are equivalent. We can also prove that they are equivalent through the properties of Boolean algebra, as follows:

$$
\begin{aligned}
F &= A\,D\,C + \overline{C}\,(\overline{A}\,B + A\,(B + D)) = \\
  &= A\,D\,C + \overline{C}\,(\overline{A}\,B + A\,B + A\,D) = \\
  &= A\,D\,C + \overline{C}\,(B\,(\overline{A} + A) + A\,D) = \\
  &= A\,D\,C + \overline{C}\,(B + A\,D) = A\,D\,C + A\,D\,\overline{C} + \overline{C}\,B = \\
  &= A\,D\,(C + \overline{C}) + \overline{C}\,B = A\,D + \overline{C}\,B.
\end{aligned}
$$

1.16.5   Example: Defining the Behavior of a Logical Network

We want to define the truth table of a combinational network with three inputs A, B, and C: output F must assume value 1 when the number of input 1s is odd.

First, we prepare the truth table (bottom left). Then, line by line, we count the 1s and write the value of F based on the given definition. For example, in the last line we count three (an odd number of) 1s so we insert a 1 in the output column. In the end, we get the table at the right.

A	B	C	F
0	0	0	.
0	0	1	.
0	1	0	.
0	1	1	.
1	0	0	.
1	0	1	.
1	1	0	.
1	1	1	.

$\rightarrow$

A	B	C	F
0	0	0	0
0	0	1	1
0	1	0	1
0	1	1	0
1	0	0	1
1	0	1	0
1	1	0	0
1	1	1	1

1.16.6   Example: Circuit Schematic from the Truth Table

Let's examine the table derived from the last example and start by writing the Boolean expression of the function using the first form (AND–OR) of Shannon's Theorem. Then, we'll draw its circuit schematic.

We obtain the following from the table:

$$F = \overline{A}\,\overline{B}\,C + \overline{A}\,B\,\overline{C} + A\,\overline{B}\,\overline{C} + A\,B\,C$$

The expression defines four three-input ANDs that merge into one single four-input OR. After the components of this schematic are designed and interconnected, we will connect inputs A, B, and C to the ANDs, taking any negations into account. In the end, we get:

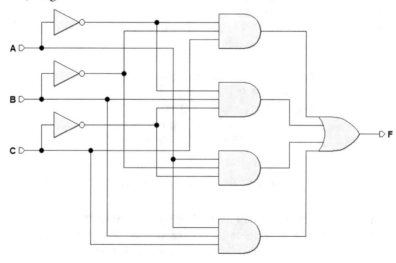

1.16.7 Example: Controlling a Heating System

Designing the control circuit of a heating system.

The system is composed of a thermostat, a heater, and a switch. T, I, and C are Boolean variables: the first two are inputs, the last, an output.

Defining the Variables

Given t_0 as a certain threshold temperature, the thermostat indicates that we are above or below it. Thus, we suppose that:

$$T = 0 \qquad \text{(if } t \geq t_0) \qquad T = 1 \qquad \text{(if } t < t_0)$$

The heater could be ON or OFF:

$$C = 0 \qquad \text{(heater OFF)} \qquad C = 1 \qquad \text{(heater ON)}$$

The same goes for the switch, which could be ON or OFF:

$$I = 0 \qquad \text{(switch OFF)} \qquad I = 1 \qquad \text{(switch ON)}$$

Defining the Network Operation

For the heater to be ON ($C = 1$) it must be $t < t_0$ ($T = 1$) and the switch must be ON ($I = 1$).

Let's translate this *verbal description* into the truth table:

		I	T	C
• if I is OFF	$\Rightarrow$ heater is OFF ($C = 0$)	0	0	0
• if I is ON but ($t \geq t_0$)	$\Rightarrow$ heater is OFF ($C = 0$) $\Rightarrow$	0	1	0
• if I is ON and ($t < t_0$)	$\Rightarrow$ heater is ON ($C = 1$)	1	0	0
		1	1	1

Synthesis

Synthesis is the process that allows us to find the logical expression (and the network schematic) from a truth table. The resulting logical expression is $C = I \cdot T$, and the circuit representation is:

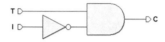

Note: the values assigned to the three variables to represent the different physical conditions are arbitrary. For example, we would be able to define I and C as before and T as follows:

$$T = 0 \text{ se } t < t_0$$
$$T = 1 \text{ se } t \geq t_0$$

gives us:

I	T	C
0	0	0
0	1	0
1	0	1
1	1	0

Notice that $C = 1$ if and only if $I = 1$ and $T = 0$. Here, the logical expression is $C = I \cdot \overline{T}$. The new circuit representation is:

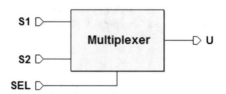

In the two cases, both $C = I \cdot T$ and $C = I \cdot \overline{T}$ are minterms. There is one only combination of I and T whose product is $C = 1$. In general, to synthesize a network with just one 1 in the output, we take the minterm of the unique combination of variables that give the output 1. We do this by inserting the variables that equal 1 in that specific combination into the *direct* form and those that equal 0 into the *negated* form.

1.16.8   Example: Two Channels Multiplexer (Selector)

This is a system that provides a Boolean output variable (U) that copies one of the two possible inputs ($S1$, $S2$), depending on the value of a control variable (SEL).

S1 ▷
Multiplexer ▷ U
S2 ▷
SEL ▷

Before defining the problem in Boolean terms, let's draw the timing diagram of the possible $S1s$, $S2s$, and $SELs$ over time and see how U should change.

Let's represent $S1$ and $S2$ as *digital signals*, that is, sequences of 0s and 1s that vary over time and encode information such as a phone call or a TV program, according to a certain standard. Without concerning ourselves with the type of code, let's use SEL to select which of the signals will be routed to output U.

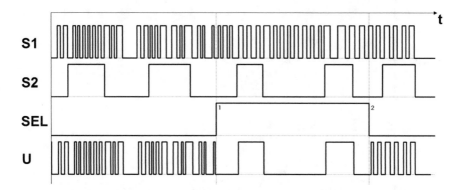

From the image above, we see that U needs to assume the values of $S1$ if $SEL = 0$, or of $S2$ if $SEL = 1$. Thus, it is possible to define the truth table. Notice that the table is valid "moment by moment," and does not describe the history of the network operations:

SEL	$S1$	$S2$	U	
0	0	0	0	
0	0	1	0	
0	1	0	1	(a)
0	1	1	1	(b)
1	0	0	0	
1	0	1	1	(c)
1	1	0	0	
1	1	1	1	(d)

Let's apply the first form of Shannon's Expansion Theorem: meaning, let's do an AND–OR canonical synthesis. For all the 1s in the output column, let's write the corresponding minterms, merging them in a OR:

$$U = \overline{SEL} \cdot S1 \cdot \overline{S2} + \text{ (a)}$$
$$\overline{SEL} \cdot S1 \cdot S2 + \text{ (b)}$$
$$SEL \cdot \overline{S1} \cdot S2 + \text{ (c)}$$
$$SEL \cdot S1 \cdot S2 \quad \text{ (d)}$$

This is the circuit schematic, including the NOTs (three suffice):

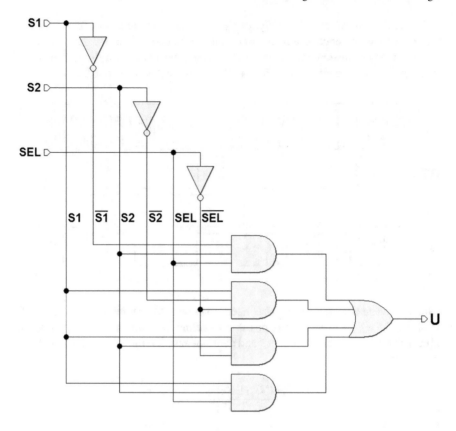

Notice that every time we see one of the four combinations that the table assigns at 1, the corresponding minterm (and only that) generates a 1. The OR makes it so that the output U goes to 1 for all four combinations above.

Minimizing

The AND/OR canonical form that we derived can be minimized by using the properties of Boolean algebra. We need to minimize the *number of literals* of the expression that describes the network. Let's see the following:

$$U = \overline{SEL} \cdot S1 \cdot \overline{S2} + \overline{SEL} \cdot S1 \cdot S2 + SEL \cdot \overline{S1} \cdot S2 + SEL \cdot S1 \cdot S2 =$$
$$= (S1 \cdot \overline{S2} + S1 \cdot S2) \cdot \overline{SEL} + (\overline{S1} \cdot S2 + S1 \cdot S2) \cdot SEL =$$
$$= ((\overline{S2} + S2) \cdot S1) \cdot \overline{SEL} + ((\overline{S1} + S1) \cdot S2) \cdot SEL =$$
$$= S1 \cdot \overline{SEL} + S2 \cdot SEL$$

Take into account that $(S1 + \overline{S1}) = 1$ and $(\overline{S2} + S2) = 1$. Originally the expression had 12 literals; now it only has four. See the logical schematic below. The network's complexity is markedly reduced:

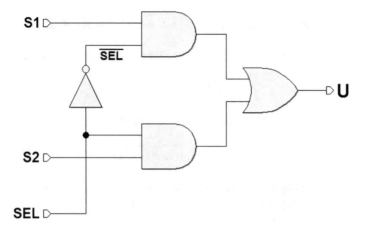

The final expression:

$$U = S1 \cdot \overline{SEL} + S2 \cdot SEL$$

can also be reinterpreted intuitively. It simply says that output U copies $S1$ when SEL is 0 (thus $\overline{SEL} = 1$), while it copies $S2$ when SEL is 1.

Note: we went through the analytical process of minimization for practice. To reduce the complexity of a Boolean function, we can use simpler methods like the "maps" that we will see in Chap. 2.

1.17 Exercises

1. Negate the following term and transform it into a four-term logical sum.

$$\overline{A}\,B\,\overline{C}\,D$$

2. Negate the following term and then transform it into a single product term.

$$\overline{A} + \overline{B} + C$$

3. Using the theorems of Boolean algebra, minimize the following logical expression:

$$A + A\,B + C\,B + C\,\overline{B}$$

4. Using the theorems of Boolean algebra, it is possible to prove that the following expression equals 1 only when A and B are contemporaneously at 1 or when D is at 1 and C is contemporaneously at 0:

$$F = \overline{A} \, C \, B + \overline{A} \, C \, \overline{B} + \overline{A} \, \overline{D} + \overline{B} \, C \, D + \overline{B} \, C + \overline{B} \, \overline{D}$$

5. Minimize the following logical function with the criteria of Boolean algebra.

$$Y = \overline{A} \, (A + B) + \overline{C} + B \, C$$

6. Using De Morgan's theorem, minimize the following logical function:

$$Y = \overline{A + A \, \overline{B} + C \, D}$$

7. Design the circuit that implements the expression $\overline{A} \, B + A \, \overline{B} \, \overline{C} \, (B + C)$, and verify that it is equivalent to the following network:

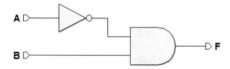

8. Design the circuits corresponding to the following logical expressions:

   (a) $C = (A + B) + \overline{A \, B}$
   (b) $D = A \, (B + C) \, \overline{C}$
   (c) $E = \overline{A} \, D \, (C + \overline{D})$
   (d) $G = \overline{A} \, B \, \overline{C} + D \, (C + A \, \overline{B} \, C)$

9. Do the following conversions between the canonical forms:

   (a) $F = \overline{A} \, B \, \overline{C} + \overline{A} \, B \, C + A \, \overline{B} \, C + A \, B \, \overline{C}$ to the OR–AND form.
   (b) $G = (\overline{A} + B + C)(\overline{A} + \overline{B} + \overline{C})(A + B + C)$ to the AND–OR form.

10. Analytically derive the logical function of the following circuit:

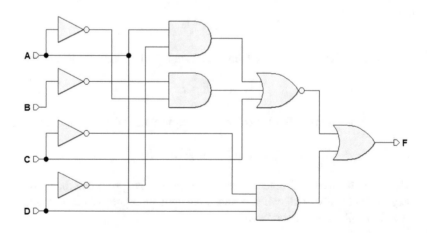

1.18 Solutions

1. $\overline{\overline{A}\,\overline{B}\,\overline{C}\,\overline{D}} = \overline{\overline{A}} + \overline{\overline{B}} + \overline{C} + \overline{D} = A + \overline{B} + C + \overline{D}$
2. $\overline{A} + \overline{B} + C = \overline{\overline{A}\,\overline{B}\,\overline{C}} = A\,B\,\overline{C}$
3. $A + AB + CB + C\overline{B} = A(1 + B) + C(B + \overline{B}) = A + C$
4. We need to demonstrate that the expression given equals the one below:
$$F = A\,B + \overline{C}\,D$$

By minimizing the expression given we obtain:

$$
\begin{aligned}
F &= \overline{\overline{A}\,C\,B + \overline{A}\,C\,\overline{B} + \overline{A}\,D + \overline{B}\,C\,D + \overline{B}\,C + \overline{B}\,\overline{D}} = \\
&= \overline{\overline{A}\,C\,(B + \overline{B}) + \overline{A}\,D + \overline{B}\,C\,(D + 1) + \overline{B}\,\overline{D}} = \\
&= \overline{\overline{A}\,C + \overline{A}\,D + \overline{B}\,C + \overline{B}\,\overline{D}} = \\
&= \overline{\overline{A}\,(C + \overline{D}) + \overline{B}\,(C + \overline{D})} = \\
&= \overline{(\overline{A} + \overline{B}) \cdot (C + \overline{D})} = \\
&= \overline{(\overline{A} + \overline{B})} + \overline{(C + \overline{D})} = A\,B + \overline{C}\,D
\end{aligned}
$$

5. We obtain:

$$
\begin{aligned}
Y &= \overline{A}\,(A + B) + \overline{C} + B\,C = \\
&= \overline{A}\,A + \overline{A}\,B + \overline{C} + B\,C = \overline{A}\,B + \overline{C} + B\,\overline{C} + B\,C = \\
&= \overline{A}\,B + \overline{C} + B\,(\overline{C} + C) = \overline{A}B + \overline{C} + B = \\
&= B + \overline{C}
\end{aligned}
$$

6. We obtain:

$$Y = \overline{A + A\overline{B} + CD} = \overline{A(1 + \overline{B}) + CD} = \overline{A + CD} = \overline{A}(\overline{C} + \overline{D})$$

7. The network is as follows:

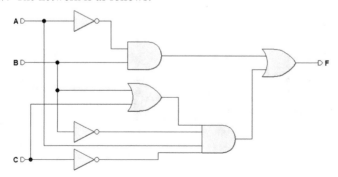

Minimizing the expression of F:

$$F = \overline{A}\,B + A\,\overline{B}\,\overline{C}\,(B + C) = \overline{A}\,B + A\,\overline{B}\,\overline{C}\,B + A\,\overline{B}\,\overline{C}\,C =$$
$$= \overline{A}\,B + A\,\overline{C}\,(\overline{B}\,B) + A\,\overline{B}\,(\overline{C}\,C) = \overline{A}\,B + A\,\overline{C}\,(0) + A\,\overline{B}\,(0) =$$
$$= \overline{A}\,B$$

8. These are the circuits corresponding to the expressions:

(a) $C = (A + B) + \overline{A}\,B$:

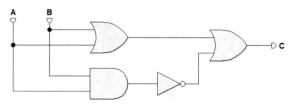

(b) $D = A\,(B + C)\,\overline{C}$:

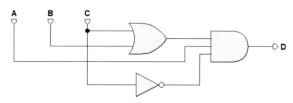

(c) $E = \overline{A}\,D\,\overline{(C + \overline{D})}$:

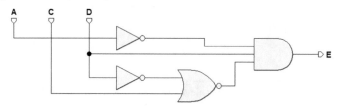

(d) $G = \overline{A}\,B\,\overline{C} + D\,(C + A\,\overline{B}\,C)$:

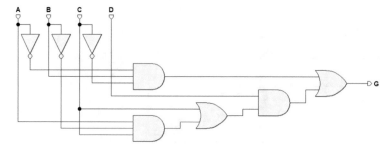

9. (a) We derive the truth table and represent the negated output.

$$
\begin{array}{ccc|c|c}
A & B & C & F & \overline{F} \\
\hline
0 & 0 & 0 & 0 & 1 \\
0 & 0 & 1 & 0 & 1 \\
0 & 1 & 0 & 1 & 0 \\
0 & 1 & 1 & 1 & 0 \\
1 & 0 & 0 & 0 & 1 \\
1 & 0 & 1 & 1 & 0 \\
1 & 1 & 0 & 1 & 0 \\
1 & 1 & 1 & 0 & 1 \\
\end{array}
$$

From this table, we derive the AND/OR expression of the negated function and we apply De Morgan's Theorem.

$$
\overline{F} = \overline{A}\,\overline{B}\,\overline{C} + \overline{A}\,\overline{B}\,C + A\,B\,C + A\,\overline{B}\,\overline{C}
$$
$$
= \overline{(A + B + C)(A + B + \overline{C})(\overline{A} + \overline{B} + \overline{C})(\overline{A} + B + C)} =
$$

Finally, we once more negate the whole expression.

$$
F = (A + B + C)(A + B + \overline{C})(\overline{A} + \overline{B} + \overline{C})(\overline{A} + B + C).
$$

(b) We complete the truth table, from which we will then directly derive the AND–OR canonical synthesis.

$$
\begin{array}{ccc|c}
A & B & C & G \\
\hline
0 & 0 & 0 & 0 \\
0 & 0 & 1 & 1 \\
0 & 1 & 0 & 1 \\
0 & 1 & 1 & 1 \\
1 & 0 & 0 & 0 \\
1 & 0 & 1 & 1 \\
1 & 1 & 0 & 1 \\
1 & 1 & 1 & 0 \\
\end{array}
$$

$$
G = \overline{A}\,\overline{B}\,C + \overline{A}\,B\,\overline{C} + \overline{A}\,B\,C + A\,\overline{B}\,C + A\,B\,\overline{C}
$$

10. $F = A\,\overline{C}\,D + \overline{(A\,\overline{D} + \overline{A}\,\overline{B} + C)}$.

Chapter 2
Combinational Network Design

Abstract This chapter deals with the transition from Boolean algebra to the implementation of combinational networks. Karnaugh maps provide a simple and intuitive method to represent and minimize functions with a few variables. The Variable-Entered Maps extend their usefulness and overcome some of their limitation. Standard networks such as decoders, multiplexers, and demultiplexers provide a wider view of combinational circuits, where the random approach of classical synthesis is enriched with an architectural one that introduces the concepts of programmable logic. Finally, this chapter deals with time behavior of non-ideal components and its implications on the synthesis.

2.1 Karnaugh Maps

In Chap. 1, we used two *"languages"* to describe a logical network: *Boolean expressions* and *truth tables*. Here, we will see a third language: *maps*.

Describing a function through its Boolean expression or truth table makes simplifying them logically an arduous task. Maps offer a way to write truth tables in a format that makes simplification easier. As we will see, maps order and highlight minterms because they have a geometric structure that makes applying the *absorption* and *logic adjacency* properties easy.

Absorption:

$$A + A B \quad = A \qquad A + \overline{A} B \quad = A + B$$
$$A \cdot (A + B) = A \qquad A \cdot (\overline{A} + B) = A B$$

Logic Adjacency:

$$A B + A \overline{B} \quad = A$$
$$(A + B) \cdot (A + \overline{B}) = A$$

Given a two-variable function $f(A, B)$, we represent in a two-dimensional space (A, B) its set definition, which consists in the four combinations (A, B):

© Springer International Publishing AG, part of Springer Nature 2019
G. Donzellini et al., *Introduction to Digital Systems Design*,
https://doi.org/10.1007/978-3-319-92804-3_2

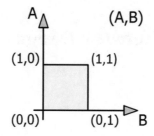

For a three-variable function $f(A, B, C)$, the set of definition can be represented in a three-dimensional space by placing each of the eight possible combinations of A, B, and C at the vertexes of a cube:

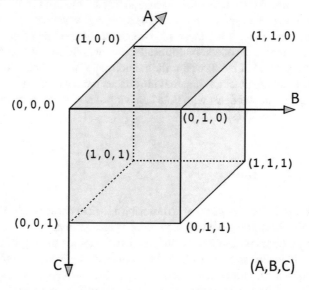

By moving along one edge of the cube from one vertex to another, *geometrically adjacent* one, we see that *only one variable* changes value.

There is *logic adjacency* between two combinations of variables when they are at distance 1; that is, they differ by only one variable.

The cube above is an order-3 cube. Within its structure, various cubes of inferior order, or subcubes, can be identified. The order is the number of geometrical dimensions: a square is an "order-2" (two-dimensional) cube; an n-order cube will have 2^n vertexes. In the cube above, we see:

- Eight vertexes (*order-0 cubes*): Every vertex has one single combination of all the variables; for example, in the vertex $(1,1,1)$ all the input variables have the value 1.

- Twelve edges (*order-1 cubes*): Each edge has two combinations of variables at distance 1. In other words, each edge is univocally located by one single pair of two-out-of-three variables; e.g., the upper left-hand edge of the figure is defined by $B = 0$ and $C = 1$.
- Six faces (*order-2 cubes*): Each face has four different combinations of variables, and it is univocally located by the value of a single variable; e.g., the farthest right face in the figure is defined by $B = 1$.

Now that, we have made these preliminary points, and let's draw the *map* corresponding to a three-dimensional cube by "cutting it out" and arranging its eight vertexes on a plane, maintaining as much as possible the same position they had in space.

The map is a lattice of cells, each containing one of the vertexes:

$(0,0,0)$	$(0,1,0)$	$(1,1,0)$	$(1,0,0)$
$(0,0,1)$	$(0,1,1)$	$(1,1,1)$	$(1,0,1)$

We see that:

- In the four lower cells, $C = 1$.
- In the four higher cells, $C = 0$.
- In the four cells to the right, $A = 1$.
- In the four cells to the left, $A = 0$.
- In the four central cells, $B = 1$.
- In the four lateral cells, $B = 0$.

So, we can represent the areas of the map by using the variables as the "coordinates" of the cells. One method is shown in the figure below (left), where the values of the variables identify the rows and columns.

The figure below (right) divides the map into geometrical areas corresponding to the values of the variables. This is the method we will use in this book. The name of each variable is drawn in correspondence with the area where it equals 1.

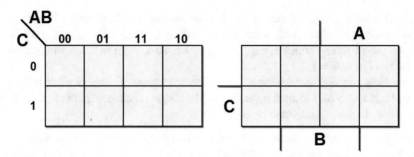

Since a map's cell corresponds univocally to one combination of variable values, it is possible to copy into the cell itself the corresponding function value $f(C, B, A)$ for that specific combination.

These types of maps are called *Karnaugh maps* (or *K-maps*).

2.2 Using Maps for AND-OR Synthesis

Here, we will look at the truth table of the multiplexer we examined in Chap. 1:

SEL	S1	S2	U
0	0	0	0
0	0	1	0
0	1	0	1
0	1	1	1
1	0	0	0
1	0	1	1
1	1	0	0
1	1	1	1

Let's copy the function values into a Karnaugh map:

	S1		
	0	0	1	1
SEL	0	1	1	0
		S2	

The expression of the AND-OR canonical form that we found was:

$$U = \overline{S1} \cdot S2 \cdot SEL + S1 \cdot \overline{S2} \cdot \overline{SEL} + S1 \cdot S2 \cdot \overline{SEL} + S1 \cdot S2 \cdot SEL$$

In the following figure, we see the minterms and their position on the map.

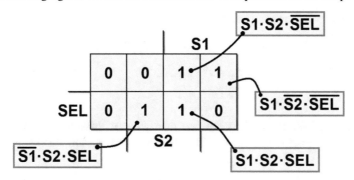

Remember that cells are adjacent if they are at distance 1, that is if they differ by one variable only. On the map, we can group together the 1s that are at distance 1 in two unidimensional subcubes (or "groupings"):

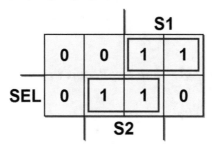

In each grouping, we have a variable whose value changes within the group, while the values of the other variables do not. For each grouping, let's write a term that contains only the variables whose value remains the same and ignores the others.

From the upper grouping, we get $S1 \cdot \overline{SEL}$: this is the area where $S1 = 1$ and $SEL = 0$, while $S2$ assumes different values.

Likewise, we have $S2 \cdot SEL$ in the lower grouping: this is the area where $S2 = 1$ and $SEL = 1$, while $S1$ varies.

In the end, by adding the terms obtained, we quickly and directly derive the minimized expression:

$$U = S1 \cdot \overline{SEL} + S2 \cdot SEL$$

In Chap. 1, we obtained the same expression by minimizing the canonical form using the properties of Boolean algebra.

Proof

In the above right grouping, the two minterms are:

$$S1 \cdot \overline{S2} \cdot \overline{SEL} \quad \text{and} \quad S1 \cdot S2 \cdot \overline{SEL}$$

In OR, these two terms can be simplified:

$$S1 \cdot \overline{S2} \cdot \overline{SEL} + S1 \cdot S2 \cdot \overline{SEL} =$$
$$S1 \cdot \overline{SEL} \cdot (S2 + \overline{S2}) =$$
$$S1 \cdot \overline{SEL}$$

The eliminated variable, $S2$, is the one that assumes different values within the grouping. The other term is treated likewise: here, the variable that changes is $S1$, thus showing:

$$S1 \cdot S2 \cdot SEL + \overline{S1} \cdot S2 \cdot SEL =$$
$$S2 \cdot SEL \cdot (S1 + \overline{S1}) =$$
$$S2 \cdot SEL$$

The function is $U = S1 \cdot \overline{SEL} + S2 \cdot SEL$, as defined above.

2.2.1 Implicants and Prime Implicants in Karnaugh Maps

Consider the following map.

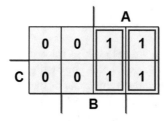

From the two unidimensional subcubes, we derive:

$$U = A B + A \overline{B}$$

Each one is an implicant because if $A \overline{B}$ or $A B$ equals 1, the function U also equals 1. Let's now consider the two-dimensional subcube that comprises all the 1s as in the following figure.

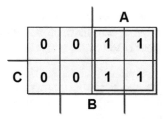

In this subcube, the only variable that does not change is A: thus, the function is reduced to the expression $U = A$. We could reach the same conclusion this way:

$$U = A\overline{B} + A B = A(\overline{B} + B) = A$$

The simplification is possible because $\overline{A}\,B$ and $A\,B$ are not prime implicants while A is. To obtain the best grouping, we must locate the largest possible subcubes. If *there is no* largest subcube that completely covers what we are considering, it is called a prime implicant.

Let's see some examples of how to derive the expression of the function from the maps. From the map on the left, we derive $U = A\,\overline{B}\,\overline{C}$, while we get $U = \overline{B}\,\overline{C}$ from the one on the right.

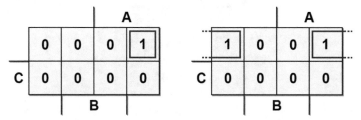

Despite appearances, the two cells are *logically adjacent*, in the map on the right. As we have seen, the map is a representation on a two-dimensional space of a multi-dimensional cube. So, we need to imagine the map as a piece of paper that can be folded over itself vertically and horizontally; the upper and lower edges and the right and left edges are adjacent to each other. As proof, note that only variable A changes between each two pair of cells.

Now, consider the following map:

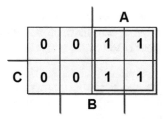

The largest subcube containing the 1s is highlighted: thus $U = \overline{B}$.

Now let's look two borderline cases. The map on the left has a 1 in every cell and represents the constant function $U = 1$; likewise, the map on the right only contains $0s$, represented by the constant $U = 0$:

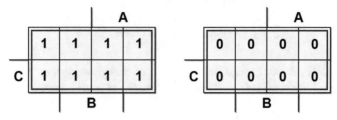

In a three-variable map, we have seen that all the minterms that differ by one variable are represented on the map by adjacent cells. This also holds for four-variable maps, as in the following example:

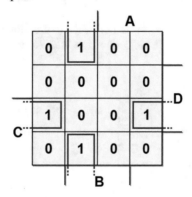

From this map, we get: $U = \overline{A} B \overline{D} + \overline{B} C D$.

Five-variable maps (order-5) can be represented by two order-4 maps, in which cells in the same position in the two maps are adjacent (imagine one map on the top of the other):

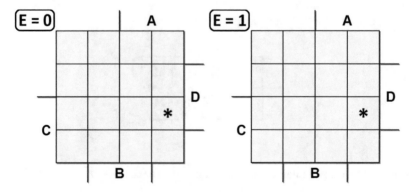

The cells with asterisks are adjacent: it is therefore possible to group them and, therefore, eliminate the variable E.

Six-variable maps are made with four order-4 maps. Higher order maps are too complicated to represent.

2.2.2 Using Maps for Minimization

Let's define the *essential prime implicant* as the only prime implicant that contains a certain 1 on the map. On the following map, for example, there are two order-1 essential prime implants and one order-2 essential prime implicant.

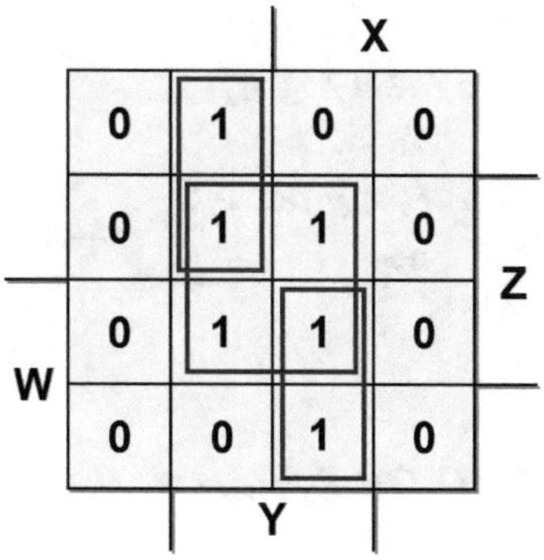

The cells with the $1s$ can be considered more than once. This is useful for finding higher order subcubes. We obtain the synthesis:

$$U = Y Z + \overline{X} Y \overline{W} + X Y W$$

An implicant completely covered by essential prime implicants is called a *redundant implicant*. Redundant implicants, as such, must be eliminated from the synthesis. Therefore, by using the maps, we obtain the minimum synthesis by taking only the essential prime implicants.

2.2.3 "Checkerboard" Maps

A *checkerboard map*, as seen in the figure below, cannot be minimized in terms of two-level AND-OR or OR-AND networks.

		A	
0	**1**	**0**	**1**
C **1**	**0**	**1**	**0**

B

It can be verified that it represents an *XOR tree*:

$$U = \overline{A}\,B\,\overline{C} + A\,\overline{B}\,\overline{C} + \overline{A}\,\overline{B}\,C + A\,B\,C =$$
$$= (\overline{A}\,B + A\,\overline{B}) \cdot \overline{C} + (\overline{A}\,\overline{B} + A\,B) \cdot C =$$
$$= (A \oplus B) \cdot \overline{C} + \overline{(A \oplus B)} \cdot C =$$
$$= A \oplus B \oplus C = (A \oplus B) \oplus C$$

$\Rightarrow$

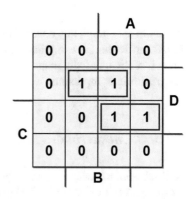

2.2.4 Examples of AND-OR Synthesis

1. Considering the truth table below, let's complete the map at the right.

A	B	C	D	F
0	0	0	0	0
0	0	0	1	0
0	0	1	0	0
0	0	1	1	0
0	1	0	0	0
0	1	0	1	1
0	1	1	0	0
0	1	1	1	0 ⇒
1	0	0	0	0
1	0	0	1	0
1	0	1	0	0
1	0	1	1	1
1	1	0	0	0
1	1	0	1	1
1	1	1	0	0
1	1	1	1	1

		A		
0	**0**	**0**	**0**
0	**1**	**1**	**0**
0	**0**	**1**	**1**
0	**0**	**0**	**0**

C

B

D

The grouping in the third row gives us the term $A\,C\,D$; the grouping in the second gives us $B\,\overline{C}\,D$, thus: $F = A\,C\,D + B\,\overline{C}\,D$. Let's verify:

$$A\,B\,C\,D + A\,\overline{B}\,C\,D = A\,C\,D\,(B + \overline{B}) = A\,C\,D$$
$$\overline{A}\,B\,\overline{C}\,D + A\,B\,\overline{C}\,D = B\,\overline{C}\,D\,(A + \overline{A}) = B\,\overline{C}\,D$$

2. Let's simplify $F = \overline{A}\,\overline{B}\,\overline{C} + A\,B\,\overline{C} + \overline{A}\,\overline{B}\,C$. We get this map:

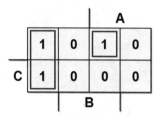

By grouping the two 1s at the left we get: $\overline{A}\,\overline{B}$. The overall function:

$$F = \overline{A}\,\overline{B} + A\,B\,\overline{C}.$$

3. Let's synthesize the following map:

The four 1s that are adjacent even though on opposide edges, can be grouped into a single product. The resulting function is:

$$F = \overline{B}\,D.$$

4. Likewise, we synthesize the other map with the 1s at the angles.

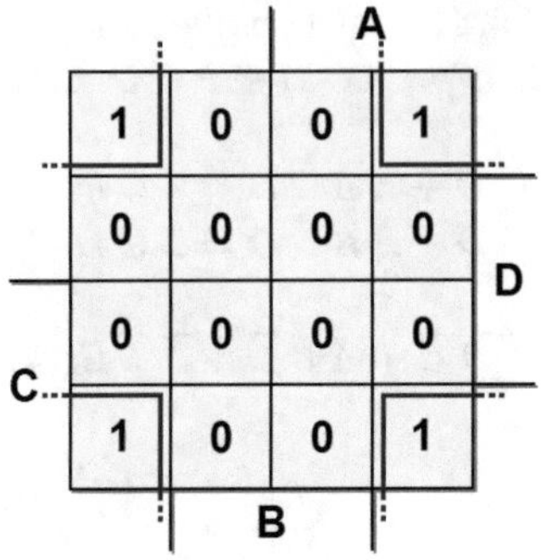

As above, we group one single term. The function is:

$$F = \overline{B}\,\overline{D}.$$

5. Let's derive the function from the map below:

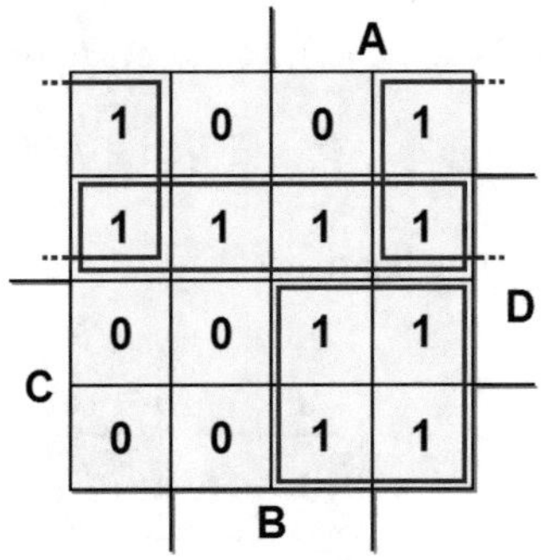

The groupings all have four bits. The upper grouping provides the term $\overline{B}\,\overline{C}$; in the second row, we have $\overline{C}\,D$; at the bottom left, we get $A\,C$. So, the desired function is:

$$F = \overline{B}\,\overline{C} + \overline{C}\,D + A\,C.$$

6. Here is another example (*cyclical map*):

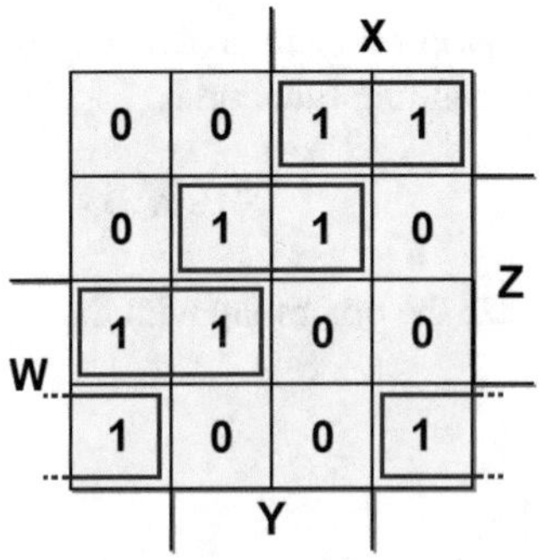

There are four order-1 essential implicants, so we get:

$$U = X\overline{Z}\,\overline{W} + Y\overline{W}Z + \overline{X}ZW + \overline{Y}\,\overline{Z}W$$

2.3 OR-AND Synthesis

In an AND-OR synthesis, we have seen that we synthesize the 1s of the function through the AND of the input variables. They are *direct* if the variable equals 1 and *negated* if it equals 0. Then, all outputs of the AND gates are OR-ed together.

In the OR-AND synthesis, according to the second form of Shannon's theorem, we synthesize the 0 of the function through the OR of the input variables (*direct* if 0 and *negated* if 1) and all the ORs end up in an AND.

To minimize the OR-AND synthesis using Karnaugh maps, the property of logic adjacency is still valid. Let's take the two-input channel multiplexer as an example, but use the OR-AND technique. Here is the truth table:

SEL	$S1$	$S2$	U
0	0	0	0
0	0	1	0
0	1	0	1
0	1	1	1
1	0	0	0
1	0	1	1
1	1	0	0
1	1	1	1

After completing the map with the corresponding values from the table, we identify the subcubes that group the 0s (the rules are the same as for the subcubes with 1s). In the map below, we see the best grouping possible for our example:

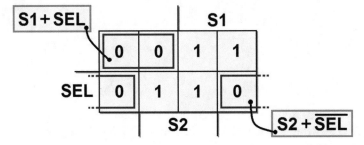

From this, we directly derive the minimized OR-AND expression:

$$U = (S1 + SEL)(S2 + \overline{SEL}).$$

2.3.1   Synthesis of the Negated Function

For an OR-AND synthesis, we can work differently, by first doing the AND-OR synthesis from the 0s as if they were 1s and as if the function were negated.

If we apply this method to the previous example, we get the following AND-OR synthesis of $\overline{U}$:

$$\overline{U} = \overline{S1} \cdot \overline{SEL} + \overline{S2} \cdot SEL$$

By applying the De Morgan Theorem, we obtain the OR-AND expression:

$$U = (S1 + SEL)(S2 + \overline{SEL})$$

Note that it is not possible to obtain the OR-AND synthesis by applying the principle of duality directly to the AND-OR expression.

2.4   NAND-NAND Synthesis

Once we have the AND-OR synthesis, we can implement the function with a network composed exclusively of NANDs (NAND-NAND synthesis) by substituting every OR or AND with a NAND, as in the figure below:

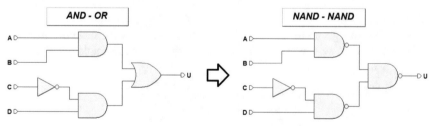

Proof:

Let's substitute the ANDs on the left with NANDs followed by NOT. Then, let's apply De Morgan's Theorem to the OR gate, transforming it into AND, whose inputs and output are negated:

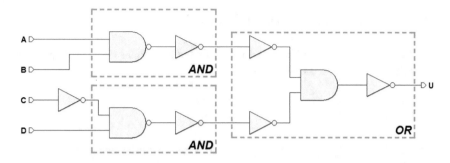

Let's merge the AND gate with the NOT that follows and obtains a NAND, and then separates the two NANDS on the left from the NOTs.

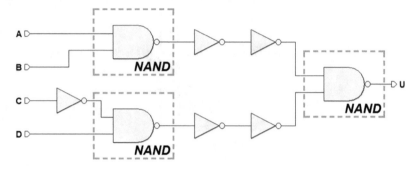

Finally, when we eliminate the double negation, we get the NAND-NAND network. If the AND-OR function has one or more inputs that go directly to the OR gate, they must be negated.

2.5 NOR-NOR Synthesis

Starting from an OR-AND network, we substitute each OR or AND with a NOR gate and get a NOR-NOR synthesis as in the example below. The proof is analogous to the NAND-NAND network.

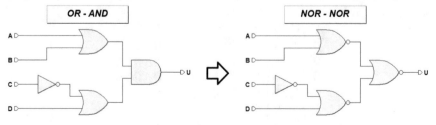

2.6   Standard Combinational Networks

These are general-use combinational networks with a regular circuit structure that are available to the user as functional blocks.

2.6.1   Decoders

Decoders have n inputs and 2^n outputs: every combination of inputs activates one and only one output.

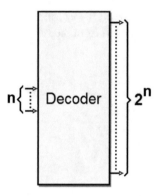

A $3 \rightarrow 8$ decoder has 3 inputs and 8 outputs:

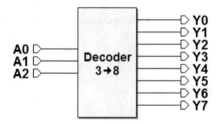

As we can see in the truth table below, output activation occurs in order. The combination $A2A1A0 =$ "000" activates the output $Y0$ and so on until $A2A1A0 =$ "111" activates output $Y7$:

A2	A1	A0	Y0	Y1	Y2	Y3	Y4	Y5	Y6	Y7
0	0	0	1	0	0	0	0	0	0	0
0	0	1	0	1	0	0	0	0	0	0
0	1	0	0	0	1	0	0	0	0	0
0	1	1	0	0	0	1	0	0	0	0
1	0	0	0	0	0	0	1	0	0	0
1	0	1	0	0	0	0	0	1	0	0
1	1	0	0	0	0	0	0	0	1	0
1	1	1	0	0	0	0	0	0	0	1

Synthesizing a truth table with eight outputs requires using a combinational network for each output. Here, each output is activated by a single combination of inputs, so maps are not useful since each output calls for only one minterm. The figure below shows the synthesis of the $3 \rightarrow 8$ decoder:

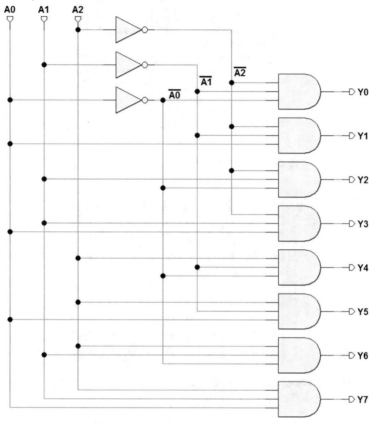

Decoders generally have one *enable input*. When the decoder is enabled, it behaves as described above; otherwise, no output is activated. In the figure below, we see the synthesis of the output $Y0$, with the enable input EN.

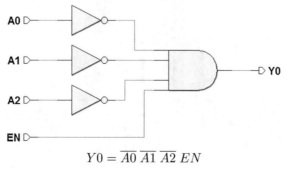

$$Y0 = \overline{A0}\ \overline{A1}\ \overline{A2}\ EN$$

The figure below shows the "Dec 3–8" component from the *Deeds* library, along with its truth table (notice the functionality of the *EN* input):

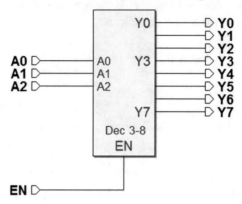

EN	$A2$	$A1$	$A0$	$Y0$	$Y1$	$Y2$	$Y3$	$Y4$	$Y5$	$Y6$	$Y7$
0	–	–	–	0	0	0	0	0	0	0	0
1	0	0	0	1	0	0	0	0	0	0	0
1	0	0	1	0	1	0	0	0	0	0	0
1	0	1	0	0	0	1	0	0	0	0	0
1	0	1	1	0	0	0	1	0	0	0	0
1	1	0	0	0	0	0	0	1	0	0	0
1	1	0	1	0	0	0	0	0	1	0	0
1	1	1	0	0	0	0	0	0	0	1	0
1	1	1	1	0	0	0	0	0	0	0	1

More than one decoder can be connected in order to form a larger one. For example, with two "Dec 3–8s" we get a 4 → 16 decoder:

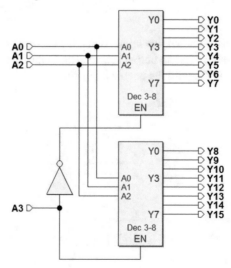

The truth table below can be divided into two parts: the upper part where the input $A3 = 0$ and the lower part where $A3 = 1$. The "Dec 3–8" that generates the first eight outputs must be enabled ($EN = 1$) when $A3 = 0$ and the other when $A3 = 1$. A simple NOT, as we see in the logic diagram, does the job.

A3	A2	A1	A0	Y0	Y1	Y2	Y3	Y4	Y5	Y6	Y7	Y8	Y9	Y10	Y11	Y12	Y13	Y14	Y15
0	0	0	0	1	0	0	0	0	0	0	0	0	0	0	0	0	0	0	0
0	0	0	1	0	1	0	0	0	0	0	0	0	0	0	0	0	0	0	0
0	0	1	0	0	0	1	0	0	0	0	0	0	0	0	0	0	0	0	0
0	0	1	1	0	0	0	1	0	0	0	0	0	0	0	0	0	0	0	0
0	1	0	0	0	0	0	0	1	0	0	0	0	0	0	0	0	0	0	0
0	1	0	1	0	0	0	0	0	1	0	0	0	0	0	0	0	0	0	0
0	1	1	0	0	0	0	0	0	0	1	0	0	0	0	0	0	0	0	0
0	1	1	1	0	0	0	0	0	0	0	1	0	0	0	0	0	0	0	0
1	0	0	0	0	0	0	0	0	0	0	0	1	0	0	0	0	0	0	0
1	0	0	1	0	0	0	0	0	0	0	0	0	1	0	0	0	0	0	0
1	0	1	0	0	0	0	0	0	0	0	0	0	0	1	0	0	0	0	0
1	0	1	1	0	0	0	0	0	0	0	0	0	0	0	1	0	0	0	0
1	1	0	0	0	0	0	0	0	0	0	0	0	0	0	0	1	0	0	0
1	1	0	1	0	0	0	0	0	0	0	0	0	0	0	0	0	1	0	0
1	1	1	0	0	0	0	0	0	0	0	0	0	0	0	0	0	0	1	0
1	1	1	1	0	0	0	0	0	0	0	0	0	0	0	0	0	0	0	1

We can extend this technique at will to obtain larger decoders than the ones available.

Decoders are very important because they are very commonly used. For example, they represent the fundamental component to implement "addressing" of digital system devices. In other words, they allow us to use an identification number to identify and/or select the elements of a system and operate on them, as with memories and microcomputer systems.

2.6.2 Multiplexer

The "multiplexer" (short form "mux," or "channel selector" or simply "selector") is a combinational network with 2^n data inputs, one single output, and n selection inputs. The output assumes the input value identified by the selection inputs.

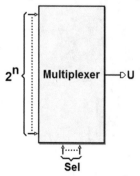

We have already seen the two-input channel selector and how it is created through logical gates. In this case, input S selects which of the two I0 and I1 inputs will be reproduced as the output.

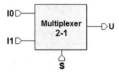

It is easy to extend this concept to multiplexers with a higher number of inputs. In the figure below, three inputs S2, S1, and S0 control which of the eight inputs (I0..I7) will be brought over to the output.

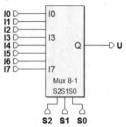

The network has eight data inputs and three selection inputs. It is impractical to do the synthesis through the truth table of a network with 11 inputs (providing 2048 rows in the table). The concept of decoding (see below) can help here.

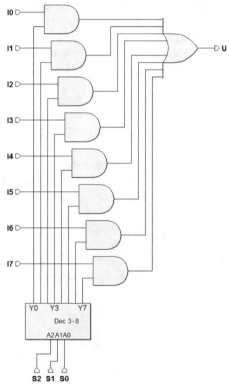

Sending the selection inputs to a decoder, we enable only one of the eight AND gates that transmit each of the eight inputs I0..I7 through the OR gate.

Considering the circuital structure of the decoder already discussed, by just adding an input to each of the AND gates that generate the decoder's outputs, we can get rid of the eight external ANDs. The result is that the network is reduced to just two levels, as in the next figure.

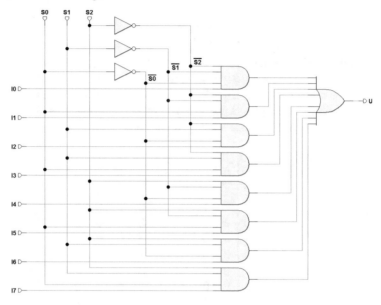

2.6.3 Demultiplexers

The demultiplexer ("deselector" or "DEMUX") is a combinational network with only one data input IN, n selection inputs, and 2^n outputs. It is a combinational network that mirrors the multiplexer. The value of the single input is transferred to one of the many outputs, which is chosen by the logical value of the selection inputs.

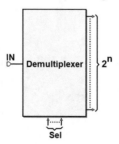

The next figure shows an example of an eight-output demultiplexer.

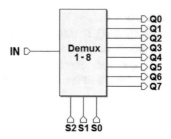

The truth table is represented below. The non-selected outputs always generate 0, while the selected one (from $S2$, $S1$, and $S0$) copies the value of input IN. Therefore, in the first row of the table, (for any $S2$, $S1$, and $S0$ with $IN = 0$), the selected and non-selected outputs all equal 0. However, on the other half of the table (for $IN = 1$), the selected output equals 1, since it copies input IN:

IN	$S2$	$S1$	$S0$	$Q0$	$Q1$	$Q2$	$Q3$	$Q4$	$Q5$	$Q6$	$Q7$
0	–	–	–	0	0	0	0	0	0	0	0
1	0	0	0	1	0	0	0	0	0	0	0
1	0	0	1	0	1	0	0	0	0	0	0
1	0	1	0	0	0	1	0	0	0	0	0
1	0	1	1	0	0	0	1	0	0	0	0
1	1	0	0	0	0	0	0	1	0	0	0
1	1	0	1	0	0	0	0	0	1	0	0
1	1	1	0	0	0	0	0	0	0	1	0
1	1	1	1	0	0	0	0	0	0	0	1

This truth table could be examined from another perspective: all the outputs are 0 when input IN equals 0, while the only output at 1 is the selected output if IN equals 1. This behavior matches that of a decoder with an enable input. It follows that a decoder with an enable input can be used as a demultiplexer, using the enable input as a data input.

The figure below shows that the "Demux 1–8" and "Dec 3–8" (*Deeds* components) are equivalent:

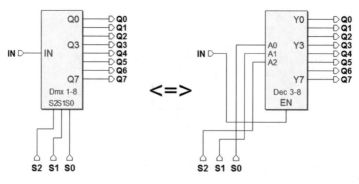

Example

When a multiplexer and a demultiplexer are connected as in the following figure, they allow data transfer from more than one source through a single line to more than one destination. This is especially useful when the transmitting system is very far from the receiving system: it can make radio or cable transmission practical.

The multiplexer's $T2$, $T1$, and $T0$ selection inputs allow us to choose the source to transmit, while $R2$, $R1$, and $R0$ select the data destination.

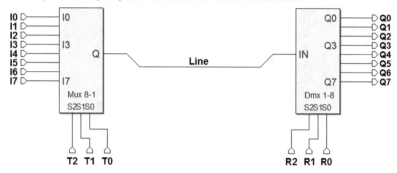

2.6.4 Seven-Segment Display Decoder

A seven-segment display is a device for visualizing numbers and letters. It has seven inputs, one for every luminous segment. Let's assume that a 1 turns on the corresponding segment, while a 0 keeps it off. The segments are denoted by the lower-case letters "a, b, c, d, e, f, g" as in the figure below.

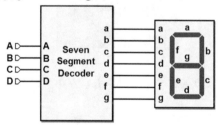

The "hexadecimal" (Hex) code will be examined in Chap. 3 but for now suffice it to say that it encodes, using 4 bits, the decimal digits from 0 to 9 followed by the first six letters of the alphabet (for a total of 16 symbols).

According to the four-bit input binary number $DCBA$, the decoder must activate the segments that make up the corresponding hexadecimal symbol, as seen in the figure below.

The decoder will have seven outputs: a, b, c, d, e, f, and g, one for each segment. Each of the outputs is controlled by an independent combinational network that switches that particular segment on or off.

The following table describes the networks that generate a, b, c, d, e, f, and g (the Hex column was inserted to facilitate understanding).

D	C	B	A	a	b	c	d	e	f	g	Hex
0	0	0	0	1	1	1	1	1	1	0	0
0	0	0	1	0	1	1	0	0	0	0	1
0	0	1	0	1	1	0	1	1	0	1	2
0	0	1	1	1	1	1	1	0	0	1	3
0	1	0	0	0	1	1	0	0	1	1	4
0	1	0	1	1	0	1	1	0	1	1	5
0	1	1	0	1	0	1	1	1	1	1	6
0	1	1	1	1	1	1	0	0	0	0	7
1	0	0	0	1	1	1	1	1	1	1	8
1	0	0	1	1	1	1	1	0	1	1	9
1	0	1	0	1	1	1	0	1	1	1	A
1	0	1	1	0	0	1	1	1	1	1	B
1	1	0	0	1	0	0	1	1	1	0	C
1	1	0	1	0	1	1	1	1	0	1	D
1	1	1	0	1	0	0	1	1	1	1	E
1	1	1	1	1	0	0	0	1	1	1	F

Below is the synthesis of the networks that drive segments "a" and "b".

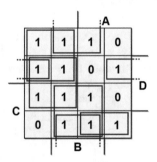

$$a = \overline{A}\,D + B\,\overline{D} + B\,C + \overline{B}\,\overline{C}\,D + A\,C\,\overline{D} + \overline{A}\,\overline{C}$$

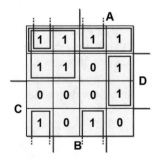

$$b = \overline{C}\,\overline{D} + \overline{A}\,\overline{C} + \overline{A}\,\overline{B}\,D + A\,\overline{B}\,D + A\,B\,\overline{D}$$

Likewise, we find the other outputs.

2.6.5 Seven-Segment BCD Decoder (using "don't-cares")

We want to design a similar device to visualize only the decimal digits from 0 to 9. Since there are only ten symbols to represent, we need only ten input combinations. On the other hand, ten combinations still require four inputs (if there were three inputs, there would only be $2^3 = 8$ combinations available), so there are six unused combinations.

When the input is a combination that does not correspond to any decimal digit, the problem of what to visualize on the screen can be resolved by two different approaches:

- Leave all the segments of the display off.
- Ignore which segments are on or off.

The purpose of the device is actually to decode a decimal number, not to operate on combinations that fall outside the expected ones. In practice, we hypothesize the inputs $DCBA$ always remain within the constraints, in this case, meaning combinations from "0000" to "1001" (from 0 to 9 in decimal figures).

If we are not interested in what we will visualize if we have unexpected combinations (the second approach), we place the symbol "-" meaning "don't-care") on the table and the maps corresponding to the outputs related to nonsignificant inputs. This symbol will then be treated in the synthesis randomly as either 0 or 1, in order to get the most economical implementation of the device. This is, therefore, the decoder's truth table:

D	C	B	A	a	b	c	d	e	f	g	Dec
0	0	0	0	1	1	1	1	1	1	0	0
0	0	0	1	0	1	1	0	0	0	0	1
0	0	1	0	1	1	0	1	1	0	1	2
0	0	1	1	1	1	1	1	0	0	1	3
0	1	0	0	0	1	1	0	0	1	1	4
0	1	0	1	1	0	1	1	0	1	1	5
0	1	1	0	1	0	1	1	1	1	1	6
0	1	1	1	1	1	1	0	0	0	0	7
1	0	0	0	1	1	1	1	1	1	1	8
1	0	0	1	1	1	1	1	0	1	1	9
1	0	1	0	–	–	–	–	–	–	–	–
1	0	1	1	–	–	–	–	–	–	–	–
1	1	0	0	–	–	–	–	–	–	–	–
1	1	0	1	–	–	–	–	–	–	–	–
1	1	1	0	–	–	–	–	–	–	–	–
1	1	1	1	–	–	–	–	–	–	–	–

The "don't-cares" allow us to minimize the network more efficiently since we can make them into 0s or 1s in order to choose the largest subcubes.

For example, we can choose among four subcubes for output "a", as below:

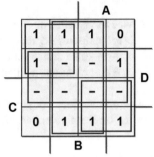

We get: $a = B + D + \overline{A}\,\overline{C} + A\,C$.

2.6.6 Using Multiplexers to Synthesize Combinational Networks

An interesting way to create combinational networks without going to the synthesis procedure is to use a multiplexer. This method has the advantage of allowing the network to be re-configured, even while it is working, when it is matched with memorization systems (which we have not seen yet).

We use multiplexers like those we studied earlier that allow us to generate the desired function by selecting the values of the function itself as if we were *reading* them from the truth table. We need to have the truth table of the function. If, however, we start from the Boolean expression, we must obtain the truth table from it.

Example 1:

Synthesize with a multiplexer a function $f(A, B, C)$, which is expressed as:

$$G = \overline{B}\,C + A\,B\,\overline{C} + \overline{A}\,B\,C.$$

Let's get the truth table:

A	B	C	G
0	0	0	0
0	0	1	1
0	1	0	0
0	1	1	1
1	0	0	0
1	0	1	1
1	1	0	1
1	1	1	0

As shown in the next figure (left), we prepare a network based on a multiplexer $8 \rightarrow 1$ connecting input variables A, B, and C to the selection inputs.

Thus (same figure, right), we connect the eight inputs $I0..I7$ of the multiplexer to constants 0 and 1 in the same order as they appear in the truth table.

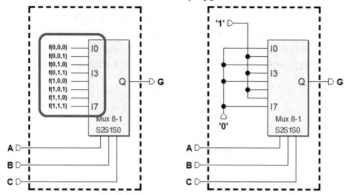

The network is ready: for each combination of inputs A, B, and C, the multiplexer will transfer the value we have set as input over to the output.

Example 2:

The function is: $H = A\,\overline{B}\,D + \overline{A}\,B\,D + B\,C\,\overline{D} + \overline{A}\,C\,\overline{D}$.

There are four inputs (A, B, C, and D), but we want to use a multiplexer with just three selection lines. The trick is to condition the eight inputs $I0..I7$ to the value of input D.

Let's derive the truth table from the expression: its farthest right column reports the value of H as a function of D taken from the same table.

The figure on the right shows the resulting network when the constants and the inputs D and $\overline{D}$ are connected:

A B C D	H	H
0 0 0 0	0	
0 0 0 1	0	0
0 0 1 0	1	
0 0 1 1	0	$\overline{D}$
0 1 0 0	0	
0 1 0 1	1	D
0 1 1 0	1	
0 1 1 1	1	$1 \Rightarrow$
1 0 0 0	0	
1 0 0 1	1	D
1 0 1 0	0	
1 0 1 1	1	D
1 1 0 0	0	
1 1 0 1	0	0
1 1 1 0	1	
1 1 1 1	0	$\overline{D}$

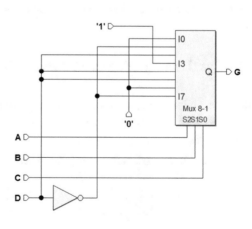

2.7   Variable-Entered Maps

Now, let's discuss a method of synthesis using Karnaugh maps that contain variables. This method will be used (in Chap. 7) to synthesize Finite State Machines (FSMs) with the ASM method, since applying the rules of synthesis will directly produce this type of map. The figure below shows an example of a map with entered variables.

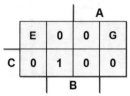

Aside from variables A, B, and C, shown around the map, let's look at variables E and G shown inside two of the cells. This map shows a five-variable function $f(A, B, C, E, G)$. Writing a certain variable (or an entire expression) inside a cell means conditioning the result of the function to that variable or expression.

In the example here, the upper left-hand cell corresponds to the minterm $\overline{A}\,\overline{B}\,\overline{C}$; however, since the variable is in the cell, the term must be conditioned to E, giving: $\overline{A}\,\overline{B}\,\overline{C}\,E$.

In the upper right-hand cell, we condition the corresponding minterm to the variable G, giving: $A\,\overline{B}\,\overline{C}\,G$. The resulting function is:

$$f = \overline{A}\,\overline{B}\,\overline{C}\,E + A\,\overline{B}\,\overline{C}\,G + \overline{A}\,B\,C$$

In this example, the function's expression cannot be minimized, but in general the problem of minimization exists. Now, let's look at one of the methods to obtain a minimum synthesis (or at least minimized as far as possible).

2.7.1   Synthesizing Maps with Entered Variables

Here, we examine a method for synthesizing maps with entered variables. In the example, we find variable $X3$ in a cell and its negated form in another. We take the following steps:

1. We reduce all the variables on the map to zero and synthesize the 1s.

We obtain the first partial result. $f' = X_1\,\overline{X_2}$.

2. We turn the 1s synthesized in step 1 into don't-care terms. We then choose the cells containing the same variable, making them equal to 1, leaving the remaining ones at 0. We consider the variable and its negated value to be two independent variables to be considered one by one in two different steps. We then synthesize this map and place the result in AND with the chosen variable.

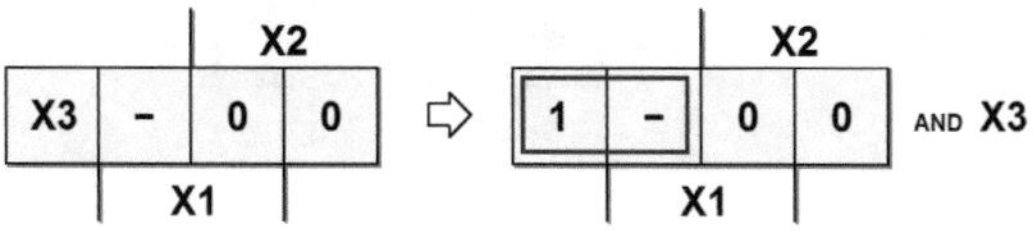

We write the second partial result as: $f'' = \overline{X_2}\, X_3$.

We repeat step 2 for all the other variables on the map. When this cycle of steps is finished, the equivalent expression is given by the OR of all the AND expressions found. In the example, step 2 is repeated once more:

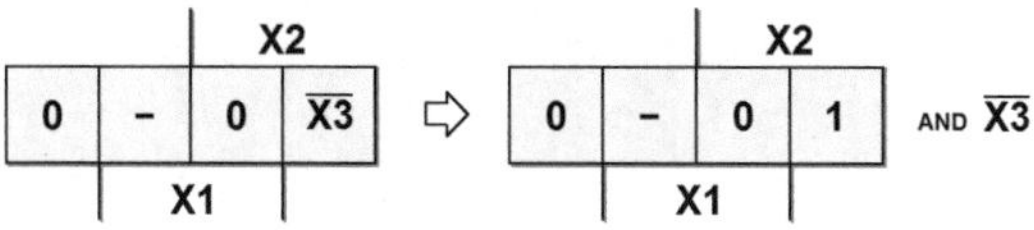

From which we derive the last partial result: $f''' = \overline{X_1}\, X_2\, \overline{X_3}$, so the final expression is:

$$f = f' + f'' + f''' = X_1\,\overline{X_2} + \overline{X_2}\,X_3 + \overline{X_1}\,X_2\,\overline{X_3}$$

Example 1: Given the following map, derive the corresponding function.

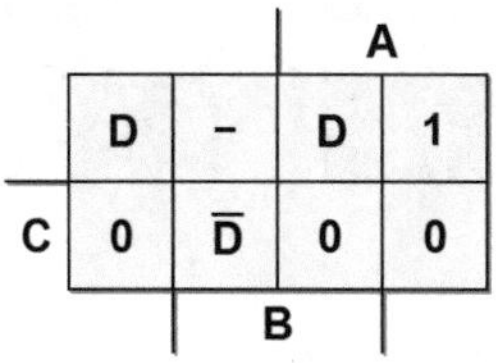

1. Set the cells containing D and $\overline{D}$ to 0 and synthesize the remaining 1s. Note that the don't-care terms remain the same and could be used to minimize the map, although it would not help in this example.

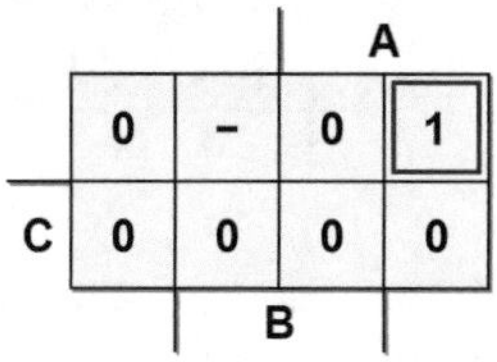

We obtain the expression: $f' = A\,\overline{B}\,\overline{C}$.

2. We substitute the only 1 on the map with a don't-care term, set the cells containing D at 1, and set the cells with $\overline{D}$ at 0.

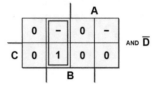

This way, we obtain the synthesis $\overline{C}$, which is placed in AND with D, so the second term of the desired expression is: $f'' = \overline{C} \, D$.

3. We repeat step 2 for the cells with $\overline{D}$, as if this were an independent variable from D, by setting the cells with $\overline{D}$ to 1 and those with D to 0.

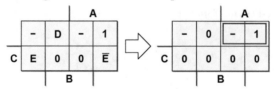

From this map, we derive the term $\overline{A} \, B$. When this is placed in AND with $\overline{D}$, it provides the last term of the desired expression: $f''' = \overline{A} \, B \, \overline{D}$.

4. The final expression is: $f = f' + f'' + f''' = A \, \overline{B} \, \overline{C} + \overline{C} \, D + \overline{A} \, B \, \overline{D}$.

Example 2: Deriving the function from a map with two entered variables:

After the first step, we get $A \, \overline{C}$. Note that an equally valid synthesis of step 1 could be $\overline{B} \, \overline{C}$. Let's start to do step 2, by inserting 1 in the cells with D, and 0 in those with E e $\overline{E}$:

The synthesis of the map gives $\overline{C}$, so the desired term is $\overline{C} \, D$. Notice how only the considered variable enters into the term, while the others appear neither in direct nor in negated form.

Let's do step 2 again, this time for E: we reduce cells D and $\overline{E}$ to zero:

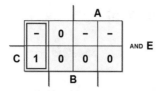

Synthesizing the map provides the term $\overline{A}\,\overline{B}$, hence $\overline{A}\,\overline{B}\,E$. Finally, we do step 1 once again for cells containing $\overline{E}$, reducing those with D and E to zero:

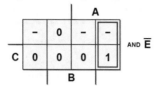

We obtain the term $A\,\overline{B}$, which gives us $A\,\overline{B}\,\overline{E}$. The final expression is:

$$f = A\,\overline{C} + \overline{C}\,D + \overline{A}\,\overline{B}\,E + A\,\overline{B}\,\overline{E}$$

2.7.2 Entered Variables and Theorems of Expansion

A map with entered variables could be considered a hybrid, a middle ground between representing logical networks through their Boolean expression and doing it through their Karnaugh map.

The "classic" K-map that contains only 0s and 1s is none other than a map with entered variables that happens to lack entered variables. If we enter all the variables of a map, what we obtain in the only remaining cell is the Boolean expression of the network. It would be possible to demonstrate, by using Shannon's expansion theorems, how maps with entered variables can result from the "compression" of ordinary Karnaugh maps and vice versa.

It is interesting to look again at the map of the two-input multiplexer and compare it with the map of the same network with entered variables, obtained by compressing the K-map by entering the variables $S1$ and $S2$.

It would have been much simpler to create the map with entered variables by means of the multiplexer's specifications: if $SEL = 0$, the output is $S1$, and if $SEL = 1$ the output is $S2$. In fact, the variable-entered map can be seen as a synthetic tool for describing combinational networks.

2.8   Time Behavior of Combinational Networks

2.8.1   *Definitions and Timing Models*

A logical gate introduces a delay between the change in value of the input and that of the output. In the figure below, the input and output signals of a NOT are represented in idealized form, since the transitions between logical values are instantaneous.

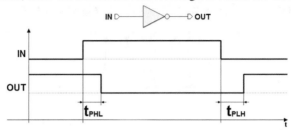

t_{PHL} :   propagation time (high to low)

t_{PLH} :   propagation time (low to high)

The delays in this form are called **transport delays**. This idealized model makes them quick and easy to read.

For simplicity's sake, we could denote propagation times generically as:

$$t_p = \max(t_{PLH}, t_{PLH})$$

When necessary, it is useful to adopt a more realistic signal model where level transitions (edges) do not occur instantly (in 0 time) but in finite times with linear progression. We use this model to describe **inertial delays**:

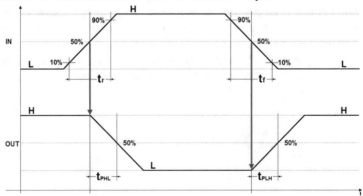

With this model, we assume that the logical level transitions occur when the signal reaches (rises to or lowers to) 50% of its excursion.

Propagation times t_{PHL} and t_{PLH} are the times between the signal-level transition at input and that at output.

The rise time t_r is the time interval between 10 and 90% of the signal excursion from low to high. In a similar way, the fall time t_f is defined as the time interval between 10 and 90% of the signal excursion from high to low. Make sure not to confuse rise and fall times with propagation times.

Representation through *transport delays* requires only the translation of the output signal over the input signal over time and does not explain the fact that in reality, signals whose duration falls under a certain threshold are not propagated.

The inertial model, however, does allow us to represent this behavior. Let's look at the following logical circuit.

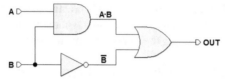

If we analyze the circuit with a digital circuit simulator set to calculate only the *transport delays*, we get:

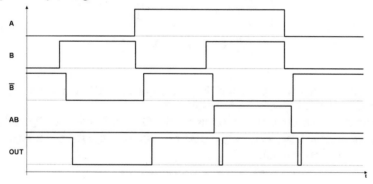

Thus, based on this model, there should be two short pulses on the OUT output due to the delay differences along the signal paths and the difference between times t_{PHL} and t_{PLH}.

If we repeat the simulation using the *inertial delay* model, we get:

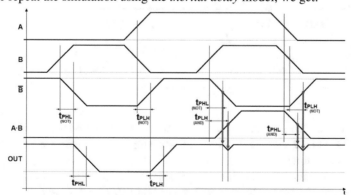

In this new, more realistic simulation, we see that the two pulses are present but will not be propagated since their amplitude fails to reach the threshold due to the finite slope of the transitions.

2.8.2   Hazards

As we have seen, combinational networks can give rise to impulsive behavior due to differences in delays. These behaviors are called ***hazards***. Depending on the physical component's technology, these behaviors might be mitigated or removed by the inertial behavior of the circuits. In any case, these phenomena are potentially damaging due to their effect on the circuits that receive them and they should be avoided when they can cause errors.

The hazards that arise due to asymmetrical delay paths, due in turn to the presence of inverters or other gates, are called ***static hazards***. They can generally be eliminated (or *masked*) through algebraic methods like the one explained later.

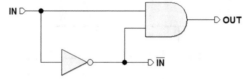

In the figure above, a simple network made up of an AND gate and a NOT gate allows us to observe the effect of different paths.

This network should generate a constant ($OUT = IN \cdot \overline{IN} = 0$). To perform a simplified time analysis, let us only consider *transport delays* so that the hazard will not be masked by the inertial delay.

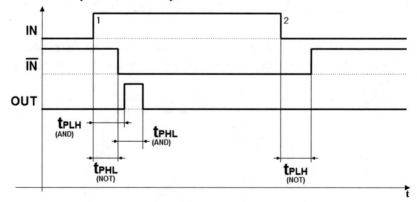

The figure above shows that the output is not always 0 but produces a pulse at 1 (the hazard). Due to the delay of the NOT, the AND inputs are together at 1 for a certain, small period of time.

To be thorough, let's examine another simple circuit based on an OR gate. The output should be constant here as well (given that $OUT = IN + \overline{IN} = 1$):

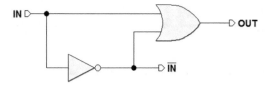

The time analysis using the same criteria as above also shows a hazard (in the opposite transition) in this case as well. Timing analysis, executed as in the case before, shows the presence of a hazard, this time on the falling edge of the input transition.

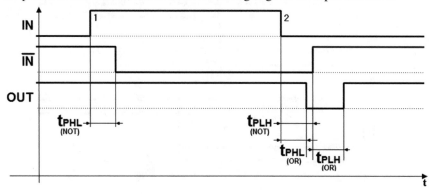

In AND-OR networks, hazards generally appear as brief transitions to 0 of signals that should be stable at 1. In OR-AND networks, we see signals that transition to 1 when they should be at 0.

2.8.3 Elimination of Static Hazards

In the following two-level AND-OR circuit (the $2 \rightarrow 1$ multiplexer), there is a hazard in the transition $1 \rightarrow 0$ of C when inputs A and B equal 1:

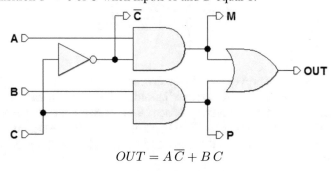

$$OUT = A\,\overline{C} + B\,C$$

Time simulation shows that the hazard occurs since, for a short time, the two logical products $A\,\overline{C}$ e $B\,C$ will be reduced to zero due to the delay induced by NOT.

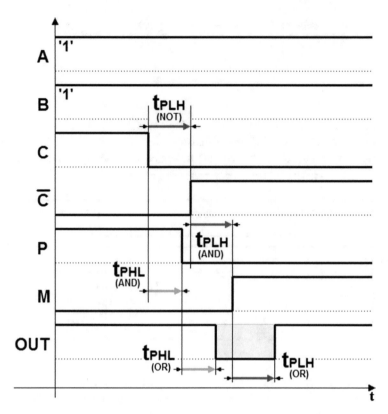

If we look at the Karnaugh map, we see that the hazard develops in the transition indicated by the arrow (i.e., when C goes from 1 to 0, while $A = 1$ and $B = 1$).

There is a theorem (not proven here) that affirms: a two-level combinational network synthesized as a sum of products is free of hazards if there are no hazards in its $1 \rightarrow 1$ transitions (from one set of inputs that produce 1 to another set of inputs that produce 1).

In a two-level AND-OR combinational network, there can be a hazard when there is a pair of 1s near each other on the map that does not belong to the same implicant, as in the map above.

Let's check for a hazard in this network since there are two adjacent 1s in different implicants. To eliminate the hazard, we must add the implicant that contains these two 1s (term $A\,B$):

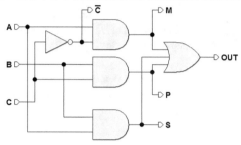

The result is: $OUT = A\,\overline{C} + B\,C + A\,B$. See the schematic of the new network below:

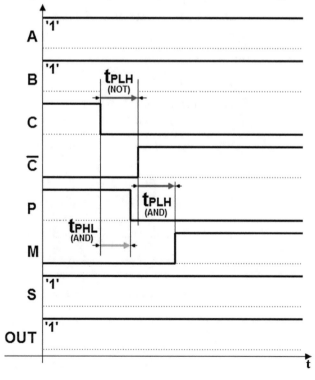

This is no longer a minimal synthesis, but the added product term masks (or *covers*) the hazard, which disappears from the time simulation.

In fact, in the short time in which the two logical products $A\,\overline{C}$ e $B\,C$ go to zero, the term $A\,B$ keeps a 1 in the input of OR, as we can see in the figure below:

In the OR-AND networks, the situation is symmetrical: the hazard occurs in $0 \rightarrow 0$ transitions and is eliminated by adding the implicants containing two nearby zeros that are not contained in the same subcube.

2.8.4 Notes on Eliminating Hazards

Generally, to eliminate static hazards, we systematically add nonessential implicants in all the situations that can produce hazards. However, let's note that eliminating hazards is necessary only when they are damaging, typically in *asynchronous sequential circuits*.

Hazards can generally be tolerated in *synchronous sequential circuits*, in which the signals are read in well-defined times where possible hazards cannot occur. We will see more on this later in the book.

2.9 Exercises

2.9.1 Maps

1. Draw the truth tables and maps for the following functions:

   (a) $G = A\,B\,C + B\,\overline{C}$
   (b) $H = (A + \overline{B})(B + \overline{C})$

2. Construct the maps of the following functions:

   (a) $F = \overline{A}\,\overline{B}\,D + \overline{A}\,B\,C + A\,B\,D + A\,\overline{B}\,\overline{C}\,\overline{D} + \overline{A}\,B\,\overline{C}\,D$
   (b) $M = (\overline{A} + C)(\overline{A} + \overline{C} + D)(\overline{A} + B)(A + B + \overline{C} + D)$

3. Minimize the logical functions in the maps below as sums of products.

   (a) F1:

	A		
1	1	1	0
1	1	1	0
0	0	1	0
0	1	1	0

with C on the left side, D on the right, B at the bottom.

(b) F2:

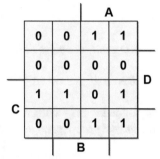

(c) F3:

(d) F4:

4. Minimize the function $F = \overline{B}\,C\,\overline{D} + \overline{A}\,B\,\overline{C}\,\overline{D} + A\,\overline{B}\,\overline{D}$ as a sum of products, keeping in mind that inputs $ABCD = $ "$11--$" and $ABCD = $ "$--11$" are never present (combinations $A = B = 1$ or $C = D = 1$ can never arise).

5. Synthesize the logical function in the map below as a sum of products.

7. Synthesize the following map, which contains don't-cares.

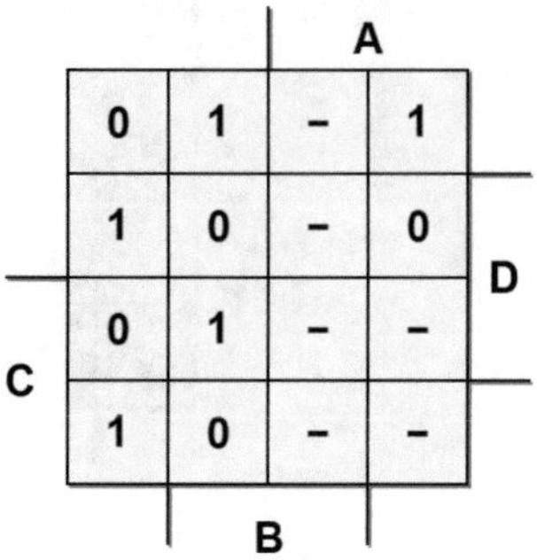

		A	
0	1	–	1
1	0	–	0
0	1	–	–
1	0	–	–

(rows labeled C on left, columns B at bottom, D on right)

8. Using only NAND gates, draw the simplest circuit that generates the function:

$$F = (A\,B + \overline{A}\,\overline{B})\,C\,D + (A\,\overline{C} + \overline{A}\,C)\,B\,D + \overline{A}\,C\,D + A\,\overline{B}\,C\,D.$$

9. Synthesize the following map with entered variables (group only the don't-cares that give us the largest groupings; ignore those that would add groups).

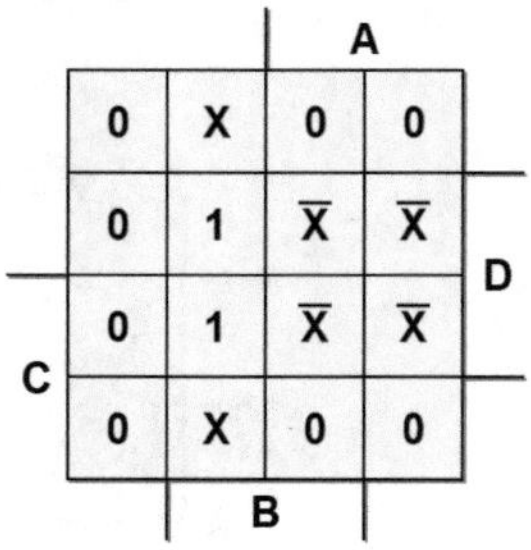

		A	
0	X	0	0
0	1	$\overline{X}$	$\overline{X}$
0	1	$\overline{X}$	$\overline{X}$
0	X	0	0

(rows labeled C on left, columns B at bottom, D on right)

2.9.2   Hazards

1. Derive the minimal synthesis from the following map:

    (a) Ignoring the hazards;
    (b) Eliminating the hazards.

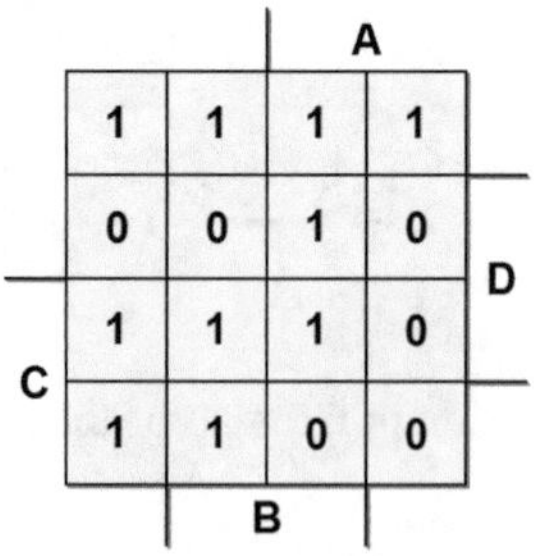

		A	
1	1	1	1
0	0	1	0
1	1	1	0
1	1	0	0

(rows labeled C on left, columns B at bottom, D on right)

2. Identify the hazards in the circuit and eliminate them.

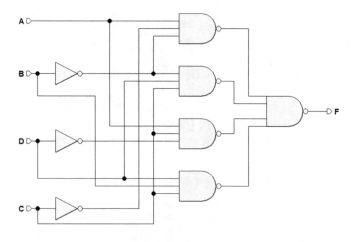

2.10   Solutions

2.10.1   *Maps*

1. (a) $G = A\,B\,C + B\,\overline{C}$

A	B	C	G
0	0	0	0
0	0	1	0
0	1	0	1
0	1	1	0
1	0	0	0
1	0	1	0
1	1	0	1
1	1	1	1

		A	
0	1	1	0
C 0	0	1	0
	B		

(b) $H = (A + \overline{B})(B + \overline{C})$

A	B	C	H
0	0	0	1
0	0	1	0
0	1	0	0
0	1	1	0
1	0	0	1
1	0	1	0
1	1	0	1
1	1	1	1

		A	
1	0	1	1
C 0	0	1	0
	B		

2. (a) $F = \overline{A}\,\overline{B}\,D + \overline{A}\,B\,C + A\,B\,D + A\,\overline{B}\,\overline{C}\,\overline{D} + \overline{A}\,B\,\overline{C}\,D$

		A		
0	0	0	1	
1	1	1	0	D
1	1	1	0	
0	1	0	0	
		B		

(b) $M = (\overline{A} + C)(\overline{A} + \overline{C} + D)(\overline{A} + B)(A + B + \overline{C} + D)$

		A		
1	1	0	0	
1	1	0	0	D
1	1	0	0	
0	1	1	0	
		B		

3. (a) $F1 = \overline{A}\,\overline{C} + A\,B + B\,\overline{D}$:

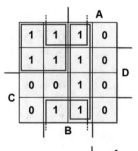

(b) $F2 = \overline{B}\,\overline{C} + A\,B\,D + B\,C\,\overline{D}$:

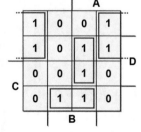

(c) $F3 = A\overline{D} + \overline{B}CD + \overline{A}CD$:

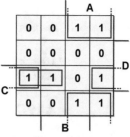

(d) $F4 = \overline{B}\,\overline{C}\,\overline{D} + AB\overline{D} + BCD$:

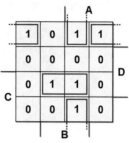

4. $F = A\overline{D} + \overline{B}C + B\overline{C}\overline{D}$:

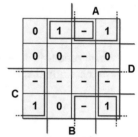

5. These are the three order-2 subcubes for AND-OR synthesis:

We obtain the expression: $F = \overline{A} + \overline{B} + \overline{C}$.

6. For OR-AND synthesis, we just need to consider the only 0.

We derive the same expression as in the previous exercise, but using only one grouping.

7. $F = \overline{A}\,\overline{B}\,\overline{C}\,D + \overline{B}\,C\,\overline{D} + B\,\overline{C}\,\overline{D} + B\,C\,D + A\,\overline{D}$

8. We expand the given expression in terms of minterms.

$$F = (AB + \overline{A}\,\overline{B})\,CD + (A\,\overline{C} + \overline{A}\,C)\,BD + \overline{A}\,CD + A\,\overline{B}\,CD =$$
$$= ABCD + \overline{A}\,\overline{B}\,CD + AB\,\overline{C}\,D +$$
$$+\overline{A}\,BCD + \overline{A}\,(B + \overline{B})\,CD + A\,\overline{B}\,CD =$$
$$= ABCD + \overline{A}\,\overline{B}\,CD + A\,B\,\overline{C}\,D + \overline{A}\,BCD + A\,\overline{B}\,CD$$

The resulting map is:

		A	
0	0	0	0
0	0	1	0
1	1	1	1
0	0	0	0

which produces the minimized expression: $F = C\,D + A\,B\,C$.
We transform this into NAND-NAND with De Morgan's Theorem.

$$F = \overline{\overline{C\,D} \cdot \overline{A\,B\,C}}$$

This is the network:

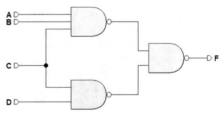

9. With the method, we studied for Variable-Entered Maps, and we have:

$$F = \overline{A}\,B\,D + \overline{A}\,B\,X + A\,D\,\overline{X}$$

2.10.2 Hazards

1. Synthesis of a function considering the hazards issue:

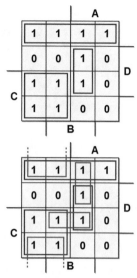

a) With hazards:
$\overline{C}\,\overline{D} + \overline{A}\,C + A\,B\,D$:

b) With hazards "*covered*":
$\overline{C}\,\overline{D} + \overline{A}\,C + A\,B\,D+$
$\overline{A}\,\overline{D} + B\,C\,D + A\,B\,\overline{C}$:

2. Keeping NAND-NAND network equivalence in mind, we derive the expression of the AND-OR network, and from this, we get the map.

$$F = A\,\overline{B}\,\overline{C} + \overline{B}\,C\,D + A\,C\,\overline{D} + B\,C\,D$$

			A
0	0	0	1
0	0	0	1
1	1	1	1
0	0	1	1

We group the map for minimal synthesis.

			A
0	0	0	1
0	0	0	1
1	1	1	1
0	0	1	1

On the map, all the adjacent 1s are masked, so the minimal synthesis is already free of hazards. See its expression and the logical network below.

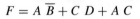

$$F = A \, \overline{B} + C \, D + A \, C$$

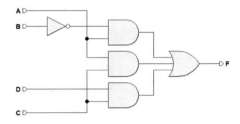

Chapter 3
Numeral Systems and Binary Arithmetic

Abstract The representation of numbers is essential for the digital logic design. In this chapter, positional number systems (decimal, binary, octal, hexadecimal), BCD and Gray codes are presented together with the rules for the conversion between numbers encoded in different bases and the representations of negative numbers. Then, the rules for the arithmetic operations and the circuits that execute them are presented. The addition of binary number is examined with particular attention, since it is the operation at the basis of all computational circuits. Alphanumeric codes and the concept of parity for error detection complete the chapter.

3.1 Binary Information

The basic unit of binary information is called a *bit* (binary digit). The bit can only assume the value of 0 or 1.

Normally, bits are grouped into meaningful sets called *nibbles* (4 bits), *bytes* (8 bits), or generically *words* if they contain more bits (e.g., 16, 32, or 64 bits):

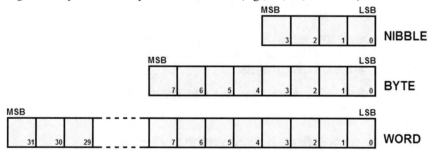

© Springer International Publishing AG, part of Springer Nature 2019
G. Donzellini et al., *Introduction to Digital Systems Design*,
https://doi.org/10.1007/978-3-319-92804-3_3

In this chapter, we will see how to use these sets of bits to represent numbers or other types of information through standard codes.

In the previous figure, the abbreviation MSB refers to the *Most Significant Bit* and LSB refers to the *Least Significant Bit*, when these sets are used to codify numbers.

3.2   Binary Numbering System (BIN)

Like the decimal system, binary numbering is *positional*. Positional notation makes it possible to express a number N as:

$$N = n_{m-1} \cdot R^{m-1} + \cdots + n_1 \cdot R^1 + n_0 \cdot R^0$$

where R is the *base* of the system; m is the number of digits the representation is composed of; the exponents denote the *position k* of each digit from 0 to $m-1$ starting from the right; the set $A = \{0, \ldots, R-1\}$ is the *alphabet* of the numbering system and is made up of *symbols* we use to represent numbers; coefficients n are the numbers that correspond to the value of the symbols in set A.

In the decimal system, the alphabet of symbols is $A = \{0, 1 \ldots, 9\}$ and, obviously, $R = 10$. When we write a number N, for example, 2017, we mean:

$$2017_{10} = 2 \cdot 10^3 + 0 \cdot 10^2 + 1 \cdot 10^1 + 7 \cdot 10^0$$

Let's refer to the *binary numbering system* as that system that has base $R = 2$ and alphabet $A = \{0, 1\}$. For example, the number 1110 in base two corresponds to the number 14 in base ten:

$$1110_2 = 1 \cdot 2^3 + 1 \cdot 2^2 + 1 \cdot 2^1 + 0 \cdot 2^0 = 8 + 4 + 2 + 0 = 14_{10}$$

Finally, let's denote *weight* as:

$$p = R^k$$

In the decimal system, weight refers to units, decimals, hundreds, thousands, etc., (powers of 10). In the binary numbering system, magnitudes have the power of base $R = 2$: 1, 2, 4, 8, 16, 32, 64, 128, etc.

Note: For a short guide about powers of 2, refer to Appendix A.

3.2.1   Converting from Binary System to Decimal

We obtain this conversion by directly applying the definition above. Here are examples for 4 and 8 digits:

$$0000_2 = 0 \cdot 2^3 + 0 \cdot 2^2 + 0 \cdot 2^1 + 0 \cdot 2^0 = 0_{10}$$

$$0011_2 = 0 \cdot 2^3 + 0 \cdot 2^2 + 1 \cdot 2^1 + 1 \cdot 2^0 = 3_{10}$$

$$1111_2 = 1 \cdot 2^3 + 1 \cdot 2^2 + 1 \cdot 2^1 + 1 \cdot 2^0 = 15_{10}$$

$$00000010_2 = 0 \cdot 2^7 + 0 \cdot 2^6 + 0 \cdot 2^5 + 0 \cdot 2^4 + 0 \cdot 2^3 + 0 \cdot 2^2 + 1 \cdot 2^1 + 0 \cdot 2^0 = 2_{10}$$

$$10001110_2 = 1 \cdot 2^7 + 0 \cdot 2^6 + 0 \cdot 2^5 + 0 \cdot 2^4 + 1 \cdot 2^3 + 1 \cdot 2^2 + 1 \cdot 2^1 + 0 \cdot 2^0 = 142_{10}$$

$$11110011_2 = 1 \cdot 2^7 + 1 \cdot 2^6 + 1 \cdot 2^5 + 1 \cdot 2^4 + 0 \cdot 2^3 + 0 \cdot 2^2 + 1 \cdot 2^1 + 1 \cdot 2^0 = 243_{10}$$

$$11111111_2 = 1 \cdot 2^7 + 1 \cdot 2^6 + 1 \cdot 2^5 + 1 \cdot 2^4 + 1 \cdot 2^3 + 1 \cdot 2^2 + 1 \cdot 2^1 + 1 \cdot 2^0 = 255_{10}$$

3.2.2 Converting from Decimal System to Binary

A natural number N (with $N \geq 2$) can be expressed as:

$$N = 2 \cdot q_0 + r_0 \quad \text{with} \quad r_0 = (0, 1)$$

where q_0 is the quotient and r_0 is the remainder of the division of N by 2. We can repeat the process if $q_0 \geq 2$, by writing that:

$$q_0 = 2 \cdot q_1 + r_1 \quad \text{with} \quad r_1 = (0, 1)$$

By substituting the latter expression in the former one, we obtain:

$$N = 2 \cdot (2 \cdot q_1 + r_1) + r_0$$

Continuing, if $q_1 \geq 2$, we derive:

$$N = 2 \cdot (2 \cdot (2 \cdot q_2 + r_2) + r_1) + r_0 = 2^3 \cdot q_2 + 2^2 \cdot r_2 + 2^1 \cdot r_1 + 2^0 \cdot r_0$$

and so on as long as $q_k \geq 2$, giving us:

$$N = 2^k \cdot q_{k-1} + 2^{k-1} \cdot r_{k-1} + \ldots + 2^1 \cdot r_1 + 2^0 \cdot r_0.$$

In other words, the remainders r_i from divisions by 2 represent the binary digits of N. Consider $N_{10} = 46$, for example. Let's divide by 2 recursively, writing the remainder at the right of the line and the result below the number.

$$
\begin{array}{r|ll}
 & 46 & 0 & (r_0) \\
(q_0) & 23 & 1 & (r_1) \\
(q_1) & 11 & 1 & (r_2) \\
(q_2) & 5 & 1 & (r_3) \\
(q_3) & 2 & 0 & (r_4) \\
(q_4) & 1 & 1 & (q_4)
\end{array}
$$

We get the result of the conversion by writing in order, from the left, q_4 followed by the remainders $r_4, r_3, ..., r_0$:

$$N = 46_{10} = 101110_2$$

Proof

$$
\begin{aligned}
N &= q_4 \cdot 2^5 + r_4 \cdot 2^4 + r_3 \cdot 2^3 + r_2 \cdot 2^2 + r_1 \cdot 2^1 + r_0 \cdot 2^0 = \\
&= 1 \cdot 32 + 0 \cdot 16 + 1 \cdot 8 + 1 \cdot 4 + 1 \cdot 2 + 0 \cdot 1 = 46_{10}
\end{aligned}
$$

As an example, let's convert some decimal numbers into the binary system.

$$
258_{10}: \quad
\begin{array}{r|l}
258 & 0 \\
129 & 1 \\
64 & 0 \\
32 & 0 \\
16 & 0 \\
8 & 0 \\
4 & 0 \\
2 & 0 \\
1 & 1 \rightarrow 100000010_2
\end{array}
\qquad\qquad
237_{10}: \quad
\begin{array}{r|l}
237 & 1 \\
118 & 0 \\
59 & 1 \\
29 & 1 \\
14 & 0 \\
7 & 1 \\
3 & 1 \\
1 & 1 \rightarrow 11101101_2
\end{array}
$$

3.2.3 Maximum Representable Number

Given a set of m bits, the largest natural number that can be represented is:

$$N_{max} = 2^m - 1$$

For example, if we consider a byte ($m = 8$), where every bit equals 1, we will add all the magnitudes present among them.

$$
\begin{aligned}
11111111_2 &= 128 + 64 + 32 + 16 + 8 + 4 + 2 + 1 = \\
&= 255 = 256 - 1 = \\
&= 2^8 - 1
\end{aligned}
$$

3.3 Octal Number System (OCT)

This is a base $R = 8$ positional numbering system with alphabet $A = \{0, 1, 2, 3, 4, 5, 6, 7\}$. It is used to represent binary numbers but it is rarely used today, being replaced by the hexadecimal system. Nevertheless, it is still useful to be acquainted with it. Below is an example:

$$123_8 = 1 \cdot 8^2 + 2 \cdot 8^1 + 3 \cdot 8^0 = 64 + 16 + 3 = 83_{10}$$

The binary–octal conversion table, for a single octal digit:

BIN	OCT
000	0
001	1
010	2
011	3
100	4
101	5
110	6
111	7

Since the octal number system's base is a power of 2, the conversion from binary to octal is immediate.

Converting any binary number into octal requires just dividing it into groups of three digits (starting from the right) and replacing each with the corresponding octal digit.

See the examples below:

$$100100_2 = |\,100_2\,|\,100_2\,| = 44_8$$
$$111111_2 = |\,111_2\,|\,111_2\,| = 77_8$$
$$11001_2 = |\,011_2\,|\,001_2\,| = 31_8$$
$$1101_2 = |\,001_2\,|\,101_2\,| = 15_8$$

An integer decimal number can be converted into octal by using the method of repeated division seen above for decimal to binary conversion, in this case dividing by 8.

In this example we convert the number 267_{10} into octal:

$$
\begin{array}{r|l}
267 & 3 \quad (267 : 8 = 33 + 3) \\
33 & 1 \quad (33 : 8 = 4 + 1) \\
4 & 4 \quad (4 : 8 = 0 + 4)
\end{array}
$$

The result is: 413_8.

3.4   Hexadecimal Number System (HEX)

This is a positional number system with base $R = 16$. Its alphabet is:

$$A = \{0, 1, 2, 3, 4, 5, 6, 7, 8, 9, A, B, C, D, E, F\}.$$

To represent digits higher than 9 we use the first six letters of the Latin alphabet since they are available on a typical keyboard. A DEC $-$ BIN $-$ HEX conversion table is as follows.

DEC	BIN	HEX
0	0000	0
1	0001	1
2	0010	2
3	0011	3
4	0100	4
5	0101	5
6	0110	6
7	0111	7
8	1000	8
9	1001	9
10	1010	A
11	1011	B
12	1100	C
13	1101	D
14	1110	E
15	1111	F

Since the hexadecimal number system is also based on the power of 2, the BIN–HEX conversion is immediate. We replace every group of four binary digits with the corresponding HEX digits. Here are some examples:

$$10001111_2 = | \, 1000_2 \, | \, 1111_2 \, | = 8F_{16} = 8F_H$$
$$11100011_2 = | \, 1110_2 \, | \, 0011_2 \, | = E3_{16} = E3_H$$
$$100101_2 \ = | \, 0010_2 \, | \, 0101_2 \, | = 25_{16} = 25_H$$
$$11100_2 \ \ = | \, 0001_2 \, | \, 1100_2 \, | = 1C_{16} = 1C_H$$

In this case the conversion can also be done with repeated division, by dividing by 16. In the following example, the number 655_{10} is converted into hexadecimal.

$$
\begin{array}{r|l l}
655 & 15 & (655 : 16 = 40 + 15) \\
40 & 8 & (40 : 16 = 2 + 8) \\
2 & 2 & (2 : 16 = 0 + 2)
\end{array}
$$

The result is: $28F_H$.

To get familiar with hexadecimal numbers, let's look at some examples of column addition with two numbers. Note that when a digit is over 15 (not 9 as in the decimal system), we carry it over one column.:

$$
\begin{array}{ccccc}
28_H + & 22_H + & 40_H + & 0F_H + & 3A_H + \\
33_H = & AA_H = & 51_H = & 0F_H = & 9B_H = \\
\hline
5B_H & CC_H & 91_H & 1E_H & D5_H
\end{array}
$$

3.5 Others Binary Codes

3.5.1 Binary Coded Decimal

The most natural way to represent a number in a binary format is to use the pure binary numbering system. However, the downside of this is that decimal-to-binary and binary-to-decimal conversion becomes more taxing as the numbers get bigger.

To make the conversion easier, we can represent the digits of a decimal number one by one, using the so-called Binary Coded Decimal (BCD) codes. They code the decimal numbers' digits one by one using groups of four bits. Obviously, their arithmetic properties are less than those of the pure binary number system.

In general, binary codes are said to be *weighted* if every digit has its own weight according to its position. Those in which every combination of digits is randomly associated to a certain number are called *non-weighted*.

Self-complementing codes are those where two numbers that add up to 9 are complements to one of each other (i.e., the 1s and 0s are interchanged).

Now, let's look at some of the most commonly used BCD codes:

- BCD 8421: It is a weighted code in which the digits' weights are 8, 4, 2, 1 from left to right; this is equal to pure binary terminated at 1001_2.
- BCD 5421: It is analogous to BCD 8421 but the weights are 5, 4, 2, 1.
- AIKEN 2421: It is a weighted, self-complementing BCD code with weights of 2, 4, 2, 1.

N	BCD 8421	BCD 5421	AIKEN 2421
0	0000	0000	0000
1	0001	0001	0001
2	0010	0010	0010
3	0011	0011	0011
4	0100	0100	0100
5	0101	1000	1011
6	0110	1001	1100
7	0111	1010	1101
8	1000	1011	1110
9	1001	1100	1111

- BCD XS3: It is a non-weighted, self-complementing code that is obtained by adding 3 (11 in binary) to BCD 8421; XS3 actually means "excess 3".

N	BCDXS3
0	0011
1	0100
2	0101
3	0110
4	0111
5	1000
6	1001
7	1010
8	1011
9	1100

3.5.2   GRAY Codes

The GRAY codes are non-weighted. They are characterized by the fact that each number differs by one single digit from the one that precedes it and the one that follows it. GRAY codes can be made with any number of bits. There are many types of GRAY codes. The one shown here is the "reflected" type.

N	GRAY
0	0000
1	0001
2	0011
3	0010
4	0110
5	0111
6	0101
7	0100
8	1100
9	1101
10	1111
11	1110
12	1010
13	1011
14	1001
15	1000

3.6 Binary Arithmetic

3.6.1 Addition

Consider the addition of two bits A and B. Let's evaluate the four possible cases:

$A +$	$0 +$	$0 +$	$1 +$	$1 +$
$B =$	$0 =$	$1 =$	$0 =$	$1 =$
sum	0	1	1	10_2

In the farthest right column, the result is obviously 2_{10} but it makes sense to interpret it as "result 0, with carry over 1". So, let's denote the bit of the sum as S and introduce the bit C_o (*Carry Out*), which represents the result. In cases where there is no carry, we will write it as 0.

A $+$	$0 +$	$0 +$	$1 +$	$1 +$
B $=$	$0 =$	$1 =$	$0 =$	$1 =$
C_o S	$0\ 0$	$0\ 1$	$0\ 1$	$1\ 0$

When we add binary numbers coded on more than one bit, the carry generated from one column must be added to the result from the column to its immediate left, as in the following examples:

$0010_2 +$	$0001_2 +$	$0011_2 +$
$0001_2 =$	$0001_2 =$	$0001_2 =$
0011_2	0010_2	0100_2

In the first example, there are no carries; in the second, there is one in the farthest right column; in the last, the two right-hand columns have carries.

Based on these observations, we define the *addition rules* for a given column by introducing the carry C_i (*Carry In*), which comes from the adjoining column.

C_i $+$	$0 +$	$0 +$	$0 +$	$0 +$
A $+$	$0 +$	$0 +$	$1 +$	$1 +$
B $=$	$0 =$	$1 =$	$0 =$	$1 =$
C_o S	$0\ 0$	$0\ 1$	$0\ 1$	$1\ 0$

$$
\begin{array}{ccccccccccc}
C_i & + & & 1 & + & & 1 & + & & 1 & + & & 1 & + \\
A & + & & 0 & + & & 0 & + & & 1 & + & & 1 & + \\
B & = & & 0 & = & & 1 & = & & 0 & = & & 1 & = \\
\hline \\
C_o\ S & & & 0\ 1 & & & 1\ 0 & & & 1\ 0 & & & 1\ 1 &
\end{array}
$$

See below some examples of binary number addition.

$$
\begin{array}{ccccc}
0001_2\ + & 0110_2\ + & 0111_2\ + & 01011111_2\ + & 01011110_2\ + \\
0111_2\ = & 0110_2\ = & 0111_2\ = & 00100101_2\ = & 00101111_2\ = \\
\hline \\
1000_2 & 1100_2 & 1110_2 & 10000100_2 & 10001101_2
\end{array}
$$

When calculating the sum of two numbers, the result could exceed the maximum representable number. If, for example, in our logical network we have only 4 bits available to code a number, the following sums generate a result that is too large.

$$
\begin{array}{cccc}
1111_2\ + & 0111_2\ + & 0101_2\ + & 1111_2\ + \\
0001_2\ = & 1011_2\ = & 1100_2\ = & 1111_2\ = \\
\hline \\
10000_2 & 10010_2 & 10001_2 & 11110_2
\end{array}
$$

To contain the result, we need 5 bits. An *overflow* error has occurred. After calculating a sum, we must always check for an overflow error, that is if anything has carried over from the column of the MSB. This rule holds for numbers without signs. Next, we will see how to check for overflow with numbers that can be positive or negative.

3.6.2 Subtraction

We define the *subtraction rules* by using criteria similar to those for addition. In this case, a column can *borrow* from the column to the left if it is necessary. Due to similarities with real circuits, we will use the same symbol used for carry. C_i is what is *borrowed* from the column on the right, while C_o is what the present column *borrows* from the left.

$$
\begin{array}{ccccccccccc}
C_i & - & & 0 & - & & 0 & - & & 0 & - & & 0 & - \\
A & - & & 0 & - & & 0 & - & & 1 & - & & 1 & - \\
B & = & & 0 & = & & 1 & = & & 0 & = & & 1 & = \\
\hline \\
C_o\ S & & & 0\ 0 & & & 1\ 1 & & & 1\ 1 & & & 1\ 0 &
\end{array}
$$

$$
\begin{array}{llllll}
C_i & - & 1 & - & 1 & - & 1 & - & 1 & - \\
A & - & 0 & - & 0 & - & 1 & - & 1 & - \\
B & = & 0 & = & 1 & = & 0 & = & 1 & = \\
\hline
C_o\,S & & 0\ 1 & & 0\ 0 & & 0\ 0 & & 1\ 1
\end{array}
$$

Here are some examples of subtraction:

$$
\begin{array}{lllll}
1001_2 - & 0111_2 - & 0110_2 - & 1111_2 - & 01011011_2 - \\
0011_2 = & 0101_2 = & 0001_2 = & 0111_2 = & 00100101_2 = \\
\hline
0110_2 & 0010_2 & 0101_2 & 1000_2 & 00110110_2
\end{array}
$$

3.6.3 Products

The *product rules* are as follows (we will not deal with division):

$$
0 \times 0 = 0
$$
$$
0 \times 1 = 0
$$
$$
1 \times 0 = 0
$$
$$
1 \times 1 = 1
$$

Here are some examples of products (they are carried out in the familiar way but with the rules seen here). Note that in the partial results of the operation, we either copy the multiplicand when it is multiplied by 1, or we write all 0s when it is multiplied by 0.

$$
\begin{array}{lll}
\begin{array}{r}
1\ 1 \times \\
1\ 1 = \\
\hline
1\ 1 \\
1\ 1 - \\
\hline
1\ 0\ 0\ 1
\end{array}
&
\begin{array}{r}
1\ 1 \times \\
1\ 0\ 1 = \\
\hline
1\ 1 \\
0\ 0 - \\
1\ 1 - - \\
\hline
1\ 1\ 1\ 1
\end{array}
&
\begin{array}{r}
1\ 1\ 0 \times \\
1\ 0\ 0 = \\
\hline
0\ 0\ 0 \\
0\ 0\ 0 - \\
1\ 1\ 0 - - \\
\hline
1\ 1\ 0\ 0\ 0
\end{array}
\end{array}
$$

3.7 BCD 8421 Arithmetic

When adding in BCD 8421 arithmetic, we keep in mind that every group of 4 bits codes a decimal number so the rules of carries should be the same as with decimal representation. In the following example, when calculating in decimal on the left, there are no carries from the units to the tens. When calculating in binary, on the right, there is no carry from the units column and the result is still made up of BCD digits (since ≤ 9), so the result is correct.

$$24_{10} + \quad 0010\ 0100 +$$
$$42_{10} = \quad 0100\ 0010 =$$

$$66_{10} \qquad 0110\ 0110$$

However, in the second example, when we add the BCD digits, we come out with a sum that is not a BCD digit. The result is not valid (see the sum in decimal on the left), so we must make adjustments.

$$27_{10} + \quad 0010\ 0111 +$$
$$35_{10} = \quad 0011\ 0101 =$$

$$62_{10} \qquad 0101\ 1100$$

In the result, the number 1100_2 is not in BCD 8421 (it is greater than $9_{10} = 1001_2$). The method to correct this consists in adding 0110_2 (i.e., decimal 6) to the non-BCD number.

$$1100 +$$
$$0110 =$$

$$1\ 0010$$

The four digits to the right are now BCD. The carry is considered to be added to the BCD digit to its immediate left. In the example:

$$0101 +$$
$$0001 =$$

$$0110$$

So, the exact result of the BCD addition, with corrections, is:

$$62_{10} = 01100010_{bcd}$$

3.8 Binary Rational Numbers

The way to deal with binary rational numbers is similar to what is used for integer numbers. Let's see two examples of conversion:

$$0.1011_2 = 1 \cdot 2^{-1} + 0 \cdot 2^{-2} + 1 \cdot 2^{-3} + 1 \cdot 2^{-4} =$$
$$= 0.5 + 0.125 + 0.0625 =$$
$$= 0.6875_{10}.$$

To convert DEC into BIN, we do the following:

$$0.6875 \cdot 2 = 1.3750$$

The integer part constitutes the binary number, and the decimal part is multiplied again by 2. Continuing:

$$0.375 \cdot 2 = 0.750$$
$$0.750 \cdot 2 = 1.500$$
$$0.500 \cdot 2 = 1.000$$

we have come back to the initial number: 0.1011_2.

3.9 Arithmetic Networks

3.9.1 Half Adders

Half adders add two single-digit binary numbers generating a result and a carry. From the *Deeds* library:

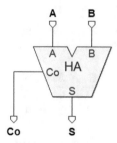

Using the addition rules seen earlier, let's complete the truth table. The addends are A and B, the sum S, and the carry C_o:

A B	C_o S
0 0	0 0
0 1	0 1
1 0	0 1
1 1	1 0

Without the use of a map, we immediately recognize the functions in the table:

$$C_o = A \cdot B \qquad\qquad e \qquad\qquad S = A \oplus B$$

So the logical network of the half adder will look like this:

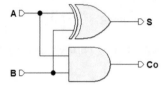

3.9.2   Full Adders

Full adders extend the possibilities of half adders by adding the carry from the preceding sum to the input.

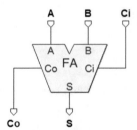

Let's derive the truth table from the addition rules. A and B are the addends, C_i is the carry from the previous addition, S is the sum, and C_o is the carry:

C_i A B	C_o S
0 0 0	0 0
0 0 1	0 1
0 1 0	0 1
0 1 1	1 0
1 0 0	0 1
1 0 1	1 0
1 1 0	1 0
1 1 1	1 1

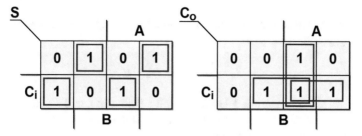

From the maps, we derive the synthesis:

$$S = \overline{A}\,\overline{B}\,C_i + \overline{A}\,B\,\overline{C_i} + A\,\overline{B}\,\overline{C_i} + A\,B\,C_i =$$
$$= \overline{A}\,(\overline{B}\,C_i + B\,\overline{C_i}) + A\,(\overline{B}\,\overline{C_i} + BC_i) =$$
$$= \overline{A}\,(B \oplus C_i) + A\,(\overline{B \oplus C_i}) =$$
$$= A \oplus (B \oplus C_{IN})$$
$$C_o = A\,B + A\,C_i + C_i\,B$$

and then the logical schematic:

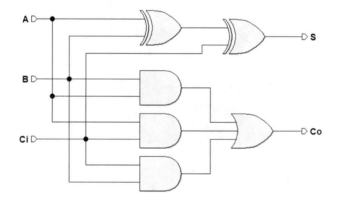

3.9.3 Ripple Carry Adder

The network in the figure below (*ripple carry adder*) allows us to calculate the sum of two 4-bit numbers, and it is extendable to any number with m bits:

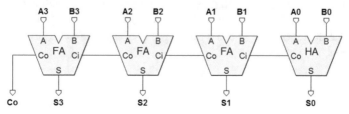

Because the carries are propagated from one stage to another, the execution time of a sum T_s is proportional to m.

$$T_s = (m - 1) \cdot t_{fa} + t_{ha}$$

where t_{fa} is the propagation delay of the full adders (FA), and t_{ha} that of the half adders (HA). The C_o, and therefore the sum, is valid only after all the intermediate carries are propagated along the network.

This is why, when we need high-speed calculation on a large number of bits, we rely on more efficient architectures where the carries are calculated, not in ripple fashion but in parallel, as in the *Carry Look Ahead* networks (not dealt with in this book).

3.9.4 Arithmetic Logic Unit (ALU)

The ALU is a combinational network that makes it possible to do different operations on two binary numbers (A and B). The figure below shows an example of an 8-bit ALU (from the *Deeds* library):

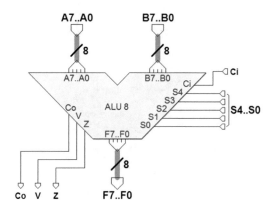

The operands are $A7..A0$ and $B7..B0$, while the result is taken from the outputs $F7..F0$. For addition and subtraction, there is one input Ci and one carry output Co.

Outputs V and Z are also available: $V = 1$ shows that the arithmetic operation gave rise to an *overflow*, while $Z = 1$ is activated when the result of the operation is zero. The operations are selected by the group of inputs $S4..S0$, according to the table below:

$S4..S0$	Function	Notes
00000	$F = 0$	
00001	$F = +1$	
00010	$F = -1$	
00011	$F = -128$	Minimum negative value
00100	$F = A$	
00101	$F = B$	
00110	$F = \overline{A}$	Ones' complement of A
00111	$F = \overline{B}$	Ones' complement of B
01000	$F = A \ and \ B$	(bitwise)
01001	$F = A \ and \ \overline{B}$	(bitwise)
01010	$F = \overline{A} \ and \ B$	(bitwise)
01011	$F = \overline{A \ or \ B}$	(bitwise)
01100	$F = A \ or \ B$	(bitwise)
01101	$F = A \ or \ \overline{B}$	(bitwise)
01110	$F = \overline{A} \ or \ B$	(bitwise)
01111	$F = \overline{A \ and \ B}$	(bitwise)
10000	$F = A \oplus B$	(bitwise)
10001	$F = \overline{A \oplus B}$	(bitwise)
10010	$F = \overline{A} + 1$	Two's complement of A
10011	$F = \overline{B} + 1$	Two's complement of B
10100	$F = A + 1$	Increment of A
10101	$F = B + 1$	Increment of B
10110	$F = A - 1$	Decrement of A
10111	$F = B - 1$	Decrement of B
11000	$F = A + B$	Addition
11001	$F = A + B + Ci$	Addition (with input carry)
11010	$F = sat \ (A + B)$	Saturating Addition
11011	$F = A - B$	Subtraction
11100	$F = A - B - Ci$	Subtraction (with input borrow)
11101	$F = sat \ (A - B)$	Saturating Subtraction
11110	$F = B - A$	Reversed Subtraction
11111	$F = B - A - Ci$	Reversed Subtraction (with input borrow)

3.10 Relative Numbers in Binary

There are many methods to represent relative numbers (signed numbers) in binary. In general, positive numbers are represented as if they had no sign whereas methods differ as to how negative numbers are coded. When the codifications we use change, so change the rules to manage the operations involving negative numbers.

3.10.1 Representation in "Module and Sign" Code

Let's consider a 4-bit packet: we use the MSB for the sign bit and the rest for the module:

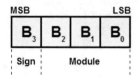

We code the sign *minus* with 1 and the sign *plus* with 0. We get the following table:

B_3	B_2	B_1	B_0	Dec	
0	1	1	1	+7	
0	1	1	0	+6	
0	1	0	1	+5	
0	1	0	0	+4	
0	0	1	1	+3	
0	0	1	0	+2	
0	0	0	1	+1	
0	0	0	0	+0	
1	0	0	0	−0	(!)
1	0	0	1	−1	
1	0	1	0	−2	
1	0	1	1	−3	
1	1	0	0	−4	
1	1	0	1	−5	
1	1	1	0	−6	
1	1	1	1	−7	

There are some downsides to this representation:

1. The zero has two different codes: 0000 and 1000.
2. The code has bad arithmetic properties.

As shown in the following examples, we cannot use a normal adder.

$$3_{10} + \quad 0011_2 + \qquad\qquad\qquad 4_{10} + \quad 0100_2 +$$
$$4_{10} = \quad 0100_2 = \qquad\qquad\qquad 4_{10} = \quad 0100_2 =$$
$$\overline{7_{10}} \quad \overline{0111_2} \quad \text{(correct)} \qquad\qquad \overline{-0_{10}} \quad \overline{1000_2} \quad \text{(overflow)}$$

$$7_{10} + \qquad\qquad 0111_2 +$$
$$-2_{10} = \qquad\qquad 1010_2 =$$
$$\overline{5_{10}} \qquad\qquad \overline{10001_2} \quad \text{(carry)}$$

So, in a *module and sign* representation, we would need to use a more complex, tailor-made adder.

3.10.2 Complementation

The complement to the base

Let the number N in base R be represented in a positional notation with m digits:

$$N = n_{m-1} \cdot R^{m-1} + n_{m-2} \cdot R^{m-2} + \cdots + n_1 \cdot R^1 + n_0 \cdot R^0$$

We define the complement to the base R of the number N:

$$C_R(N) = R^m - N$$

Let's now consider another number Q, represented with the same base and the same number of digits. By adding the number Q to $C_R(N)$, we obtain the difference between the two plus the term R^m, which represents a carry outside the number's format.

$$Q + C_R(N) = Q + (R^m - N) = (Q - N) + R^m$$

Bringing the difference $Q - N$ to the left:

$$Q - N = Q + C_R(N) - R^m$$

Let's now look at an example with decimal numbers. $R = 10$, so the complement is "to ten"; we use only two digits ($m = 2$), and assume $Q = 48_{10}$ e $N = 12_{10}$.

$$48_{10} - 12_{10} = 48_{10} + C_{10}(12_{10}) - 10^2$$

By applying the definition of the complement of the base to our example:

$$C_{10}(12_{10}) = 10^2 - 12 = 88$$

We get:

$$48_{10} - 12_{10} = 48_{10} + 88_{10} - 10^2 = 136_{10} - 10^2 = 36_{10}$$

Despite the apparent complication, we have a great advantage: the negative number has been substituted by its complement to ten, which is positive. Then taking away 10^2 from the result is very simple, as we will soon see, since it does not require subtraction.

In an arithmetic network, we can do without subtractors and use only adders as long as we know an easy way to calculate the complement to the base of a number. Note that the complement operation is, from the point of view of the calculation, the same as changing the sign of a number.

Let's go back to our example and do the addition in column:

$$
\begin{array}{r}
|48\,+ \\
|88\,= \\
\hline
\text{carry} \quad 1\,|36
\end{array}
$$

At this point, taking away 10^2 is simple: we only need to ignore the carry. The result is what we expect: $36_{10} = 48_{10} - 12_{10}$.

Note that it is necessary to represent all the numbers with the same m. For example, let's use the same quantities as before but represented on eight digits ($m = 8$):

$$C_{10}(12_{10}) = 10^8 - 12_{10} = 99999988_{10}$$

$$
\begin{array}{r}
|00000048\,+ \\
|99999988\,= \\
\hline
\text{carry} \quad 1\,|00000036
\end{array}
$$

The carry exceeds the m digits: taking away R^m from the result means just ignoring it.

Now let's look at binary numbers with $R = 2$. We'll use the 4-bit format ($m = 4$), and we'll evaluate the complement *to two* of $N = 0101_2 = 5_{10}$ from the definition:

$$C_2(0101_2) = 2^m - N = 10000_2 - 0101_2 = 1011_2$$

As we'll see below, the $C_2(N)$ can be calculated more quickly using the complement of "base minus one".

The complement to "base minus one"

We define *the complement to base minus one (complement to one)* of the number N:

$$C_{R-1}(N) = (R^m - 1) - N$$

Let's compare this definition with that of the complement of the base. The result is:

$$C_R(N) = C_{R-1}(N) + 1$$

Calculating the complement to the base minus one is simpler than calculating the complement to the base. Therefore, to obtain the complement to the base, we prefer to calculate the complement to the base minus one and then add 1.

Let's look at an example in decimal with $N = 12_{10}$. Let's represent the number with eight digits and calculate the complement to nine (base ten minus one), by first evaluating the term:

$$(R^m - 1) = 10^8 - 1 = 99999999$$

This number is composed only of the digit "9" repeated m times. The complement to nine of 12 is:

$$C_9(12) = 99999999 - 00000012 = 99999987$$

It is simpler to calculate the complement to nine because when we subtract the number we never need to borrow. Then, by adding 1 to the result, we get the complement to ten.

Le's take the example of the binary number $N = 0101_2 = 5_{10}$ seen above ($M = 4$). To evaluate ones' complement (base two minus one), we first calculate the term:

$$(R^m - 1) = 2^m - 1 = 10000_2 - 1 = 1111_2$$

The situation is similar to the one before: this number is composed of the digit "1" repeated m times. So, ones' complement of 0101_2 is:

$$C_1(0101_2) = 1111_2 - 0101_2$$

To make this clearer, let's do the subtraction in column.

$$
\begin{array}{rr}
 & 1111_2 - \\
 & 0101_2 = \\
\hline
C_1(0101_2) & 1010_2
\end{array}
$$

We see that when we subtract the number from a number made only of "1", there is no need to borrow. To calculate the result, all we need to do is replace all the "1" with "0" and vice versa. From a circuital perspective, this means negating all the bits the number N is composed of.

$$C_1(N) = \overline{N}$$

From what we have seen, by adding $+1$ to the number obtained, we get the two's complement:

$$C_2(N) = C_1(N) + 1 = \overline{N} + 1$$

3.10.3   Representation in "Ones' Complement" Code

We can represent negative binary numbers through ones' complement (C_1). In the following table, we see an example for 4-bit numbers. The positive numbers are coded in pure binary, leaving the most significant digit at 0. The negatives are the C_1 of the corresponding positive number calculated according to the criteria outlined above.

To generate the code of a negative number, as above, we just invert all the bits of the corresponding positive number, for example:

B_3	B_2	B_1	B_0	Dec	
0	1	1	1	+7	
0	1	1	0	+6	
0	1	0	1	+5	
0	1	0	0	+4	
0	0	1	1	+3	
0	0	1	0	+2	
0	0	0	1	+1	
0	0	0	0	+0	
1	1	1	1	−0	(!)
1	1	1	0	−1	
1	1	0	1	−2	
1	1	0	0	−3	
1	0	1	1	−4	
1	0	1	0	−5	
1	0	0	1	−6	
1	0	0	0	−7	

$$-3_{10} = C_1(3_{10}) = C_1(0011_2) = 1100_2$$

Arithmetic Properties

- The sum of two positives:

$+2$ +	0010 +
$+5$ =	0101 =
$+7$	0111

- The sum of a positive and a negative with a positive result:

$+5$ +	0101 +
-2 =	1101 =
$+3$	10010 ("End Around Carry") $\rightarrow 0010 + 1 = 0011$

- The sum of a positive and a negative with a negative result:

-5 +	1010 +
$+2$ =	0010 =
-3	1100

- The sum of two negatives:

-3 +	1100 +
-3 =	1100 =
-6	11000 ("End Around Carry") $\rightarrow 1000 + 1 = 1001$

The carry "outside" the number format should be added to the LSB of the result. C_P denotes the carry to the MSB and C_S the carry outside the MSB.

Let's consider the sum of two numbers $Q + N$; the table below shows us the values of C_P and C_S for all the combinations of the signs of addends and result, when the sum is correct, i.e., when the number's format is able to contain the result:

Q	N	Sum	C_P	C_S
+	+	+	0	0
+	−	+	1	1
+	−	−	0	0
−	−	−	1	1

If C_P and C_S are different, there is an *overflow* error, and we can detect it as *OVF* $= C_P \oplus C_S$. Here is an example of an overflow error:

$$
\begin{array}{ll}
+7\,+ & 0111\,+ \\
+1\,= & 0001\,= \\
\hline
+8 & 1000 \quad C_P = 1, C_S = 0, \rightarrow OVF
\end{array}
$$

3.10.4 Representation in "Two's Complement" Code

The most commonly used method for representing negative binary numbers relies on two's complement (C_2). In the table on the right, we find an example of code for four-bit numbers.

The positive numbers here are coded in pure binary as before, leaving the most significant digit at 0, while the negative numbers are calculated as two's complement of the corresponding positive number.

B_3	B_2	B_1	B_0	Dec
0	1	1	1	+7
0	1	1	0	+6
0	1	0	1	+5
0	1	0	0	+4
0	0	1	1	+3
0	0	1	0	+2
0	0	0	1	+1
0	0	0	0	+0
1	1	1	1	−1
1	1	1	0	−2
1	1	0	1	−3
1	1	0	0	−4
1	0	1	1	−5
1	0	1	0	−6
1	0	0	1	−7
1	0	0	0	−8

For practice, let's calculate the number -1_{10} starting from $+1_{10}$:

$$C_2(0001) = C_1(0001) + 1 = 1110 + 1 = 1111$$

If we calculate C_2 of zero, we get zero again (the zero in C_2 has a *univocal representation*). Note that the operation C_2 changes the number's sign: if N is positive, $C_2(N)$ is negative and vice versa.

Also, if we perform $C_2(C_2(N))$, we get the number N again:

$$C_2(C_2(N)) = 2^m - (C_2(N)) = 2^m - (2^m - N) = N$$

Arithmetic Properties

Let's look again at the carry into the MSB (C_P) and the carry outside the number (C_S). The same considerations for representation in C_1 also hold for C_2. When adding two numbers Q and N, the table used for checking the carries for *overflow* errors still holds:

Q	N	Sum	C_P	C_S
$+$	$+$	$+$	0	0
$+$	$-$	$+$	1	1
$+$	$-$	$-$	0	0
$-$	$-$	$-$	1	1

As before, there is overflow when C_P and C_S are different: $OVF = C_P \oplus C_S$.

C_2 code allows us to execute additions regardless of the sign of the addends by using a normal binary adder with no need to make corrections to the result (unlike with C_1).

3.10.5 Sign Extension

A signed binary number represented in C_2 code (or in C_1), over m bits, can be extended to a larger number of bits $v > m$, provided that the sign and value are preserved. Let's consider a positive number, for example 6_{10}, represented on four bits and its corresponding negative in C_2:

$$6_{10} = 0110_2 \qquad -6_{10} = 1010_2$$

Let's examine the positive number and consider the sign as an integral part of the number. In positional notation, it is:

$$0110_2 = 0 \cdot 2^3 + 1 \cdot 2^2 + 1 \cdot 2^1 + 0 \cdot 2^0$$

If we represent it with 8 bits, the positional notation will be:

$$0 \cdot 2^7 + 0 \cdot 2^6 + 0 \cdot 2^5 + 0 \cdot 2^4 + 0 \cdot 2^3 + 1 \cdot 2^2 + 1 \cdot 2^1 + 0 \cdot 2^0 = 00000110_2$$

We have actually added non-significant zeroes to the left of the number, as you might expect. Still, this method cannot work for negative numbers, first of all because we will change the sign to positive but also because we will also change the value.

So, let's evaluate $C_2(6_{10})$, represented with 8 bits.

$$C_2(00000110_2) = C_1(00000110_2) + 1 = 11111001_2 + 1 = 11111010_2$$

As we can see, to extend a negative number to a larger number of bits, we only need to add 1s (rather than 0s) to the left. In other words, in either case, we must add digits with a value equal to the sign on the left. From a circuital perspective, the extension of the sign is translated into a very simple network.

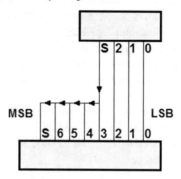

3.11 Representation of Real Numbers

There are essentially two methods to deal with real numbers in binary arithmetic.

- *Fixed point.* To assign a certain number of bits for the integer part of the number and the others for the fractional part. For example:

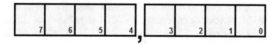

- *Floating point.* With this method, the bits available (32, in the example) are divided as follows:

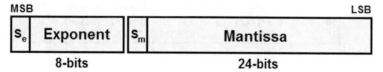

In the figure, S_e is the sign of the exponent, and S_m is the sign of the mantissa. There is a "normalized" representation with a 0, . . .-type mantissa and a 2^{Exp}-type exponent, for example: $-0, 1011000101101 1_2 * 2^{001011111_2}$.

3.12   Alphanumeric Codes

Alphanumeric codes allow us to represent uppercase and lowercase letters, the ten decimal numbers, punctuation marks, and the so-called special symbols. There are codes that allow the writer to use the characters of almost all the written languages in the world, including Chinese and Japanese. The most common among these is Unicode.[1] This type of code is very complex and is not organized only on simple correspondence tables but on libraries of software supported by modern environments for applications development. An explanation of Unicode goes beyond the scope of this book.

One code that is still rather commonly used and relatively simple is *ASCII*,[2] which has 7 significant bits in the standard version. It codes the 26 uppercase and lowercase letters of the English language, the 10 decimal numbers, punctuation, and the symbols used in that language.

It also includes a certain number of communication codes, the so-called non-printable characters, which are used only sparingly in modern systems. Below are two tables of ASCII code characters: "non-printable" and printable.

ASCII: Not Printable Characters				
Dec	Code	Description	Dec	Code Description
0	NULL	(Null character)	16	DLE (Data link escape)
1	SOH	(Start of Header)	17	DC1 (Device control 1)
2	STX	(Start of Text)	18	DC2 (Device control 2)
3	ETX	(End of Text)	19	DC3 (Device control 3)
4	EOT	(End of Transmiss	20	DC4 (Device control 4)ion)
5	ENQ	(Enquiry)	21	NAK (Negative acknowledgement)
6	ACK	(Acknowledgemen	22	SYN (Synchronous idle)t)
7	BEL	(Bell)	23	ETB (End of transmission block)
8	BS	(Backspace)	24	CAN (Cancel)
9	HT	(Horizontal Tab)	25	EM (End of medium)
10	LF	(Line feed)	26	SUB (Substitute)
11	VT	(Vertical Tab)	27	ESC (Escape)
12	FF	(Form feed)	28	FS (File separator)
13	CR	(Carriage return)	29	GS (Group separator)
14	SO	(Shift Out)	30	RS (Record separator)
15	SI	(Shift In)	31	US (Unit separator)

---

[1] http://www.unicode.org

[2] American Standard Code for Information Interchange

ASCII: Printable Characters					
Dec	Code	Description	Dec	Code	Description
32		Space	80	P	Capital P
33	!	Exclamation mark	81	Q	Capital Q
34	"	Quotation mark	82	R	Capital R
35	#	Number sign	83	S	Capital S
36	$	Dollar sign	84	T	Capital T
37	%	Percent sign	85	U	Capital U
38	&	Ampersand	86	V	Capital V
39	'	Apostrophe	87	W	Capital W
40	(	round brackets	88	X	Capital X
41	)	round brackets	89	Y	Capital Y
42	*	Asterisk	90	Z	Capital Z
43	+	Plus sign	91	[	square brackets
44	,	Comma	92	\	Backslash
45	-	Hyphen	93	]	square brackets
46	.	Dot	94	⌒	Circumflex accent
47	/	Slash	95	_	underscore
48	0	number zero	96	'	Grave accent
49	1	number one	97	a	Lowercase a
50	2	number two	98	b	Lowercase b
51	3	number three	99	c	Lowercase c
52	4	number four	100	d	Lowercase d
53	5	number five	101	e	Lowercase e
54	6	number six	102	f	Lowercase f
55	7	number seven	103	g	Lowercase g
56	8	number eight	104	h	Lowercase h
57	9	number nine	105	i	Lowercase i
58	:	Colon	106	j	Lowercase j
59	;	Semicolon	107	k	Lowercase k
60	<	Less-than sign	108	l	Lowercase l
61	=	Equals sign	109	m	Lowercase m
62	>	Greater-than sign	110	n	Lowercase n
63	?	Question mark	111	o	Lowercase o
64	@	At sign	112	p	Lowercase p
65	A	Capital A	113	q	Lowercase q
66	B	Capital B	114	r	Lowercase r
67	C	Capital C	115	s	Lowercase s
68	D	Capital D	116	t	Lowercase t
69	E	Capital E	117	u	Lowercase u
70	F	Capital F	118	v	Lowercase v
71	G	Capital G	119	w	Lowercase w
72	H	Capital H	120	x	Lowercase x
73	I	Capital I	121	y	Lowercase y
74	J	Capital J	122	z	Lowercase z
75	K	Capital K	123	{	curly brackets
76	L	Capital L	124			vertical-bar
77	M	Capital M	125	}	curly brackets
78	N	Capital N	126	~	Tilde ; swung dash
79	O	Capital O	127	DEL	Delete

3.13 Error Detection Codes: Parity Generator and Checker

Assume that in an *8-bit* system, we are using *ASCII standard code* (that only has *7 significant bits*, denoted here as $C6 \ldots C0$). So, we can use the eighth bit to create an *"error detection"* code by using it as a *"parity bit"*. In this way, we add information which is redundant for coding ASCII, but useful to control *the integrity of the data*.

With $P = 1$ at the output, an XOR gate signals the presence of an odd number of ones in the input.

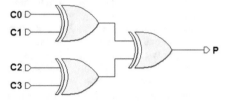

Thus, operation is called *"parity check"*. Through a tree structure, we can extend the parity check to any number of inputs.

The truth table of the 4-input XOR tree will show a 1 for each input combination with an odd number of ones:

C3	C2	C1	C0	P
0	0	0	0	0
0	0	0	1	1
0	0	1	0	1
0	0	1	1	0
0	1	0	0	1
0	1	0	1	0
0	1	1	0	0
0	1	1	1	1
1	0	0	0	1
1	0	0	1	0
1	0	1	0	0
1	0	1	1	1
1	1	0	0	0
1	1	0	1	1
1	1	1	0	1
1	1	1	1	0

In the figure below, a 7-bit parity check processes bits $C6 \ldots C0$ of the input data word, producing a 1 in output P if the set has an odd number of ones (*"odd parity"*). This structure is called a *"parity generator"*:

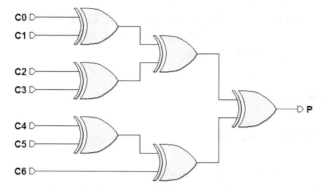

The 8-bit data word, composed of P and $C6 \ldots C0$ has an even number of ones (*"even parity"*). In the set of bits of the data word, P is called the *"parity bit"*.

P	C6	C5	C4	C3	C2	C1	C0

Associating a *redundancy* (in this case, a *parity bit*) to an original data word allows us to perform a quality check on it when data is moved from one system to another. The distance between systems, the characteristics of the communication channel, and the presence of noise all degrade the quality of the signal along its path, to the point that might damage its contents. Some bits can be received wrong. In real systems, errors are always present. They cannot be completely avoided but, with the right techniques, the probability of their happening can be greatly reduced. The error rate can be studied in statistical terms.

We introduce an analogous structure on the receiver side to verify that data parity has been preserved. We recalculate the parity of the received bits $C0 \ldots C7$ and also include bit P (see the figure below).

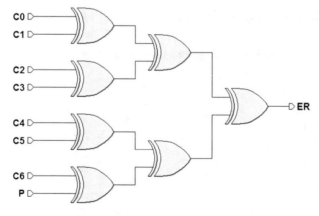

If no error has been detected, output ER (*Error*) will be zero. If ER is 1, it means that parity has not been preserved due to an error. Note that the check is accurate only if the error involves *one bit*: if there are *two* affected bits, for example, ER does not show it.

Thus, this is only useful if the probability of error is low. If we assume the probability of one bit error out of 10^4 transmitted, the probability of an error on two bits will be 10^8 bits (product of the two).

3.14   Exercises

3.14.1   Binary Numbers

1. Write 168_{10} and 143_{10} as binary numbers.
2. Calculate the decimal value of the following binary numbers:

    (a) 1110111_2
    (b) 101011101001011_2

3. Do the following calculations:

    (a) $11001101_2 + 1010101_2$
    (b) $1011011_2 - 101110_2$
    (c) $1001001_2 - 10101_2$

4. Multiply these binary numbers:

    (a) 1110_2 e 1011_2
    (b) 1011_2 e 101_2
    (c) 11100_2 e 011_2
    (d) 110_2 e 1111_2

3.14.2   Signed Binary Numbers

1. Do these calculations in two's complement with 7 bits:

    (a) $15 + 11$
    (b) $15 + (-11)$
    (c) $15 - 11$
    (d) $4 - 11$
    (e) $29 + 17$
    (f) $29 - 17$
    (g) $29 + (-17)$
    (h) $(-11) - 29$
    (i) $8 + (-18)$
    (j) $7 - 17$

2. Do these calculations in two's complement using the minimum number of bits to avoid overflow.

    (a) $3 + (-7)$

  (b) $-3 + 7$
  (c) $11 + (-11)$
  (d) $18 - (-3)$

3.14.3 Octal and Hexadecimal Numbers

1. Convert the following numbers from octal to hexadecimal.

  (a) 276534_8
  (b) 22017555724_8

2. Convert these decimals into binary and hexadecimals.

  (a) 7722_{10}
  (b) 1435_{10}

3. Convert these decimals into octals and hexadecimals.

  (a) 36625_{10}
  (b) 124_{10}

4. Calculate the decimal values of these hexadecimal numbers.

  (a) $1A2B07_{16}$
  (b) 11047_{16}

5. Calculate the decimal values of these octal numbers.

  (a) 3111_8
  (b) 276534_8

6. Solve the equations of these hexadecimal numbers.

  (a) $41AB7_{16} + C2D6F_{16}$
  (b) $A23CE_{16} + 363E6_{16}$

3.15 Solutions

3.15.1 Binary Numbers

	168	0			143	1	
	84	0			71	1	
	42	0			35	1	
1.	21	1			17	1	
	10	0			8	0	
	5	1			4	0	
	2	0			2	0	
	1	1	$\rightarrow 10101000_2$		1	1	$\rightarrow 10001111_2$

2. (a) $1110111_2 = 1 \cdot 2^6 + 1 \cdot 2^5 + 1 \cdot 2^4 + 0 \cdot 2^3 + 1 \cdot 2^2 + 1 \cdot 2^1 + 1 \cdot 2^0 +$
$= 119_2$

  (b) $101011101001011_2 = 22347_{10}$

3. (a)
$$\begin{array}{r} 11001101\ + \\ 1010101\ = \\ \hline 100100010 \end{array}$$

  (b)
$$\begin{array}{r} 1011011\ - \\ 101110\ = \\ \hline 0101101 \end{array}$$

  (c)
$$\begin{array}{r} 1001001\ - \\ 10101\ = \\ \hline 110100 \end{array}$$

4. (a)
$$\begin{array}{r} 1110\ \times \\ 1011\ = \\ \hline 1110 \\ 11100 \\ 000000 \\ 1110000 \\ \hline 10011010 \end{array}$$

  (b)
$$\begin{array}{r} 1011\ \times \\ 101\ = \\ \hline 1011 \\ 00000 \\ 101100 \\ \hline 110111 \end{array}$$

  (c)
$$\begin{array}{r} 11100\ \times \\ 00011\ = \\ \hline 11100 \\ 111000 \\ 0000000 \\ 00000000 \\ 000000000 \\ \hline 1010100 \end{array}$$

  (d)
$$\begin{array}{r} 110\ \times \\ 1111\ = \\ \hline 110 \\ 1100 \\ 11000 \\ 110000 \\ \hline 1011010 \end{array}$$

3.15.2 Signed Binary Numbers

1. (a)
$$\begin{array}{r} 15 + \\ 11 = \\ \hline 26 \end{array} \qquad \begin{array}{r} 0001111 + \\ 0001011 = \\ \hline 0011010 \end{array}$$

(b)
$$\begin{array}{r} 15 + \\ -11 = \\ \hline 4 \end{array} \qquad \begin{array}{r} 0001111 + \\ 1110101 = \\ \hline 0000100 \end{array}$$

(c)
$$\begin{array}{r} 15 - \\ 11 = \\ \hline 4 \end{array} \qquad \begin{array}{r} 0001111 - \\ 0001011 = \\ \hline 0000100 \end{array}$$

(d)
$$\begin{array}{r} 4 - \\ 11 = \\ \hline -7 \end{array} \qquad \begin{array}{r} 0000100 - \\ 0001011 = \\ \hline 1111001 \end{array}$$

(e)
$$\begin{array}{r} 29 + \\ 17 = \\ \hline 46 \end{array} \qquad \begin{array}{r} 0011101 + \\ 0010001 = \\ \hline 0101110 \end{array}$$

(f)
$$\begin{array}{r} 29 - \\ 17 = \\ \hline 12 \end{array} \qquad \begin{array}{r} 0011101 - \\ 0010001 = \\ \hline 0001100 \end{array}$$

(g)
$$\begin{array}{r} 29 + \\ -17 = \\ \hline 12 \end{array} \qquad \begin{array}{r} 0011101 + \\ 1101111 = \\ \hline 0001100 \end{array}$$

(h)
$$\begin{array}{r} -11 - \\ 29 = \\ \hline -40 \end{array} \qquad \begin{array}{r} 1110101 - \\ 0011101 = \\ \hline 1011000 \end{array}$$

(i)
$$\begin{array}{r} 8 + \\ -18 = \\ \hline -10 \end{array} \qquad \begin{array}{r} 0001000 + \\ 1101110 = \\ \hline 1110110 \end{array}$$

(j)
$$\begin{array}{r} 7 - \\ 17 = \\ \hline -10 \end{array} \qquad \begin{array}{r} 0000111 - \\ 0010001 = \\ \hline 1110110 \end{array}$$

2. (a) 4 bits are needed:
$$\begin{array}{r} 3 + \\ -7 = \\ \hline -4 \end{array} \qquad \begin{array}{r} 0011 + \\ 1001 = \\ \hline 1100 \end{array}$$

(b) 4 bits are needed:

$$
\begin{array}{rr}
-3\ + & 1101\ + \\
7\ = & 0111\ = \\
\hline
4 & 0100
\end{array}
$$

(c) 5 bits are needed:

$$
\begin{array}{rr}
11\ + & 01011\ + \\
-11\ = & 10101\ = \\
\hline
0 & 00000
\end{array}
$$

(d) 6 bits are needed:

$$
\begin{array}{rr}
18\ - & 010010\ - \\
-3\ = & 111101\ = \\
\hline
21 & 010101
\end{array}
$$

3.15.3   Octal and Hexadecimal Numbers

1. The quickest solution is to first write the number in binary:

   (a) 276534_8 in binary is 010111110101011110_2.
   The hexadecimal representation is: $17D5C_H$.
   (b) 220175557248 is $010010000001111101101101111010100_2$ in binary.
   Hexadecimal: $903EDBD4_H$.

2. (a) Converting 7722_{10}:

7722	0
3861	1
1930	0
965	1
482	0
241	1
120	0
60	0
30	0
15	1
7	1
3	1
1	1

7722	10	(A) (ex: $7722 : 16 = 482 + 10$)
482	2	
30	14	(E)
1	1	

   In binary: 1111000101010_2, in hexadecimal: $1E2A_H$.

(b) Converting 1435_{10}:

$$
\begin{array}{r|l}
1435 & 1 \\
717 & 1 \\
358 & 0 \\
179 & 1 \\
89 & 1 \\
44 & 0 \\
22 & 0 \\
11 & 1 \\
5 & 1 \\
2 & 0 \\
1 & 1 \\
\end{array}
$$

$$
\begin{array}{r|l}
1435 & 11 \\
89 & 9 \\
5 & 5 \\
\end{array}
\quad \text{(B)} \quad \text{(ex: } 1435 : 16 = 89 + 11)
$$

In binary: 10110011011_2, in hexadecimal: $59B_H$.

3. (a) Converting 36625_{10}:

$$
\begin{array}{r|l}
36625 & 1 \quad (36625 : 8 = 4578 + 1) \\
4578 & 2 \quad (4578 : 8 = 572 + 2) \\
572 & 4 \quad (572 : 8 = 71 + 4) \\
71 & 7 \quad (71 : 8 = 8 + 7) \\
8 & 0 \quad (8 : 8 = 1 + 0) \\
1 & 1 \\
\end{array}
$$

$$
\begin{array}{r|l}
36625 & 1 \quad (36625 : 16 = 2289 + 1) \\
2289 & 1 \quad (2289 : 16 = 143 + 1) \\
143 & F \quad (143 : 16 = 8 + 15) \\
8 & 8 \\
\end{array}
$$

In octal: 107421_8, in hexadecimal: $8F11_H$.

(b) Converting 124_{10}:

$$
\begin{array}{r|l}
124 & 4 \quad (124 : 8 = 15 + 4) \\
15 & 7 \quad (15 : 8 = 1 + 7) \\
1 & 1 \\
\end{array}
$$

$$
\begin{array}{r|l}
124 & 12 \\
7 & 7 \\
\end{array}
\quad \text{(C)} \quad (124 : 16 = 7 + 12)
$$

In octal: 174_8, in hexadecimal: $7C_H$.

4. First, we convert the number into binary and then into decimal:

(a) $110100010101100001111_2 = 1714951_{10}$
(b) $10001000001000111_2 = 69703_{10}$

5. From octal to binary and then to decimal:

(a) $11001001001_2 = 1609_{10}$
(b) $10111110101011100_2 = 97628_{10}$

6. (a)
$$
\begin{array}{r}
4\ 1\ A\ B\ 7\ + \\
C\ 2\ D\ 6\ F\ = \\
\hline
1\ 0\ 4\ 8\ 2\ 6 \\
\end{array}
$$

(b)
$$
\begin{array}{r}
A\ 2\ 3\ C\ E\ + \\
3\ 6\ 3\ E\ 6\ = \\
\hline
D\ 8\ 7\ B\ 4 \\
\end{array}
$$

Chapter 4
Complements in Combinational Network Design

Abstract In this chapters, we overcome the limitations of the Karnaugh maps, whose application becomes impractical when applied to expressions with more than four/five variables. We present the Quine–McCluskey method, the first algorithms for minimizing Boolean expressions developed by Willard V. Quine and improved by Edward J. McCluskey. We present both the methods for synthesize single and multiple functions at the same time.

In past chapters, we addressed minimizing Boolean expressions, i.e., using Karnaugh maps to find an expression equivalent to the original but with fewer prime implicants. Unfortunately, the map method can only be applied to expressions with a maximum of four variables, extendable to five variables through 3-D maps or, in limited cases, even more through entered variables. For more than four variables, we cannot use "manual" methods but rather algorithmic methods implemented on a computer.

4.1 Minimizing Boolean Expressions with the Quine–McCluskey Method

One of the first algorithms for minimizing Boolean expressions was developed by Willard V. Quine (1908–2000) and improved by Edward J. McCluskey (1929–2016)[1] and is known as the "Quine–McCluskey Method" (hereinafter referred to as "QM–M").

The QM–M is an algorithm that translates the manual procedure of the Karnaugh maps, and it is made up of two phases. The first is the "expansion" phase where all the implicants of the function to be minimized (Karnaugh's "cubes") are generated. The prime implicants are identified, and the others are eliminated. The

---

[1] E.J.McCluskey, Minimization of Boolean Functions, The Bell System Technical Journal, November 1956.

© Springer International Publishing AG, part of Springer Nature 2019   115
G. Donzellini et al., *Introduction to Digital Systems Design*,
https://doi.org/10.1007/978-3-319-92804-3_4

second phase known as "covering" is where the smallest number of prime implicants needed to make the function equivalent to the starting function is chosen. That is, all the *minterms* of the function are "covered." There are tables for these two phases that help keep track of the steps in the algorithm and are easy to calculate.

4.1.1 The Expansion Phase

In the preparatory phase, the QM–M uses a simple approach to identify *minterms*: an *n* variable *minterm* is identified by an *n*-bit binary number where a direct variable is denoted with the value 1 and a negated variable, with 0. Let's look at this three-variable function as an example:

$$F(X, Y, Z) = \overline{X}\,\overline{Y}\,\overline{Z} + \overline{X}\,Y\,\overline{Z} + \overline{X}\,Y\,Z + X\,\overline{Y}\,\overline{Z} + X\,Y\,\overline{Z} + X\,Y\,Z$$

Minterms are identified by the binary numbers $(000, 010, 011, 100, 110, 111)$, which in decimal are $(0, 2, 3, 4, 6, 7)$, so we can use this encoding to write the compact form of the function:

$$F(X, Y, Z) = \Sigma(0, 2, 3, 4, 6, 7)$$

The latter can be described through the map below.

From here, we can easily derive the minimal expression:

$$F(X, Y, Z) = Y + \overline{Z}$$

At the beginning of the expansion phase, the QM–M lists all the *minterms* in a table, respecting the order that we will outline below. It then proceeds to pair them to obtain all the possible implicants with one variable less than the starting *minterm*.

3 Variables		2 Variables		1 Variable	
Terms	$X\ Y\ Z\|P$	Terms	$X\ Y\ Z\ \|\ P$	Terms	$X\ Y\ Z\ \|\ P$
0	000				
2	010				
4	100				
3	011				
6	110				
7	111				

For example, *minterms* $m0$ and $m2$ can be combined, thus

$$\overline{X}\ \overline{Y}\ \overline{Z} + \overline{X}\ Y\ \overline{Z} = \overline{X}\ \overline{Z}\ (\overline{Y} + Y) = \overline{X}\ \overline{Z}$$

In QM–M notation, that would be:

$$000 + 010 \rightarrow 0{-}0$$
$$0\ +\ 2\ \ \rightarrow (0, 2)$$

The symbol "$-$" is used to show that the variable was eliminated through simplification.

When we combine all the possible *minterm* pairs, we get a new table with all the two-variable implicants (one variable less than originally).

3 Variables			2 Variables			1 Variable		
Terms	$X\ Y\ Z$	P	Terms	$X\ Y\ Z$	P	Terms	$X\ Y\ Z$	P
0	000	V	0, 2	0–0				
2	010	V	0, 4	–00				
4	100	V	2, 3	01–				
3	011	V	2, 6	–10				
6	110	V	4, 6	1–0				
7	111	V	3, 7	–11				
			6, 7	11–				

At this point, we should analyze two expedients that have been used in the first phase of the algorithm.

The first expedient refers to the initial order of the *minterms* that are grouped by their number of negated variables, that is, the number of 0s in the corresponding binary number.[2] Group 1 has the *minterms* with all the negated variables (000). Group 2 has those with two negated variables (010, 100) and so on. It is actually impossible to combine *minterms* that differ by two or more variables, for example $\overline{X}\ \overline{Y}\ \overline{Z}$ (000) and $\overline{X}\ Y\ Z$ (011), because we would obtain no simplification. It is only possible to

---

[2]In the table, the groups are separated by one continuous line.

combine copies of *minterms* that differ by just one variable which is negative in one of them and direct in the other, for example: $X\ Y\ \overline{Z}$ (110) and $X\ Y\ Z$ (111). In other words, two *minterms* can be combined only if the corresponding binary numbers have a Hamming distance of 1. By grouping the numbers that contain the same number of 0s, we get homogeneous groups and can reduce the number of comparisons. Instead of comparing all the possible copies of *minterms*, we can check only those belonging to adjacent groups, those with smaller Hamming distances. The *minterm* of group 1 (000) can actually only be combined with that of group 2 (010, 100), which in turn can only be combined with that of group 3 (011, 110) and so on.[3] Note that not all combinations are possible: the term (100) belonging to group 2 can be combined with (110) but not with (011) from group 3 because their Hamming distance is two.

The second expedient will keep track of terms (implicants) that have been combined. If two implicants are combined to obtain an implicant with one less variable it means that those implicants were not prime and should not be considered in the covering phase. We know that a non-prime implicant can be substituted with a prime implicant if the prime covers it. Column P in the table does just this: there is an indication that the term has been combined so it is non-prime and can be overlooked in the covering phase.

By applying the same procedure to two-variable implicants, we get implicants where an additional variable has been eliminated.

3 Variables				2 Variables				1 Variable		
Terms	$X\ Y\ Z$	P		Terms	$X\ Y\ Z$	P		Terms	$X\ Y\ Z$	P
0	000	V		0, 2	0–0	V		0, 2, 4, 6	––0	P_0
2	010	V		0, 4	–00	V		2, 3, 6, 7	–1–	P_1
4	100	V		2, 3	01–	V				
3	011	V		2, 6	–10	V				
6	110	V		4, 6	1–0	V				
7	111	V		3, 7	–11	V				
				6, 7	11–	V				

Note that when combining two-variable implicants, we have two different ways to get the same one-variable implicant:

$$(0, 2) + (4, 6) \rightarrow (0, 2, 4, 6)$$
$$0{-}0 + 1{-}0 \rightarrow \quad {-}{-}0$$

or

$$(0, 4) + (2, 6) \rightarrow (0, 4, 2, 6)$$
$$-00 + -10 \rightarrow \quad {-}{-}0$$

---

[3] If we had to compare all the possible pairs among n terms to see if they could be combined, we would have to make more than $n(n-1)/2$ comparisons, i.e., each term with every other bidirectionally.

Clearly, implicant $(0, 2, 4, 6)$ and implicant $(0, 4, 2, 6)$ identify the same term because they correspond to $(--0)$, which is the term $\overline{Z}$. This is why only one of these will be brought over to the right-most table.

When we can combine the implicants no further, as in this case where there are only two uncombined terms altogether, the expansion phase ends and all the unflagged terms are prime implicants.

We can now list the starting function's prime implicants, which are:

$$
\begin{array}{rcccccc}
P0 & = & (0, 2, 4, 6) & = & --0 & = & \overline{Z} \\
P1 & = & (2, 3, 6, 7) & = & -1- & = & Y.
\end{array}
$$

4.1.2   The Covering Phase

If the goal in the expansion phase is to find all the prime implicants, the goal of the covering phase is to identify the lowest number of prime implicants that can cover the starting function. We must therefore be sure that all the *minterms* that defined the function to be minimized are covered by at least one of the identified prime implicants. We must also be sure to use as few implicants as possible to obtain the coverage.

To reach these goals, the QM–M used a "covering table" where the columns show all the *minterms* and the rows show all the prime implicants identified in the expansion phase. The "Xs" in the m_i column and in the P_j row show that m_i is covered by P_j:

	m_0	m_2	m_3	m_4	m_6	m_7
P_0	X	X		X	X	
P_1		X	X		X	X

In this case, it is easy to deduce that both the prime implicants are needed to cover the function, so the minimized function is:

$$
F(X, Y, Z) = P_0 + P_1 = \overline{Z} + Y
$$

Note that if a column contains just one "X," it means that there is only one prime implicant that can cover the corresponding *minterm*. In this case, the implicant is a essential prime implicant because without it, the function could not be completely covered. In this last example, both prime implicants are essential because *minterms* m_0 and m_4 are covered only by P_0, while *minterms* m_3 and m_7 are only covered by P_1. Further on, we will see that covering tables can be much more complex than this, so after selecting the essential prime implicants, we will need to use an algorithm to select the remaining ones to achieve minimum coverage.

4.1.3 Incompletely Specified Functions

We often need to deal with incompletely specified functions like the one represented in the map below:

which is written in QM–M notation as:

$$F(X, Y, Z) = \Sigma(0, 2, 4, 6) + d(3, 7)$$

where $d()$ groups all the *minterms* corresponding to don't-care.

Here, the QM–M could be applied by simply treating the don't-cares as 1s in the expansion phase and as 0s in the covering phase. The basic idea is this: in the expansion phase whenever larger cubes are constructed, (implicants with ever fewer variables) it makes sense to use as many *minterms* as possible to raise the possibilities of simplification. In the covering phase, we want to avoid covering a *minterm* if it is not strictly necessary. Avoiding to use it in the coverage phase, we try to prevent the *minterm* from making any prime implicant superfluous.

For the function above, the expansion phase is identical to the previous case while the covering phase uses a table with no m_3 or m_7 don't-care *minterms*.

	m_0	m_2	m_4	m_6
P_0	X	X	X	X
P_1		X		X

From this table, we can immediately see that P_0 is an essential prime implicant (due to m_0 and m_4) and it is also able to cover the function, giving us:

$$F(X, Y, Z) = P_0 = \overline{Z}$$

as expected.

4.1.4 Optimizing the Covering Phase

The covering phase can be particularly complicated when there is a large number of prime implicants.

Let's assume that after the expansion phase and after identifying the essential prime implicants, k prime implicants remain $\{P_0, \ldots, P_{k-1}\}$ from which we will choose the minimum coverage.

To obtain minimum coverage, we must check through $2^k - 1$ different cases, all the possible combinations with 1, 2, 3, or 4 prime implicants, and so on until finding the right one: $\{P_0\}$, $\{P_1\}$, ..., $\{P_k\}$, $\{P_0, P_1\}$, $\{P_0, P_2\}$, $\{P_0, P_3\}$, and so on until $\{P_0, \ldots, P_{k-1}\}$.

The QM–M offers a more efficient alternative that usually requires far fewer comparisons. Let's assume we want to minimize the following function:

$$F(X, Y, W, Z) = \Sigma(1, 2, 3, 6, 9, 10, 11, 12) + d(5, 13, 14)$$

The expansion phase begins with grouping terms, as explained previously. Group 1 contains *minterms* with three negated variables, group 2 with two and group 3 with one, as shown in the following tables.

4 Variables		
Terms	$XYWZ$	P
1	0001	
2	0010	
3	0011	
5	0101	
6	0110	
9	1001	
10	1010	
12	1100	
11	1011	
13	1101	
14	1110	

3 Variables		
Terms	$XYWZ$	P

2 Variables		
Terms	$XYWZ$	P

Now we can begin comparing all the group 1 and group 2 terms. When it is possible to combine two four-variable terms, we get three-variable implicants.

4 Variables			3 Variables			2 Variables		
Terms	$XYWZ$	P	Terms	$XYWZ$	P	Terms	$XYWZ$	P
1	0001	✓	1, 3	00−1				
2	0010	✓	1, 5	0−01				
3	0011	✓	1, 9	−001				
5	0101	✓	2, 3	001−				
6	0110	✓	2, 6	0−10				
9	1001	✓	2, 10	−010				
10	1010	✓						
12	1100							
11	1011							
13	1101							
14	1110							

Now we proceed to compare the terms of group 2 and group 3. See the tables below.

4 Variables			3 Variables			2 Variables		
Terms	$XYWZ$	P	Terms	$XYWZ$	P	Terms	$XYWZ$	P
1	0001	✓	1, 3	00−1				
2	0010	✓	1, 5	0−01				
3	0011	✓	1, 9	−001				
5	0101	✓	2, 3	001−				
6	0110	✓	2, 6	0−10				
9	1001	✓	2, 10	−010				
10	1010	✓	3, 11	−011				
12	1100	✓	5, 13	−101				
11	1011	✓	6, 14	−110				
13	1101	✓	9, 11	10−1				
14	1110	✓	9, 13	1−01				
			10, 11	101−				
			10, 14	1−10				
			12, 13	110−				
			12, 14	11−0				

A tic in column P regarding all the four-variable implicants (the *minterms*) indicates that none are prime.

The algorithm continues comparing the two groups of three-variable implicants to obtain two-variable implicants.

4 Variables			3 Variables			2 Variables		
Terms	$XYWZ$	P	Terms	$XYWZ$	P	Terms	$XYWZ$	P
1	0001	√	1, 3	00−1	√	1, 3, 9, 11	−0−1	P_2
2	0010	√	1, 5	0−01	√	1, 5, 9, 13	−−01	P_3
3	0011	√	1, 9	−001	√	2, 3, 10, 11	−01−	P_4
5	0101	√	2, 3	001−	√	2, 6, 10, 14	−−10	P_5
6	0110	√	2, 6	0−10	√			
9	1001	√	2, 10	−010	√			
10	1010	√	3, 11	−011	√			
12	1100	√	5, 13	−101	√			
11	1011	√	6, 14	−110	√			
13	1101	√	9, 11	10−1	√			
14	1110	√	9, 13	1−01	√			
			10, 11	101−	√			
			10, 14	1−10	√			
			12, 13	110−	P_0			
			12, 14	11−0	P_1			

Note that we can combine both $(1, 5) + (9, 13) \to (1, 5, 9, 13)$ and $(1, 9) + (5, 13) \to (1, 9, 5, 13)$ but they produce the same term $(−−01)$, $\overline{W}\, Z$, which is reported only once in the two-variable table. The pairs $(1, 5)$ and $(9, 13)$ and $(1, 9), (5, 13)$ both receive the tic because neither is a prime implicant.

In the end, there are six prime implicants: P_0 and P_1 with three variables (represented on the Karnaugh map as two cubes with two cells) and P_2, P_3, P_4, P_5 with two variables (represented on the Karnaugh map as four cubes with four cells).

The QM–M proceeds with the covering table to find the lowest number of implicants to cover the function.

	m_1	m_2	m_3	m_6	m_9	m_{10}	m_{11}	m_{12}
P_0								X
P_1								X
P_2	X		X		X		X	
P_3	X				X			
P_4		X	X			X	X	
P_5		X		X		X		

The table clearly indicates that P_5 is an essential prime implicant because it is the only one that covers m_6, so the minimum expression will certainly include it.

$$F(X, Y, W, Z) = P_5 + \cdots = W\,\overline{Z} + \cdots$$

So we can write a new table and eliminate row P_5, which has already been selected as well as columns m_2, m_6, m_{10}, which are covered by P_5:

	m_1	m_3	m_9	m_{11}	m_{12}
P_0					X
P_1					X
P_2	X	X	X	X	
P_3	X		X		
P_4		X		X	

The resulting table can be further simplified by analyzing coverage by the rows and columns and eliminating some of the prime implicants (the rows) or some of the *minterms* (the columns). Column m_i can be eliminated if it covers column m_j; that is, if for every X in column m_j there is an X in the corresponding row of column m_i. In this configuration, we can eliminate because the *minterm* m_i would be covered by one of the implicants that covers m_j, so there is no need to treat it.

In the table above, the columns m_9 and m_{11} can be eliminated because they cover m_1 and m_3, respectively, (in this case, they are actually equal):

	m_1	m_3	m_{12}
P_0			X
P_1			X
P_2	X	X	
P_3	X		
P_4		X	

Likewise, row P_i can be eliminated if it is covered by another implicant P_j; that is, if for every X in row P_i there is an X in the corresponding column of row P_j. We can eliminate P_i because the prime implicant P_j covers all the *minterms* covered by P_i (and possibly more).

In the table above, we can immediately see that P_1 covers P_0 and vice versa while P_2 covers P_3 and P_4. Thus, the final table is:

	m_1	m_3	m_{12}
P_0			X
P_2	X	X	

Now it is very easy to identify the optimal coverage, composed here by P_0 and P_2. By adding these two prime implicants to the already identified (P_5), we get the final result:

$$F(X, Y, W, Z) = P_0 + P_2 + P_5 = X\,Y\,\overline{W} + \overline{Y}\,Z + W\,\overline{Z}$$

In rare cases, not all the columns or rows can be eliminated in a coverage table. In this case, the table is referred to as "cyclic" and all the possible combinations of prime implicants must be checked to obtain optimal coverage. There are various ways (some optimal, some less so) to deal with this, but they go beyond the scope of this book.

4.1.5 Simultaneous Optimization of Multiple Functions

In real digital systems, we must often create different combinational networks, each corresponding to a Boolean function in the same project. In a case like this, we can benefit in terms of circuit complexity from jointly optimizing functions to reuse some parts of circuits shared by multiple networks. This is why the QM–M was extended to optimize more than one Boolean function. The expansion and covering phases were changed to identify the prime implicants that could be used to cover the *minterms* of multiple functions.

To understand how the multiple function method is extended, see the example below. Supposing we must optimize the following three functions:

$$F_1(X, Y, W, Z) = \Sigma(2, 3, 6, 10, 11, 12) + d(14)$$

$$F_2(X, Y, W, Z) = \Sigma(1, 3, 9, 11) + d(5, 13)$$

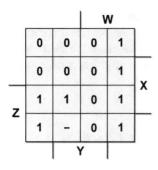

$$F_3(X, Y, W, Z) = \Sigma(1, 2, 3, 9, 10, 11) + d(5)$$

In the expansion phase, all the *minterms* of the three functions are reported in the first table. The difference from the case of only one function is that here, there is an added column that uses multi-bit masks to indicate how many functions to optimize and which ones contain the corresponding *minterm*.

4 Variables				3 Variables				2 Variables			
Terms	$XYWZ$	$F_1F_2F_3$	P	Terms	$XYWZ$	$F_1F_2F_3$	P	Terms	$XYWZ$	$F_1F_2F_3$	P
1	0001	011									
2	0010	101									
3	0011	111									
5	0101	011									
6	0110	100									
9	1001	011									
10	1010	101									
12	1100	100									
11	1011	111									
13	1101	010									
14	1110	100									

The *minterm* m_1, for example, appears in functions F_2 and F_3 but not in F_1. This is indicated by mask "011". Likewise, the *minterm* m_{14} appears in function F_1 but not in F_2 or F_3, and so the corresponding mask is "100".

In the expansion phase, the four-variable terms are combined to form three-variable terms. We must keep in mind that two terms can be combined only if they appear in the same function. For example, consider the combination of m_1 and m_3, i.e., (0001) and (0011), to obtain (00−1). This is possible only for functions F_2 and F_3, which contain both the terms. Function F_1 contains the term m_3 but not m_1, so it is impossible to simplify in this case. Thus, we will use mask "011" with (1, 3) to indicate that this term is present in only two out of the three functions: F_2 and F_3.

When applying the tic that indicates prime implicants, we must take into account which terms have been combined relative to which functions. The *minterm* m_1 has actually been simplified in all cases, so it is not a prime implicant, and the tic can be inserted in column P. The *minterm* m_3, however, has not been simplified with any other term in the function F_1, so the tic cannot be inserted because m_3 is still a potential prime implicant for this function at least during the expansion phase.

4 Variables				3 Variables				2 Variables			
Terms	$XYWZ$	$F_1F_2F_3$	P	Terms	$XYWZ$	$F_1F_2F_3$	P	Terms	$XYWZ$	$F_1F_2F_3$	P
1	0001	011	V	1, 3	00−1	011					
2	0010	101									
3	0011	111									
5	0101	011									
6	0110	100									
9	1001	011									
10	1010	101									
12	1100	100									
11	1011	111									
13	1101	010									
14	1110	100									

We can derive two simple rules from these considerations about combining terms and inserting the tic for implicants.

Rule 1

Two terms can be combined if we compute the bit-wise AND from the corresponding masks and at least one bit from the resulting mask differs from 0. We actually obtain this when there is a 1 in the same position in both masks; i.e., the terms being considered are both present in the same function. The resulting mask will be reported in the column of the simplified term. The previous case, where m_1 is combined with m_3 to obtain (1, 3): "011" AND "111" → "011".

Rule 2

If a term's mask is identical to the one resulting from the simplification, the corresponding *minterm* is definitely not a prime implicant, so the tic can be inserted. In the previous case, the resulting mask is "011" so the term m_1 is definitely not a prime implicant and we can insert the tic "∨." The masks of m_3 and (1, 3) are not the same (they are "111" and "011", respectively), so m_3 is still a potential prime implicant.

By continuing the simplification, we get the following table with three-variable terms.

4 Variables			
Terms	XYWZ	$F_1F_2F_3$	P
1	0001	011	V
2	0010	101	V
3	0011	111	V
5	0101	011	V
6	0110	100	V
9	1001	011	V
10	1010	101	V
12	1100	100	V
11	1011	111	V
13	1101	010	V
14	1110	100	V

3 Variabili			
Terms	XYWZ	$F_1F_2F_3$	P
1, 3	00−1	011	
1, 5	0−01	011	
1, 9	−001	011	
2, 3	001−	101	
2, 6	0−10	100	
2, 10	−010	101	
3, 11	−011	111	
5, 13	−101	010	
6, 14	−110	100	
9, 11	10−1	011	
9, 13	1−01	010	
10, 11	101−	101	
10, 14	1−10	100	
12, 14	11−0	100	

2 Variables			
Terms	XYWZ	$F_1F_2F_3$	P

Note that the *minterms* m_{12} and m_{13} were not combined because m_{12} is only present in function F_1 ("100") and m_{13}, only in function F_2 ("010").

Now we can combine the three-variable terms to obtain two-variable terms if possible.

4 Variables			
Terms	XYWZ	$F_1F_2F_3$	P
1	0001	011	V
2	0010	101	V
3	0011	111	V
5	0101	011	V
6	0110	100	V
9	1001	011	V
10	1010	101	V
12	1100	100	V
11	1011	111	V
13	1101	010	V
14	1110	100	V

3 Variables			
Terms	XYWZ	$F_1F_2F_3$	P
1, 3	00−1	011	V
1, 5	0−01	011	P_0
1, 9	−001	011	V
2, 3	001−	101	V
2, 6	0−10	100	V
2, 10	−010	101	V
3, 11	−011	111	P_1
5, 13	−101	010	V
6, 14	−110	100	V
9, 11	10−1	011	V
9, 13	1−01	010	V
10, 11	101−	101	V
10, 14	1−10	100	V
12, 14	11−0	100	P_2

2 Variables			
Terms	XYWZ	$F_1F_2F_3$	P
1, 3, 9, 11	−0−1	011	P_3
1, 5, 9, 13	−−01	010	P_4
2, 3, 10, 11	−01−	101	P_5
2, 6, 10, 14	−−10	100	P_6

Here as well, we must pay attention to the masks that indicate which functions we can continue simplifying and which are the prime implicants.

For example, when we combine (1, 3) with (9, 11) we get (1, 3, 9, 11) for functions F_2 and F_3 ("011"), and both starting terms are non-prime implicants.

However, when we combine (1, 9) with (3, 11) we get the same term (1, 3, 9, 11) for the same functions (F_2 and F_3) but (3, 11) is still a prime implicant because it was impossible to simplify it with another term in function F_1. The two masks are actually "011" and "111".

When the expansion phase is finished, we can move on to the covering phase, considering each function separately.

	F_1						F_2				F_3					
	m_2	m_3	m_6	m_{10}	m_{11}	m_{12}	m_1	m_3	m_9	m_{11}	m_1	m_2	m_3	m_9	m_{10}	m_{11}
P_0							X				X					
P_1		X		X				X		X		X				X
P_2						X										
P_3							X	X	X	X	X	X	X			X
P_4							X		X							
P_5	X	X		X	X							X	X		X	X
P_6	X		X	X												

Masks are useful here as well because they indicate what functions the prime implicant should be associated with. The prime implicant P_4, for example, covers the minterms m_1, m_5, m_9, and m_{13} only for function F_2, because the corresponding mask is "010", so Xs are not inserted for the other functions (F_1 and F_3).

As before, we go on to identify the essential prime implicants, in this case: P_2 (because it is the only one that covers the minterm m_{12} in F_1), P_3 (because it is the only one that covers the minterm m_9 of F_3), P_5 (because it is the only one that covers the minterms m_2 and m_{10} in F_3) and P_6 (because it is the only one that covers the minterm m_6 in F_1). Note that when a prime implicant is selected, it is selected for all the functions. The implicant P_5, for example, is selected because it is essential for F_3 but at that point it is also used to cover the minterms of F_1. In other words P_5, which corresponds to the term (2, 3, 10, 11) i.e., $\overline{Y}W$, certainly appears in the optimal expression of F_3 but can also be used for F_1. This way, the same logical network is used twice economizing on the circuit level.

By selecting P_2, P_3, P_5 and P_6 we cover all the minterms in all the functions and we now can write the resulting functions:

$$F_1(X, Y, W, Z) = P_2 + P_5 + P_6 = XY\overline{Z} + \overline{Y}W + W\overline{Z}$$
$$F_2(X, Y, W, Z) = P_3 = \overline{Y}Z$$
$$F_3(X, Y, W, Z) = P_3 + P_5 = \overline{Y}Z + \overline{Y}W$$

Note that the P_3 and P_5 combinational networks are used twice in different functions providing some savings in the overall complexity of the circuit.

4.2 Exercises

4.2.1 Quine–McCluskey: Single Function Synthesis

1. Synthesize the following Boolean function with the QM–M

$$F(A, B, C, D) = \Sigma(5, 7, 8, 9, 12, 13, 15) + d(4)$$

4 Variables		
Terms	$ABCD$	P

3 Variables		
Terms	$ABCD$	P

2 Variables		
Terms	$ABCD$	P

	m_5	m_7	m_8	m_9	m_{12}	m_{13}	m_{15}
P_0							
$\bar{P}_1$							
$\bar{P}_2$							
$\bar{P}_3$							
$\bar{P}_4$							
$\bar{P}_5$							
$\bar{P}_6$							
$\bar{P}_7$							
$\bar{P}_8$							
$\bar{P}_9$							
$\bar{P}_{10}$							
$\bar{P}_{11}$							
$\bar{P}_{12}$							
$\bar{P}_{13}$							
$\bar{P}_{14}$							
$\bar{P}_{15}$							

2. Synthesize the following Boolean functions with the QM–M

$$F(A, B, C, D) = \Sigma(0, 1, 2, 3, 4, 5, 15) + d(10, 14)$$

4 Variables		
Terms	$ABCD$	P

3 Variables		
Terms	$ABCD$	P

2 Variables		
Terms	$ABCD$	P

	m_0	m_1	m_2	m_3	m_4	m_5	m_{15}
$\bar{P}_0$							
$\bar{P}_1$							
$\bar{P}_2$							
$\bar{P}_3$							
$\bar{P}_4$							
$\bar{P}_5$							
$\bar{P}_6$							
$\bar{P}_7$							
$\bar{P}_8$							
$\bar{P}_9$							
$\bar{P}_{10}$							
$\bar{P}_{11}$							
$\bar{P}_{12}$							
$\bar{P}_{13}$							
$\bar{P}_{14}$							
$\bar{P}_{15}$							

3. Synthesize the following Boolean function with the QM–M

$$F(A, B, C, D) = \Sigma(1, 9, 11, 13, 15) + d(0, 4, 5)$$

4 Variables				3 Variables				2 Variables		
Terms	$ABCD$	P		Terms	$ABCD$	P		Terms	$ABCD$	P

	m_1	m_9	m_{11}	m_{13}	m_{15}
P_0					
P_1					
P_2					
P_3					
P_4					
P_5					
P_6					
P_7					
P_8					
P_9					
P_{10}					
P_{11}					
P_{12}					
P_{13}					
P_{14}					
P_{15}					

4.2.2 Quine–McCluskey: Jointly Synthesis of Multiple Functions

1. Jointly synthesize the following Boolean functions with the QM–M

$$F_1(A, B, C, D) = \Sigma(5, 7, 13, 15) + d(4, 12)$$
$$F_2(A, B, C, D) = \Sigma(8, 9, 12, 13) + d(4, 5)$$
$$F_3(A, B, C, D) = \Sigma(4, 5, 12, 13)$$

4 Variables				3 Variables				2 Variables			
Terms	$ABCD$	$F_1F_2F_3$	P	Terms	$ABCD$	$F_1F_2F_3$	P	Terms	$ABCD$	$F_1F_2F_3$	P

	F_1				F_2				F_3			
	m_5	m_7	m_{13}	m_{15}	m_8	m_9	m_{12}	m_{13}	m_4	m_5	m_{12}	m_{13}
P_0												
P_1												
P_2												
P_3												
P_4												
P_5												
P_6												
P_7												
P_8												
P_9												
P_{10}												
P_{11}												
P_{12}												
P_{13}												
P_{14}												
P_{15}												

2. Jointly synthesize the following Boolean functions with the QM–M

$$F_1(A, B, C, D) = \Sigma(0, 1, 2, 15) + d(3)$$
$$F_2(A, B, C, D) = \Sigma(2, 3, 4, 5) + d(0, 1)$$
$$F_3(A, B, C, D) = \Sigma(0, 1, 4, 5)$$

4 Variables			
Terms	$ABCD$	$F_1 F_2 F_3$	P

3 Variables			
Terms	$ABCD$	$F_1 F_2 F_3$	P

2 Variables			
Terms	$ABCD$	$F_1 F_2 F_3$	P

	F_1				F_2				F_3			
	m_0	m_1	m_2	m_{15}	m_2	m_3	m_4	m_5	m_0	m_1	m_4	m_5
P_0												
P_1												
P_2												
P_3												
P_4												
P_5												
P_6												
P_7												
P_8												
P_9												
P_{10}												
P_{11}												
P_{12}												
P_{13}												
P_{14}												
P_{15}												

3. Jointly synthesize the following Boolean functions with the QM–M

$$F_1(A, B, C, D) = \Sigma(1, 4, 5) + d(13)$$
$$F_2(A, B, C, D) = \Sigma(1, 9, 13) + d(0, 5)$$
$$F_3(A, B, C, D) = \Sigma(9, 11, 13, 15) + d(1, 5)$$

4 Variables			
Terms	$ABCD$	$F_1 F_2 F_3$	P

3 Variables			
Terms	$ABCD$	$F_1 F_2 F_3$	P

2 Variables			
Terms	$ABCD$	$F_1 F_2 F_3$	P

	F_1			F_2			F_3			
	m_1	m_4	m_5	m_1	m_9	m_{13}	m_9	m_{11}	m_{13}	m_{15}
P_0										
P_1										
P_2										
P_3										
P_4										
P_5										
P_6										
P_7										
P_8										
P_9										
P_{10}										
P_{11}										
P_{12}										
P_{13}										
P_{14}										
P_{15}										

4.3   Solutions

4.3.1   Quine–McCluskey: Synthesis of a Single Function

1. $F(A, B, C, D) = B\,D + A\,\overline{C}$

4 Variables			3 Variables			2 Variables		
Terms	$ABCD$	P	Terms	$ABCD$	P	Terms	$ABCD$	P
4	0100	√	5, 4	010–	√	13, 12, 9, 8	1–0–	P_0
8	1000	√	9, 8	100–	√	13, 12, 5, 4	–10–	$\bar{P_1}$
5	0101	√	12, 4	–100	√	15, 13, 7, 5	–1–1	P_2
9	1001	√	12, 8	1–00	√			
12	1100	√	7, 5	01–1	√			
7	0111	√	13, 5	–101	√			
13	1101	√	13, 9	1–01	√			
15	1111	√	13, 12	110–	√			
			15, 7	–111	√			
			15, 13	11–1	√			

	m_5	m_7	m_8	m_9	m_{12}	m_{13}	m_{15}
P_0			X	X	X	X	
$\bar{P_1}$	X̄				X̄	X̄	
P_2	X	X				X	X

2. $F(A, B, C, D) = A\,B\,C + \overline{A}\,\overline{C} + \overline{A}\,\overline{B}$

4 Variables			3 Variables			2 Variables		
Terms	$ABCD$	P	Terms	$ABCD$	P	Terms	$ABCD$	P
0	0000	√	1, 0	000–	√	3, 2, 1, 0	00– –	P_3
1	0001	√	2, 0	00–0	√	5, 4, 1, 0	0–0–	$\bar{P_4}$
2	0010	√	4, 0	0–00	√			
4	0100	√	3, 1	00–1	√			
3	0011	√	3, 2	001–	√			
5	0101	√	5, 1	0–01	√			
10	1010	√	5, 4	010–	√			
14	1110	√	10, 2	–010	P_0			
15	1111	√	14, 10	1–10	P_1			
			15, 14	111–	P_2			

	m_0	m_1	m_2	m_3	m_4	m_5	m_{15}
P_0			X				
P_1							
P_2							X
P_3	X	X	X	X			
$\bar{P_4}$	X̄	X̄			X̄	X̄	

3. $F(A, B, C, D) = A\,D + \overline{C}\,D$

4 Variables Terms	ABCD	P
0	0000	V
1	0001	V
4	0100	V
9	1001	V
5	0101	V
11	1011	V
13	1101	V
15	1111	V

3 Variables Terms	ABCD	P
1, 0	000–	V
4, 0	0–00	V
5, 1	0–01	V
5, 4	010–	V
9, 1	–001	V
11, 9	10–1	V
13, 5	–101	V
13, 9	1–01	V
15, 11	1–11	V
15, 13	11–1	V

2 Variables Terms	ABCD	P
5, 4, 1, 0	0–0–	P_0
13, 5, 9, 1	––01	P_1
15, 13, 11, 9	1––1	P_2

	m_1	m_9	m_{11}	m_{13}	m_{15}
P_0	X				
P_1	X	X		X	
P_2		X	X	X	X

4.3.2 Quine–McCluskey: Joint Synthesis of Multiple Functions

1. The solution is

$$F_1(A, B, C, D) = B\,\overline{C}$$

$$F_2(A, B, C, D) = A\,\overline{C}$$

$$F_3(A, B, C, D) = B\,D$$

4 Variables Terms	ABCD	$F_1 F_2 F_3$	P
4	0100	111	V
8	1000	010	V
5	0101	111	V
9	1001	010	V
12	1100	111	V
7	0111	100	V
13	1101	111	V
15	1111	100	V

3 Variables Terms	ABCD	$F_1 F_2 F_3$	P
5, 4	010–	111	V
4, 12	–100	111	V
8, 9	100–	010	V
8, 12	1–00	010	V
5, 7	01–1	100	V
5, 13	–101	111	V
9, 13	1–01	010	V
12, 13	110–	111	V
7, 15	–111	100	V
13, 15	11–1	100	V

2 Variables Terms	ABCD	$F_1 F_2 F_3$	P
4, 5, 12, 13	–10–	111	P_0
8, 9, 12, 13	1–0–	010	P_1
5, 7, 13, 15	–1–1	100	P_2

	F_1				F_2				F_3			
	m_5	m_7	m_{13}	m_{15}	m_8	m_9	m_{12}	m_{13}	m_4	m_5	m_{12}	m_{13}
P_0	X		X				X	X	X	X	X	X
P_1					X	X	X	X				
P_2	X	X	X	X								

2. The solution is

$$F_1(A, B, C, D) = A\,B\,C\,D + \overline{A}\,\overline{B}$$
$$F_2(A, B, C, D) = \overline{A}\,\overline{B} + \overline{A}\,\overline{C}$$
$$F_3(A, B, C, D) = \overline{A}\,\overline{C}$$

4 Variables			
Terms	$ABCD$	$F_1F_2F_3$	P
0	0000	111	✓
1	0001	111	✓
2	0010	110	✓
4	0100	011	✓
3	0011	110	✓
5	0101	011	✓
15	1111	100	P_0

3 Variables			
Terms	$ABCD$	$F_1F_2F_3$	P
0, 1	000–	111	P_1
0, 2	00–0	110	✓
0, 4	0–00	011	✓
1, 3	00–1	110	✓
1, 5	0–01	011	✓
2, 3	001–	110	✓
4, 5	010–	011	✓

2 Variables			
Terms	$ABCD$	$F_1F_2F_3$	P
0, 1, 2, 3	00– –	110	P_2
0, 1, 4, 5	0–0–	011	P_3

		F_1				F_2				F_3			
	m_0	m_1	m_2	m_{15}	m_2	m_3	m_4	m_5	m_0	m_1	m_4	m_5
P_0				X								
P_1	X	X							X	X		
P_2	X	X	X		X	X						
P_3							X	X	X	X	X	X

3. The solution is

$$F_1(A, B, C, D) = \overline{A}\,\overline{C}\,D + \overline{A}\,B\,\overline{C}$$
$$F_2(A, B, C, D) = \overline{C}\,D$$
$$F_3(A, B, C, D) = A\,D$$

4 Variables			
Terms	$ABCD$	$F_1F_2F_3$	P
0	0000	010	✓
1	0001	111	✓
4	0100	100	✓
5	0101	111	✓
9	1001	011	✓
11	1011	001	✓
13	1101	111	✓
15	1111	001	✓

3 Variables			
Terms	$ABCD$	$F_1F_2F_3$	P
0, 1	000–	010	P_0
1, 5	0–01	111	P_1
1, 9	–001	011	✓
4, 5	010–	100	P_2
5, 13	–101	111	P_3
9, 11	10–1	001	✓
9, 13	1–01	011	✓
11, 15	1–11	001	✓
13, 15	11–1	001	✓

2 Variables			
Terms	$ABCD$	$F_1F_2F_3$	P
1, 5, 9, 13	– –01	011	P_4
9, 11, 13, 15	1– –1	001	P_5

		F_1				F_2				F_3			
	m_1	m_4	m_5	m_1	m_9	m_{13}	m_9	m_{11}	m_{13}	m_{15}
P_0				X						
P_1	X		X	X						
P_2		X	X							
P_3			X			X			X	
P_4				X	X	X	X		X	
P_5							X	X	X	X

Chapter 5
Introduction to Sequential Networks

Abstract The transition from combinational to sequential networks is explained step by step, starting from a simple gate with feedback and arriving to the structure and behavior of the principal types of flip-flops. They are classified according to their temporal response (direct command, level enabled, master–slave, and edge triggered) and the logical operation (SR, D, JK). The timing parameters of physically implemented devices are considered. The chapter introduces the concept and techniques for synchronization that will be further examined in the following ones.

It is very rare for a digital device to be based only on combinational networks. In a real situation, it is important to have devices that can memorize data, generate sequences, and respond to conditions that change over time.

5.1 From Combinational Networks to Sequential Networks

We have seen combinational networks (see RC in the figure below left) where, at any given moment, the output U is the function of only the I inputs. Combinational networks are identified precisely by the fact that every input combination *always* produces the same output $U = f(I1, I2, \ldots In)$.

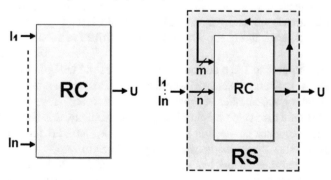

© Springer International Publishing AG, part of Springer Nature 2019
G. Donzellini et al., *Introduction to Digital Systems Design*,
https://doi.org/10.1007/978-3-319-92804-3_5

A *sequential network* (RS, below right) does not follow this rule. The same input combination can generate different outputs when applied at different moments.

We can obtain a sequential network by starting with a combinational network and bringing one or more of its outputs into the inputs, as seen in the figure above. This type of connection is called *feedback*.

Generally speaking, connecting *m* outputs of a combinational network to as many inputs makes it so that the outputs' behavior as a function of the inputs depends *in part on the outputs themselves*. If we consider recursively that each set of current outputs was produced by the inputs plus the preceding outputs, we can say that the outputs depend not only on the *current* inputs but on their *history*.

In other words, the functioning of the network depends on the *sequence of inputs* that produce a *sequence of outputs*. A logical network that behaves this way is unsurprisingly called a *sequential network*.

Next we will see that in sequential networks, the outputs, the inputs, and the special conditions of the network at that moment (called *"state"*) can all be expressed analytically. As we will see, the concept of *state* will allow us to concisely express the *history* of the network.

5.1.1   Introductory Example

Let's look at this simple example of a sequential network that uses the OR function, which as we know is a combinational network. Let's construct a *feedback* by bringing the output *U* to one of the two inputs (*B*), as in the following figure.

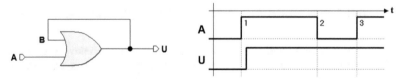

Let's try to understand how this network acts, operating on the only available input *A*, since *B* is already driven by *U*, so it is unavailable. Let's suppose that *A* and *U* (and therefore *B*) are initially equal to 0. This is described in the timing diagram to the right.

If we force input *A* to 1 at a certain moment, the output will go to 1. If this were a combinational network this change would have exhausted the number of possible cases (two), taking the number of inputs (one) into account. If we reduce input *A* to zero, this should also force output *U* to zero, but actually, this does not happen. Because of the feedback connection between *U* and *B*, output *U* remains *forced* at 1 for any input value *A* so it is impossible to get it back to zero.

We can say that the network has *memorized* the value 1. Note that the diagram shows in an approximate way the propagation time of the OR gate.

Evidently, this is no longer a combinational network but a *sequential* one, where the output value depends on the *history* of the inputs. We can no longer describe how it works through a truth table like we did with combinational networks.

5.1.2 *Memorizing an Information Bit: Flip-Flops*

If we find a way to force the output to 0, the simple sequential network we have just seen can be used to store an *information bit*, i.e., to memorize both the value 0 and 1.

A potential change is shown in the figure below, where an AND gate was inserted in the feedback loop and the input C was added.

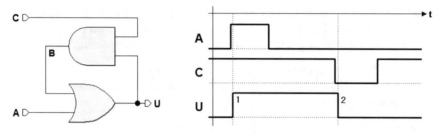

The AND gate and input C were added to establish or remove the connection between U and B. If $C = 1$ the network will behave as before since every variation of U is transferred through the AND gate on B ($B = U \cdot 1 = U$). If $C = 0$, then B is at 0, regardless of the value of U ($B = U \cdot 0 = 0$).

In the timing diagram in the figure, we assume that we start with input A and output U at 0 and with input C at 1. Then, the activation of the input A forces the output to 1. The new value $U = 1$, brought from the AND gate on input B of the OR, makes it so that further variations of A can no longer change the output. We have memorized a bit at 1.

To force the output U to 0, we will have to *open* the feedback loop by applying the value 0 to input C, as shown in the timing diagram. Forcing the output to zero memorizes a bit at 0. Further variations of C do not change the situation. As before, in the figure the propagation times are represented in a approximated way.

Note that each input can be considered in two ways:

- the purpose that each command has in the network;
- the logical modality that produces an effect.

In this case, input A produces the effect of memorizing 1 in the output, which happens when it is brought to 1. This is called an *active-high* input, that is, it does its job when it is *brought to* 1 and is *inactive* when it is at 0.

Input C has the job of memorizing a 0, which happens when it is *brought to* 0. We say that this input at 0 is *active-low* and when it is *inactive* it is at 1.

With this sequential network, we are able to *memorize a bit* whose value is maintained until a new value is memorized. This is one of the ways to create an *elementary memory cell*, also called *bistable element*, *one-bit register* or, more commonly, "*flip-flop.*" Flip-flops are the basic logical elements generally used to build sequential digital systems.

Now let's try to slightly modify the network, making the commands symmetrical and easier to operate. We want to gradually build the network analyzed in the paragraph below, the classic elementary memory cell called *Set-Reset flip-flop*.

To achieve this, we make the following changes. Let's

- add a NOT before input C, to make *active-high* the command memorizing the zero.
- call the new input we obtain $RESET$ (because it brings the output to 0).
- change the name of input A to SET (because it brings the output to 1).
- change the name of output U to Q (to follow an established naming tradition).

The network with these changes appears in the figure below left. To continue transforming it, we apply De Morgan's theorem and substitute the OR with an AND (in the figure at the right).

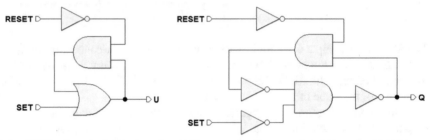

The NOTs directly following the ANDs suggest that we should use the NANDs to get the network below left. The transformation is now complete; it is convenient to re-draw the network to highlight its symmetry without altering the connections (in the figure below right).

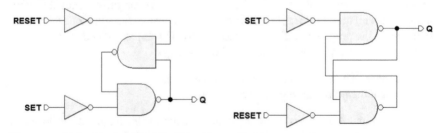

The network in the right is called *Set-Reset flip-flop*.

5.1.3   Flip-Flop Classification: Logical Type and Command Type

In the last section, we saw how to derive the sequential structure called the *Set-Reset flip-flop*. This is one of many flip-flops used to memorize a bit of information. Flip-flops are fundamental blocks used to build more complex sequential networks. We classify the flip-flops according to their *"logical type"* and to their *"command type."*

The Logical Type

The type of flip-flop whose function is to activate the output (*Set* it) or deactivate it (*Reset* it) is the logical *Set-Reset*. Further on, we will examine other logical types, such as *D* (*Delay*), which memorizes the output value of the unique input *D* and the variant *JK* of the *Set-Reset*. This variant substitutes inputs *J* and *K* with *S* and *R*, respectively, and allows the output's state to reverse. We will also see logical types *T* and *E* that derive from *JK* and *D*, respectively.

As we will see, the logical type is described by its *function table*, which indicates the output values in function of the *logical inputs*, that is, the inputs that characterize the logical type of flip-flop and that give it its name.

The Command Type

The command type describes input behavior, which comes in three forms: *direct*, *level-enabled*, or *edge-triggered*.

In the case of direct command, logical input action is not subordinated to any other enable or synchronization input but it directly controls the behavior of the flip-flop which, in this case, is called *"asynchronous."*

In the other two cases, there is an added input that enables/disables logical inputs. Enabling can happen simply as a function of the logical level of this added input (*level-enabled* command), or when it presents a logical level transition (*edge-triggered* command).[1] In this case, the flip-flop is called *"synchronous."*

Let's turn our attention to direct command types. Before we begin to deal with other types of commands, we will need to examine some important concepts like initialization and synchronization of sequential networks.

5.2   Direct Command Flip-Flops

Now, let's look at the three types of direct commands: *SR*, *D*, and *JK*. The *SR* type has already been introduced but some of its possible variants will be dealt with here.

5.2.1   SR Flip-Flop

SR Flip-Flop (Active-High Commands, NAND Version)

The schematic in the figure below represents a *Set-Reset* (SR) flip-flop built with NAND (and NOT) gates and *active-high* inputs. This is like the version examined above but with the added output $\overline{Q}$, which takes the opposite value of *Q* under normal operating conditions, as we will soon see. This is the structure that implements the

---

[1] The transition can be *"positive"* (from 0 to 1, or *"rising edge"*) or *"negative"* (from 1 to 0, the *"falling edge"*).

Set-Reset logical type with *direct command*. Drawn on the right side of the figure is the logical symbol that represents this flip-flop in schematics:

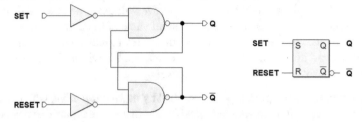

As we have seen before, this is a sequential network that can *memorize a bit*. The activation of the input SET memorizes a 1 in the cell, while the input $RESET$ imposes a 0. Given their opposite actions, it would naturally make little sense to activate SET and $RESET$ at the same time. Below, its *function table* shows what we have just described in English. The table gives both the outputs Q and $\overline{Q}$.

Set-Reset Flip-flop (Active-high Commands)				
SET	$RESET$	Q	$\overline{Q}$	
1	1	Q_p	$\overline{Q_p}$	Previous state
0	1	1	0	SET command
1	0	0	1	$RESET$ command
0	0	1	1	Invalid

The table shows that if the inputs are kept *idle* (at 0), the flip-flop stays in its previous state (Q_p, $\overline{Q_p}$), that is it maintains the previously memorized bit in the output. The next two rows describe what the SET and $RESET$ commands do. To be thorough, the last row in the table shows the *invalid* case, where both command inputs are activated.

In this network, the *invalid* combination simultaneously activates the two outputs: Q and $\overline{Q}$. The invalid configuration deserves to be treated separately later on because this limit condition has technically interesting aspects that depend in part on the specific configuration of the circuit.

The following timing diagram, obtained with the *Deeds* simulator, shows how the flip-flop behaves under various driving conditions. The duration of the signals and the visual scale of the diagram were chosen in order to show the *delay times* between the activation of SET, $RESET$, and the outputs Q and $\overline{Q}$, as evaluated by the simulator:

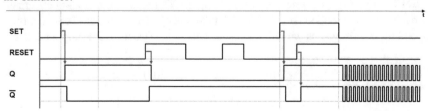

Let's analyze the behavior of the network point by point.

- In the simulated interval, we assume $Q = 0$ and $\overline{Q} = 1$ at the beginning.
- The activation of SET forces output Q to 1 and, accordingly, $\overline{Q}$ to 0.
- After SET is deactivated, $Q = 1$ remains memorized in the flip-flop.
- Activating $RESET$ forces output Q to zero.
- After $RESET$ is deactivated, $Q = 0$ remains memorized.
- Further activations and deactivations of the $RESET$ input produce no changes because output Q is already at zero.
- Output Q changes to 1 only on the next activation of SET.

In the final part of the diagram, we see the effects of applying the *invalid* input configuration. When SET and $RESET$ are both active, outputs Q and $\overline{Q}$ are both forced to 1. By examining the logical network of the flip-flop, it is easy to verify that the feedback has no effect on the NAND inputs under this condition since there is a 0 on the other input.

The network has lost both the feedback and the data memory! We can get out of this anomalous situation with no problem if we first deactivate one of the inputs and then the other. We will then get to an input configuration that would force the memorization of a known value that, at that moment, is coherent with the value of SET and $RESET$.

The timing diagram shows the *limit condition* where the inputs SET and $RESET$ are simultaneously deactivated. Normally, if SET and $RESET$ are both inactive, we get the memorization of a value. In this case, however, the information has just been lost. The simulator shows that the outputs oscillate. That is, Q and $\overline{Q}$ periodically switch between the two levels, at the same time and with the same logical value.

The simulation produces this result because the model of the components is simplified and idealized; the logical gates have the same delay time (transport). When the inputs switch at the same time, they cause the outputs to switch at the same time, too. The feedback brings them to the inputs, which then causes a further change in the level of both the outputs together and on the same instant, and so on.

This behavior would be very unlikely in a real network. In reality, the logical components are *nonlinear amplifiers*. An in-depth study of their behavior in the limit case examined here could become very complex and involve concepts like *metastability*, which will be dealt with further on.

For our purposes, it suffices to point out that the real gates' propagation times are similar but not identical. One of the gates will be faster than the other, and the feedback quickly ends up stabilizing the outputs by forcing the flip-flop into the memorization condition (although to an a priori *unknown value*).

SR Flip-Flop (Active-Low Commands, NAND Version)

If the two inverters are eliminated, we get the same flip-flop with active-low commands. Because of its simplicity, the SR flip-flop with this type of command forms the *base structure* on witch other logical types are built on.

Hereafter, we will refer to this network as *"base cell."*

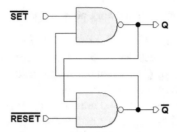

Obviously, this type is not substantially different from the type with active-high commands. Only the logical level of the command changes. Below, we have its *function table*. As we can see, this is just like the previous one except that the inputs are active-low.

Flip-flop *Set-Reset* (active-low commands)				
$\overline{SET}$	$\overline{RESET}$	Q	$\overline{Q}$	
1	1	Q_p	$\overline{Q_p}$	Previous state
0	1	1	0	*SET* command
1	0	0	1	*RESET* command
0	0	1	1	Invalid

The timing diagram below shows a simulation sequence that is identical to the previous one but with complemented input signals.

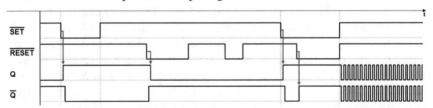

SR Flip-Flop (Active-High Commands, NOR Version)

To be thorough, let's now outline a version of the *Set-Reset* flip-flop with NOR gates. To obtain that, we go back to the network composed only of an AND and an OR, seen previously, but this time let's transform it into a network with only NOR gates. Ignoring the intermediate steps where the NOT gates are eliminated, we obtain the following network:

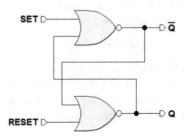

This cell has active-high inputs. Note that, unlike the NAND version, here, output Q is generated by the gate that receives the $RESET$ signal and output $\overline{Q}$ by the gate that receives SET.

Despite its simplicity, this NOR version is not commonly used since it is more convenient to use NAND gates in many technologies. Here is its function table.

Set-Reset Flip-flop (NOR gate version)				
SET	$RESET$	Q	$\overline{Q}$	
0	0	Q_p	$\overline{Q_p}$	Previous state
1	0	1	0	SET command
0	1	0	1	$RESET$ command
1	1	0	0	Invalid

This is similar to the table for the flip-flop with NAND gates except for its behavior in the invalid condition. Here, when both inputs are activated, the network responds with both outputs at 0. The analysis made before on simultaneously deactivating the inputs is still valid.

5.2.2 D Flip-Flop

Here, we introduce the concept of the logical type D flip-flop. We want one single data input D, rather than two separate SET and $RESET$ controls. The idea is to simply memorize the bit by submitting it at the input and to avoid the problems related to the *invalid* configuration.

In the figure below, on the left, we get rid of one input by adding a NOT to the SR flip-flop structure with active-high inputs. The data in input D is applied to the SET, while we attach $\overline{D}$ to the $RESET$. Once the two cascading NOTs are eliminated, we get the network below right:

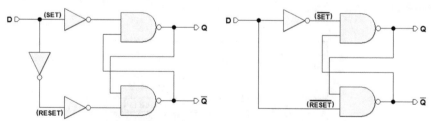

The NOT also assures that the inputs $\overline{SET}$ and $\overline{RESET}$ of the base cell will never have the same logical level. This prevents the critical condition previously discussed.

However, as seen in the timing diagram, output Q always reproduces input D. There is no command configuration that memorizes the data so this circuit is useless in the direct command version.

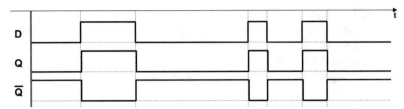

The D flip-flop is an important functional element in the *enabled command* versions, as we will see further on.

5.2.3   JK Flip-Flop

An efficient approach to eliminating the invalid configuration is the JK flip-flop with direct commands. This circuit derives from the *Set-Reset* flip-flop with active-low inputs plus two NAND gates, as shown in the next figure. The new inputs are assigned the name J and K (hence the name JK, in honor of *Jack Kilby* for his contribution to the birth of integrated electronics).

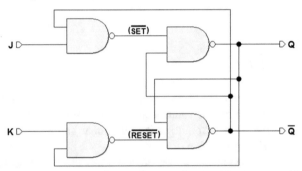

Here, we see that through one of the two new NANDs the input J drives the $\overline{SET}$ conditioned by $\overline{Q}$. Likewise, the input K drives the $\overline{RESET}$ conditioned by Q. This way, $\overline{SET}$ and $\overline{RESET}$ can never be activated at the same time. Here, the function table describes how inputs J and K act on the outputs:

JK Flip-flop				
J	K	Q	$\overline{Q}$	
0	0	Q_p	$\overline{Q_p}$	Previous state
1	0	1	0	SET command
0	1	0	1	$RESET$ command
1	1	$\overline{Q_p}$	Q_p	Toggle

In the first row of the table, J and K are both at 0, so $\overline{SET}$ and $\overline{RESET}$ are at 1, and the outputs keep the previous value. In the timing simulation below, we find this input configuration in the "MEM" time intervals:

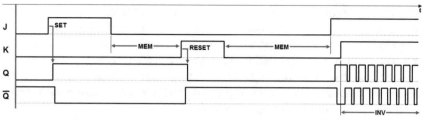

In the second row of the table $J = 1$ and $K = 0$, so the input $\overline{SET}$ is activated and output Q is forced to 1 (operation identified as "SET" in the timing simulation). In the third row, we have the opposite case and output Q is at zero ("RESET" in the same figure).

The JK flip-flop is different from the SR type because the invalid condition in SR is used to accomplish a useful function, the *inversion* of the output values (as shown in the last line of the table), also known as "*toggle*." When both inputs J and K are at 1, $\overline{SET}$ is activated if $Q = 0$, or $\overline{RESET}$ is activated if $Q = 1$. This reverses the outputs (time interval indicated as "INV" in the timing diagram).

Further on, we will see that the *toggle* function is only completely usable in *edge-triggered* JK flip-flops. Under the conditions in the timing diagram above, that is with J and K kept at 1 for a long enough time, outputs Q and $\overline{Q}$ reverse their values *continually*. At every change of outputs Q and $\overline{Q}$, the feedback reverses the values in $\overline{SET}$ and $\overline{RESET}$, in turn creating another inversion, thus giving rise to a cyclical behavior.

5.3 Initialization of a Sequential Network

Before we continue, a small digression: in the timing diagrams that we have studied, the sequences always purposefully begin with $Q = 0$ and $\overline{Q} = 1$, as in the following example:

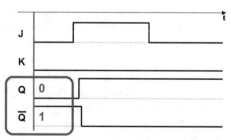

A specific moment in the normal functioning of the network has been chosen as the time to *begin to observe* the network. In the previous section, we willfully hid an important aspect of all sequential networks. To simplify things, we avoided mentioning the fact that we must launch the network with a *known configuration*. This problem involves all sequential networks and will be dealt with further when we study aspects of their design.

At the launch of a sequential network, we say in technical terms, that it must be *"initialized"* so that it can work coherently in normal operation. There will be many flip-flops in a real network: when the system is activated, in absence of specific measures, every one of these will tend to take on a *random value.*

We can see that this is unacceptable for a complex network so we need to turn to circuit techniques that allow us to supply every flip-flop with a known value *before* launching the normal functioning of the network.

In simpler networks, initialization could be irrelevant. For example, for a circuit that makes an LED light flash on a panel, it doesn't make much difference if the LED lights immediately when turned on or whether it lights half a cycle later. In almost all real cases, however, it is unacceptable for the network to start at *random* values.

Think of a digital network that controls something *intrinsically dangerous*, like opening the bulkheads in the dyke of a hydroelectric basin. We could not afford for the flip-flops to be positioned randomly at the start of a system that opens the bulkheads. Rather, we need to take every precaution so that they stay *rigorously closed*, opening only after an explicit command.

Let's return our attention to the JK flip-flop with direct commands. With a real component, when powering on the circuit, output Q of the elementary memory cell will be forced to an *a priori unknown* value. From that moment, the network will start to work (even though it starts from a random value, 0 or 1) and will follow the path imposed by the evolution of the inputs.

Still, if we try to *simulate* the network we see that the simulator cannot *resolve* the network's behavior and it gives us back the result in the figure below, where the bands represent an *unknown* value.

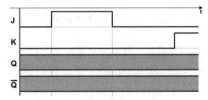

The simulator's behavior is *formally correct*: since it does not know the initial value of outputs Q and $\overline{Q}$, the simulator cannot calculate the following values. As we have seen, they depend on the inputs and also on the previous value of the outputs, which is unknown.

The simulator is a *tool for development and verification*, and for it to be efficient, the developer must be able to see errors and oversights. Here, the simulator tells us that the network *is not formally able* to begin working from a *known configuration.*

If the simulator resolved these situations by "hypothesizing" starting values, this would not do us any great favor because it could mask possible design errors.

5.3.1   Flip-Flop Initialization Inputs

So how was it possible to do a JK flip-flop network simulation? It was actually another network that was simulated, as you see in the next figure. Here an *auxiliary* initialization input called $\overline{Clear}$ acts on two of the network's NANDs (the *negation bar* highlights the fact that it is *active-low*):

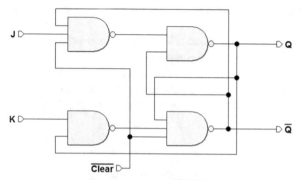

As we learn from the logical schematic, when $\overline{Clear} = 1$, the network behaves as if there were no $\overline{Clear}$ input. When it is brought to 0, however, the configuration that brings output Q to zero is *forced* on the elementary cell.

It is important to note that input $\overline{Clear}$ is *prioritized* over J and K. In fact, for all the time that it is active, $\overline{Clear}$ *keeps the output at zero* and prevents J and K from influencing the elementary cell.

In the figure below, we see the network simulation with the activation of input $\overline{Clear}$ highlighted.

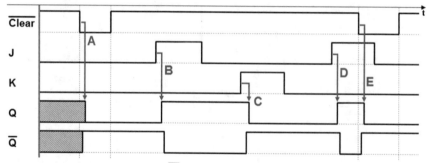

As we can see, outputs Q and $\overline{Q}$ are initially indeterminate but the activation of input $\overline{Clear}$ forces (A) a definition (the delays are due to propagation times). Then deactivating $\overline{Clear}$ allows the flip-flop to work freely (B)(C)(D), under the control of inputs J and K. Finally, on the right, we see $\overline{Clear}$ activated once more (E), which forces the flip-flop to return to zero.

To be thorough and versatile, flip-flops generally have two initialization inputs: $\overline{Clear}$ and $\overline{Preset}$. The action of $\overline{Preset}$ is perfectly symmetrical to $\overline{Clear}$: it forces the output to 1.

Notice that the function of inputs $\overline{Clear}$ and $\overline{Preset}$ is functionally equivalent to that of the $\overline{RESET}$ and $\overline{SET}$ of an RS flip-flop with active-low commands. Nevertheless, the purpose of initialization inputs remains and they should not be used differently.

For this reason, $\overline{Clear}$ and $\overline{Preset}$ should be considered *mutually exclusive*. In order to initialize the flip-flop, we use only one of them based on the project needs, while the other will be connected to a constant high logic level, so as to remain inactive.

Regardless of logical type, inputs $\overline{Clear}$ and $\overline{Preset}$ exist in all flip-flops, whether they be physical or CAD system component.

In the figure below, we see the terminations common among all the logical types: inputs $\overline{Clear}$ and $\overline{Preset}$ and outputs Q and $\overline{Q}$. The inputs are shown here in a generic form seeing that they vary depending on the logical type.

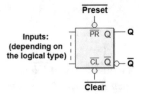

For completeness, the entire logical schematic of the JK flip-flop with direct commands, inputs J and K, outputs Q and $\overline{Q}$, and inputs $\overline{Clear}$ and $\overline{Preset}$ appears below left. The one on the right is the corresponding schematic symbol.

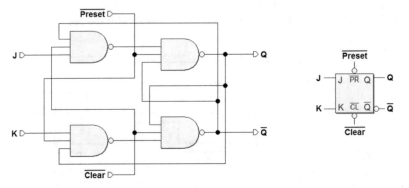

5.3.2 *Generating an Initialization Signal*

After discussing the need to initialize sequential networks, we must now examine how the signal that is fed to inputs $\overline{Clear}$ (or $\overline{Preset}$) of the individual flip-flops is produced when the system is powered up.

The next figure shows what is called a *Reset Generator*: a mixed analog and digital circuit, that we will study only from the functional point of view.

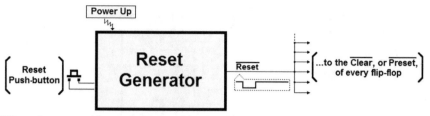

When the system is powered up, this circuit *automatically* generates a pulse of a determined duration on the $\overline{Reset}$ line, that is linked to all the flip-flops in the network (either to $\overline{Clear}$ or to $\overline{Preset}$ according to need).

There is often a push-button that the user may press to manually re-initialize the system (as with PCs for example).

Activating the $\overline{Reset}$ keeps all the system's flip-flops blocked during the initial *power up transient*. The power supply takes a certain amount of time (a few tens mS) to bring the circuit voltages from zero to nominal values. The generator keeps the $\overline{Reset}$ active for the time necessary and then deactivates it, allowing the system to start working.

5.4 Level-Enabled Flip-Flops

All the flip-flops discussed previously share the commonality that the action of the logical inputs is not subordinated to any other input. When designing a digital system, however, it is often necessary that the outputs of the flip-flops change at the *same instant*, namely that they be *synchronized*.

In this section, we will deal with *level-enabled* flip-flops, where a specific input conditions the action of the inputs. This type of flip-flop is also called *Latch*.

Take note: to make the explanation clear, the logical networks of the flip-flops presented here are simplified and do not include initializing circuits (inputs $\overline{Clear}$ and $\overline{Preset}$ will not appear in the descriptions). Remember that real flip-flops of course have these inputs, as described before.

5.4.1 SR-Latch Flip-Flop

In the following schematic, the base structure of the flip-flop is SR type with active-low inputs and has been integrated with two NAND gates.

The two new gates make the SET and $RESET$ act upon the base cell only if the *Enable* input $EN = 1$. If so, the network's overall behavior is identical to that of the SR flip-flop with direct command and active-high inputs.

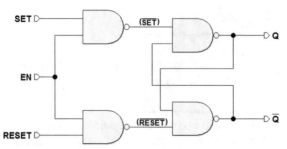

If $EN = 0$, the two NANDs that condition the input generate a 1, regardless of the value of inputs SET and $RESET$, so the base cell maintains the value of the outputs. The function table shown below summarizes this information.

Set-Reset Flip-flop (Level-enabled)					
EN	SET	$RESET$	Q	$\overline{Q}$	
0	–	–	Q_p	$\overline{Q_p}$	Previous state
1	0	0	Q_p	$\overline{Q_p}$	Previous state
1	1	0	1	0	SET command
1	0	1	0	1	$RESET$ command
1	1	1	1	1	Invalid

In the timing simulation below, we observe the behavior of the SR flip-flop as the inputs vary (for simplicity's sake, the invalid configuration was omitted).

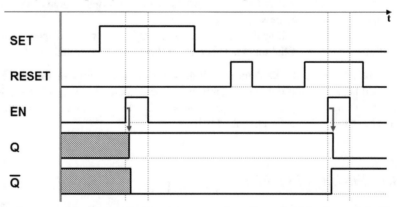

In the first part of the image, the simulator cannot predict an initial value so it indicates the outputs as *undefined*. The first activation of input EN allows input SET to force the output Q to 1. Afterward, the flip-flop keeps the value of the outputs since input EN is inactive. Then EN is active again, while input $RESET$ is active, forcing output Q to zero.

As we can see, input EN is used to restrict the changes of the outputs of the flip-flop within the time intervals it is active. By shortening these intervals, we get the first form of *synchronization*, as we will see further on.

5.4.2 D-Latch Flip-Flop

In the schematic below, a NOT was added to the structure of the SR latch flip-flop. Input SET of the SR flip-flop was renamed D. Negated, it drives the input $RESET$.

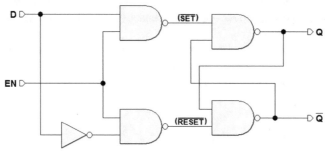

This way, we obtained a D flip-flop but this time the enable input EN is present (this flip-flop is usually called *D-Latch*). The invalid configuration is automatically eliminated by the NOT, as we saw with the direct command D flip-flop. Thanks to the enable input, however, in this case we can *memorize a bit*. The *function table* for this flip-flop is shown below:

EN	D	Q	$\overline{Q}$	D-Latch Flip-flop
0	–	Q_p	Q_p	Previous state
1	1	1	0	SET command
1	0	0	1	$RESET$ command

When EN is at 1 (active), output Q copies the value of input D, whereas when $EN = 0$, there is no transmission between input D and the base cell of memory, which keeps its value.

Due to its simplicity and economy, the D-Latch flip-flop is commonly used to make many types of *registers* and *semiconductor memory devices* that have applications in sequential networks and in computers in general. To store the information, we need first to submit the bit to be memorized at input D, and then activate and release input EN. The timing diagram below shows a typical sequence of use.

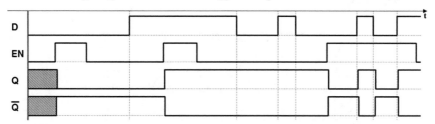

As in the other cases, we do not know the value of the outputs at the start of the simulation. First, input D of the data is set to 0 and, during this interval, input EN

is activated for a certain time. The data at input D is *transferred* to output Q. When EN is deactivated, the flip-flop *captures the value* at the output at that moment (it is said to *memorize the last transited value*).

The same activation sequence is repeated immediately after, but with input data $D = 1$: the flip-flop memorizes 1 on the output. The diagram shows that in the interval when $EN = 0$, input D does not produce changes in the outputs even though it changes many times.

Lastly, the diagram shows that EN is maintained at 1 for a certain period of time. We see that the output repeats the value of the input D (when it is enabled, the flip-flop is said to be *transparent*).

5.4.3   JK-Latch Flip-Flop

The schematic below shows a variant of the direct command JK flip-flop with an added enable input EN:

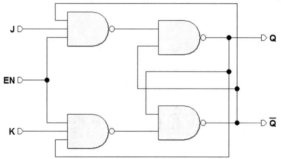

As with the direct command JK flip-flop, the input J is conditioned by output $\overline{Q}$, and input K by output Q, making it impossible to simultaneously activate the inputs of the base cell.

Input EN conditions both the inputs J and K. When EN is active the network behaves exactly like a direct command JK. Otherwise it keeps the previously memorized value (the condition $EN = 0$ is equivalent to having both J and K at 0). The function table summarizes the relationship among the inputs J, K, and EN, and the outputs Q and $\overline{Q}$.

JK Flip-flop (Level-enabled, or JK-Latch)					
EN	J	K	Q	$\overline{Q}$	
0	–	–	Q_p	$\overline{Q}_p$	Previous state
1	0	0	Q_p	$\overline{Q}_p$	Previous state
1	1	0	1	0	*SET* command
1	0	1	0	1	*RESET* command
1	1	1	$\overline{Q}_p$	Q_p	Toggle

This type of *level-enabled* JK flip-flop is not commonly used because the *toggle* configuration ($J = 1$ and $K = 1$) can only be used if the duration of EN is shorter than the network propagation time. As mentioned, the *toggle* function is used in *edge-triggered* JK flip-flops.

5.5 Synchronization of Sequential Networks

Level-enabled structures only partially satisfy the requirements of modern digital systems where it is required that the outputs of flip-flops change *periodically* and *simultaneously*. This type of system is called *synchronous*. To design a synchronous system, we must use more elaborate sequential components than those we have seen so far. Above all it is important to completely understand what *synchronization* of sequential networks really means and what issues are involved. Before introducing the flip-flops used in synchronous systems in the following sections, we will first introduce the concept of *synchronicity* using familiar elements.

5.5.1 The Synchronization Signal

Consider a network that uses level-enabled flip-flops. To satisfy the requirement of *periodicity*, the enabling command EN must take on a cyclical form as seen in part (*a*) of the figure below.

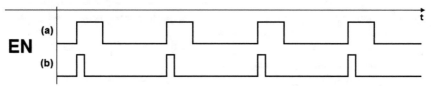

Flip-flops will change their outputs only in response to the *periodic activation* of the EN enable command. This becomes the *synchronization signal* meaning the *time reference* for the time evolution of the network. Level-enabled flip-flops, however, only partly guarantee simultaneous changes.

Their outputs can only change when EN is active and since the activation interval is finite, transitions can still occur at different times given that outputs can change the whole time EN is active.

If the duration of EN is restricted (part (*b*) in the figure above), it reduces the interval in which the outputs can change. This brings us closer to the outputs' *simultaneity* requirement.

5.5.2   Pulse Command in Level-Enabled Flip-Flops

Let's consider a level-enabled JK flip-flop identical to the one examined in previous sections. Let's drive its EN enable input with a periodic pulse sequence.

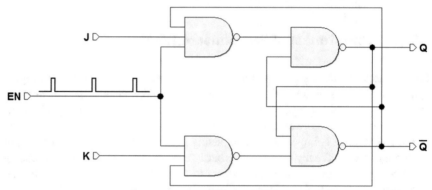

The flip-flop will change outputs Q and $\overline{Q}$ when the pulses on EN occur, as seen in the timing diagram below, which describes typical usage. Note that the flip-flop responds to inputs and only changes its outputs (after the network's propagation times) when EN is activated (see the vertical dotted lines).

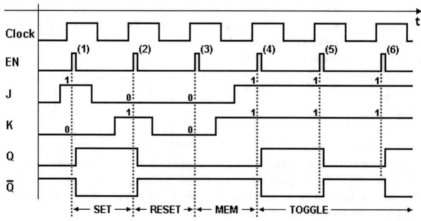

In the previous figure, the pulse (1) on EN where $J = 1$ and $K = 0$, forces output Q to 1. Pulse (2) where $J = 0$ and $K = 1$ forces Q to zero. Pulse (3) where $J = 0$ and $K = 0$ keeps the previous output value. From pulse (4) on, where $J = 1$ and $K = 1$, the output value is inverted at every cycle (*toggle*).

Note that when EN is active and $J = 1$ and $K = 1$, the short duration of the enable command does not allow for the continuous switching of the outputs typical of the implementation of the same flip-flop with direct command. The duration of the pulse must be carefully assessed at the project level in relation to network timing.

The two figures below describe two abnormal situations, where the excessive duration of the EN pulse allows for two output inversions in case (a) and three in case (b).

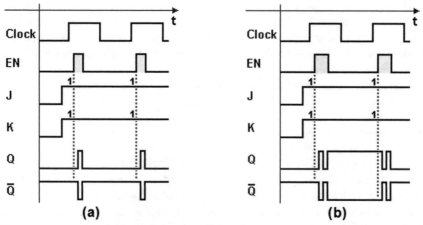

(a)   **(b)**

The function table below summarizes the logical behavior of the level-enabled JK flip-flop that is driven impulsively.

JK Flip-flop (Level-enabled), *Impulsively Driven*					
J	K	EN	Q	$\overline{Q}$	
0	0	⊓⊔	Q_p	$\overline{Q_p}$	Previous state
1	0	⊓⊔	1	0	*SET* command
0	1	⊓⊔	0	1	*RESET* command
1	1	⊓⊔	$\overline{Q_p}$	Q_p	Toggle

The symbols in the EN column represent the *pulse command*. Keep in mind while reading the table that a given combination of inputs J and K corresponds to the indicated outputs only after input EN is *pulse activated*.

5.5.3   The "Clock" and the "Edge-Triggered Command"

This example has shown us that level-enabled flip-flops make it possible to create synchronous systems. As we have seen, however, defining the length of the pulse poses a downside. In the example above, the inputs of the flip-flop are supposed to remain stable during EN activation, but it is not always possible to fulfill this condition, necessary to guarantee *simultaneity* of output changes.

To resolve these problems, we need to shorten the length of the activation pulse as much as possible in level-enabled flip-flops. Technically, however, we cannot go lower than a certain minimum value. Historically, level-enabled flip-flops were mainly used in the first logical circuits using *discrete components*, because of their simplicity.

With integrated circuits, which began to substitute discrete components as of the 1970s, the situation has evolved. Since microelectronics greatly lowered the cost of a single logical function, more dependable and economical, although more circuitally complex structures have been developed. Current digital systems no longer employ *level-enabled* flip-flops but rather *edge-triggered*, i.e., from a *level transition* of the synchronization signal.

The synchronization signal in digital systems is traditionally called *"clock."* A clock is a two-level periodic signal generated by a dedicated circuit, the *Clock Generator*:

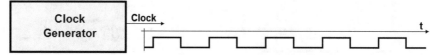

The timing evolution of a periodic signal is called a *"waveform."* In the example below, the signal is a symmetrical *square wave* with a *duty cycle* (percentage of the time the signal is high in its period) of 50%. The designer chooses the clock's *oscillation frequency* based on the system specifications.

In the figure below, the clock signal's *transitions* from 0 to 1 are highlighted by arrows pointing to the top:

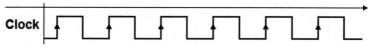

This type of transition is called the *positive (or rising) edge*; its opposite, the *negative (or falling) edge*. As we will soon see, this type of device will render the timing evolution of digital systems rigorously *synchronized* by the edge of the clock (there are positive-edge-triggered components and negative-edge-triggered components).

5.5.4 Master–Slave Structure

The first general-use structure where the change in outputs is synchronized by an edge was the *master–slave* structure. It is no longer commonly used but it is useful to examine it before going on to more modern structures.

The master–slave configuration uses two level-enabled RS flip-flops. The *Clock* signal controls the triggering of the two flip-flops and thus the data transfer from the input to the output. Here we refer to a master–slave structure that creates a JK flip-flop (the logic type it was developed for).

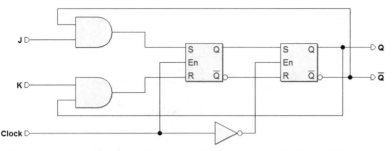

The RS flip-flop that the input data is applied to (*master*) is driven directly by the *Clock*, while the other (*slave*) is controlled by $\overline{Clock}$.

When $Clock = 1$, the data on J and K is applied to the master flip-flop but this has no effect on the slave since its inputs are disabled ($\overline{Clock} = 0$). When the *Clock* goes back to 0, the master input is disabled and the data is transferred to the slave.

The function table that describes the JK master–slave is similar to that of the level-enabled JK flip-flop that is driven impulsively, but note that the output changes occur on the falling edge of the *Clock*.

JK Flip-Flop (*Master-slave*)					
J	K	$Clock$	Q	$\overline{Q}$	
0	0	⌐⌐	Q_p	$\overline{Q_p}$	Previous state
1	0	⌐⌐	1	0	SET command
0	1	⌐⌐	0	1	$RESET$ command
1	1	⌐⌐	$\overline{Q_p}$	Q_p	Toggle

The timing diagram in the next figure shows the behavior of the master–slave JK flip-flop. The *semiperiods* of the *Clock* when it is at 1 are shown by dotted lines to demonstrate that a master–slave flip-flop acquires inputs the whole time the *Clock* is at 1 while it changes its outputs according to its *falling edges*.

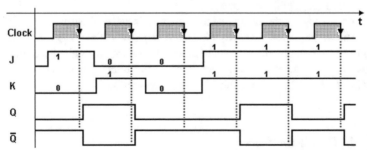

In the figure above, the triggering sequence is the same as the one for the JK flip-flop that is driven impulsively. All the possible combinations for J and K are represented including the *toggle* mode.

The master–slave structure eliminates the limitation on the duration of triggering pulses typical of the level-enabled structures. Note, however, that inputs J and K must be *stable* when the *Clock* is at 1 to make the JK master–slave flip-flop work correctly. One drawback of this structure is that it is sensitive to the changes in *J* and *K* during this interval.

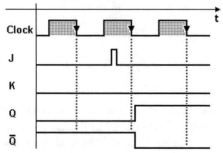

In the figure above, for example, a brief pulse on *J* (if $Q = 0$) is memorized in the master flip-flop causing a change in output on the falling edge.

Thus, we can say this structure is edge-triggered, provided that the inputs for the time when the *Clock* is at 1 are guaranteed to be stable. This problem was resolved by a variant called *data lock-out (DLO)*, which takes on the values of *J* and *K* on the *Clock's rising edge* and changes its outputs on the *falling edge*, while remaining impervious to the changes when the *Clock* is at 1.

Master–slave and DLO triggered flip-flops are now obsolete. They have been replaced by flip-flops that work *on only one edge*. We examine them in the next section.

5.6   Edge-Triggered Flip-Flops

Edge-triggered flip-flops are widely used in the current implementations of digital system. In this structure, both the acquisition of data and the change of outputs occur on a transition (*edge*) of the *Clock*.

If the active edge is the rising one the structure is called *Positive-Edge-Triggered* (PET); if the falling edge is active, it is called *Negative-Edge-Triggered* (NET).

The flip-flop remains impervious to inputs throughout the rest of the time. Thus, the inputs can change at any time except for a brief interval around the active edge of the *Clock*. This aspect is examined in the next section.

5.6.1   D-PET Flip-Flop

In the figure below left, we observe a logical network that creates the D-PET type *positive-edge-triggered* flip-flop. The structure is reminiscent of the master–slave flip-flop but it uses two level-enabled D flip-flops.

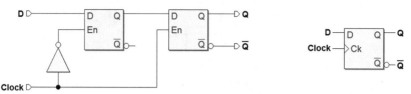

For the whole time that the *Clock* is at 0, the first flip-flop, having an active *En*, follows the variations of input *D*. In the same time interval, the second flip-flop with an inactive *En* keeps the previous value. When the *Clock* changes to 1 the first flip-flop memorizes the last value in D, while the data in its output is transferred to the second flip-flop. The figure at the right shows the logical symbol of the D-PET flip-flop. The small triangle at the input of the *Clock* indicates the sensitivity to the edge.

In the function table below, the symbol in the *Clock* column indicates the *rising edge* to which the changes in the outputs are associated.

D-PET Flip-flop				
D	*Clock*	*Q*	$\overline{Q}$	
0	⌐	0	1	Memorizes 0
1	⌐	1	0	Memorizes 1
–	0	Q	$\overline{Q_p}$	Previous state
–	1	Q	$\overline{Q_p}$	Previous state

The last two rows on the table are superfluous but they highlight the fact that the flip-flop keeps the previous outputs in the absence of an edge. Further on, these two rows will not appear in the other tables.

The timing diagram in the next figure shows that the output *Q* of the flip-flop copies the value of input *D* on the *Clock's* rising edges and keeps it stable for the length of one clock period.

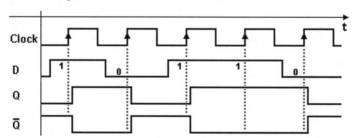

Note that *Q* changes only in response to the *Clock* rising fronts. A signal that keeps a *fixed timing relation* with the *Clock* is defined as "*synchronous.*"

The figure below shows the circuit of the D-PET flip-flop, which is commonly used in many commercial products. It is fast and economical, and it is built using three *base cells*.

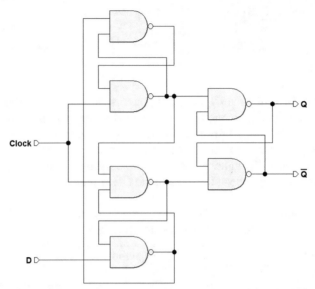

Its function table and timing diagram coincide with those of the D-PET structure we have just seen before.

The figure below shows the schematic of a D-PET flip-flop like the previous one but with $\overline{Clear}$ and $\overline{Preset}$ networks. On the right, we see the corresponding circuit symbol.

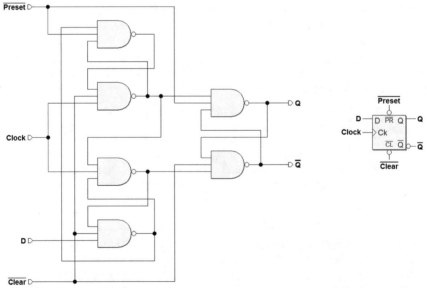

The top rows on the function table of the D-PET flip-flop with $\overline{Clear}$ and $\overline{Preset}$ (seen below) describe how the initialization inputs function.

D-PET Flip-Flop (with $\overline{Clear}$ and $\overline{Preset}$)						
$\overline{Clear}$	$\overline{Preset}$	D	$Clock$	Q	$\overline{Q}$	
0	1	–	–	0	1	Action of Clear
1	0	–	–	1	0	Action of Preset
0	0	–	–	1	1	(invalid)
1	1	0	⌐	0	1	Memorizes 0
1	1	1	⌐	1	0	Memorizes 1

Note that D and $Clock$ have no effect on the component when either $\overline{Clear}$ or $\overline{Preset}$ is active.

The behavior of the D-PET flip-flop with inactive $\overline{Clear}$ and $\overline{Preset}$ (the bottom two rows of the table) corresponds to the previous table of the flip-flop with no initialization inputs.

For structures based on the NAND cell, simultaneously activating $\overline{Clear}$ and $\overline{Preset}$ forces outputs Q and $\overline{Q}$ to 1 creating a parallel situation to that of the base cell of the SR flip-flop, with the same problems. However, this combination of values is easily avoidable because we would normally choose to initialize the flip-flop by using $\overline{Clear}$ or $\overline{Preset}$ and connecting the other to the constant 1.

Now let's analyze the time behavior of the D-PET flip-flop, including its response to the activation of inputs $\overline{Clear}$ and $\overline{Preset}$.

To make this analysis, we must focus not only on the level of input D signal when the active edge of the $Clock$ occurs but also on the level of the asynchronous $\overline{Clear}$ and $\overline{Preset}$.

As we have seen, $\overline{Clear}$ and $\overline{Preset}$ are *prioritized* and act *independently* of the $Clock$. The figure below shows the example of $\overline{Clear}$ activation and the other, the activation of $\overline{Preset}$.

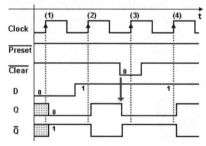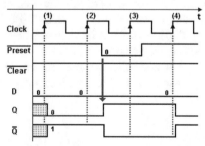

In the example on the left, we initially do not know the value of Q but with the rising edge (1) of the $Clock$ the value 0 on D is memorized by the flip-flop and appears in output Q. At the rising edge (2), the flip-flop takes on the new value of D, forcing Q to 1.

Activating $\overline{Clear}$ involves asynchronously forcing the output to zero and the insensitivity to edge (3) of the $Clock$. Note that deactivating $\overline{Clear}$ produces no changes in the output.

In the example on the right, the $\overline{Preset}$ input is activated, which *forces* the output to 1. Note that the flip-flop is insensitive to the edges of the *Clock* during the activation interval of $\overline{Preset}$ as well.

5.6.1.1   Example (D-PET as "synchronizer")

Finally, let's observe the behavior of output Q in the figure below in relation to input D.

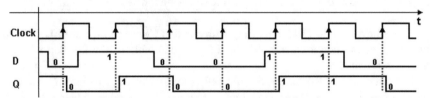

As we can see, input D changes irregularly, without respect to the edges of the *Clock*. At each occurrence of the active edge of the *Clock*, memorization of the signal at input D creates a *"synchronized"* copy in the output, i.e., with a *fixed timing relation* to the *Clock*.

A typical application of the D-PET flip-flop is the *synchronization of signals*. Further on, we will see how important signal synchronization is and the issues involved.

5.6.2   E-PET Flip-Flop

Digital networks, especially those based on large-scale integration devices, employ another type of flip-flop called "E-PET," an extension of the D type. The E-PET flip-flop has the capacity to store an input value only upon request. This differs from the D type, which stores a new value at each active edge of the clock.

We would be wrong to try to achieve this by using a D flip-flop and condition its *Clock* to an E enable signal, as in the following network:

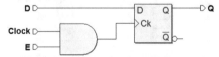

The logical gate allows us to block the *Clock* when input E is at 0. This technique is not used, however, because the gate's propagation time defeats the requirement that the outputs should be simultaneous to other flip-flops in the system.[2]

---

[2] In large-scale integration devices that use a high number of flip-flops, the designer takes great care with the physical connections of the clock to avoid time misalignments among the various elements of the network. Logical gates along the clock path are also invalid.

Thus, we make use of a different structure where the input of the D-PET flip-flop is driven by a multiplexer, as in the figure below left. At right is the circuit symbol of the E-PET flip-flop.

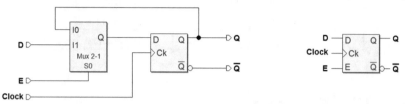

The multiplexer's input E ("Enable") allows us to copy the value of output Q to the input of the D flip-flop if $E = 0$, or the value of the external input D if $E = 1$. When the rising edge of $Clock$ arrives, in the first case the memorized data stays the same. In the second case, output Q takes on the value of external input D. Note that the $Clock$ is not affected.

See below the function table of the E-PET flip-flop with no initialization inputs.

E-PET Flip-flop					
E	D	$Clock$	Q	$\overline{Q}$	
0	–	⌐	Q_p	$\overline{Q_p}$	Previous value
1	0	⌐	0	1	Memorizes 0
1	1	⌐	1	0	Memorizes 1

In the timing diagram below, we see the effect of input E, in response to the active edges of the $Clock$. The rising edges of the $Clock$ where the flip-flop memorizes the new value are: (2), (4) and (7).

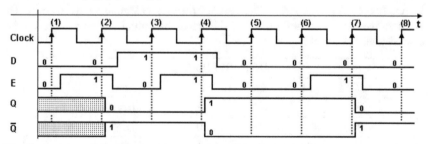

Finally, let's consider the complete version of the E-PET flip-flop with initialization inputs. Its logical symbol and function table are found below.

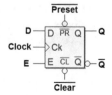

E-PET Flip-flop					
E	D	$Clock$	Q	$\overline{Q}$	
0	–	⤒	Q_p	$\overline{Q_p}$	Previous value
1	0	⤒	0	1	Memorizes 0
1	1	⤒	1	0	Memorizes 1

5.6.3 JK-PET Flip-Flop

In the figure below left, we see a logical network that creates a *positive-edge-triggered* JK-PET flip-flop. We derive this structure from that of the D-PET where input D is controlled by an AND-OR combinational network, which calculates D's value based on inputs J and K and outputs Q and $\overline{Q}$.

The image on the right shows the corresponding logical symbol. The triangle at the input of $Clock$ indicates it is sensitive to the edge.

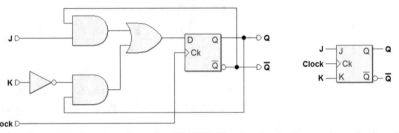

The function table that describes the JK-PET flip-flop is similar to that of other JK structures we have seen before, but in this case, the output changes occur on the rising edge of the $Clock$.

JK-PET Flip-flop					
J	K	$Clock$	Q	$\overline{Q}$	
0	0	⤒	Q_p	$\overline{Q_p}$	Previous state
1	0	⤒	1	0	SET command
0	1	⤒	0	1	$RESET$ command
1	1	⤒	$\overline{Q_p}$	Q_p	Toggle

The AND-OR combinational network that generates the value of D is derived by analyzing the function table. This analysis provides us with the table below, which allows us to go directly to the synthesis of the network.

J	K	Q	D
0	0	0	0
0	0	1	1
0	1	0	0
0	1	1	0
1	0	0	1
1	0	1	1
1	1	0	1
1	1	1	0

This gives us: $D = (J\,\overline{Q} + \overline{K}\,Q)$

The timing diagram in the figure describes the typical behavior of a JK-PET flip-flop. At the rising edge of the *Clock*, the flip-flop responds to the inputs and then updates the outputs after the network's propagation time. The input activation sequence is the same as that for previous versions of flip-flops.

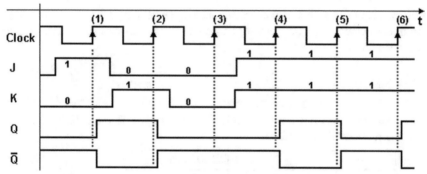

On edge (1) Q is activated in response to the acquisition of J. On (2), it is deactivated (because $K = 1$), while on (3) its value is maintained ($J = K = 0$). Finally, on the following edges (4, 5, 6) the outputs switch their value (given that $J = K = 1$).

Like the other logical types, the JK-PET flip-flop often includes inputs $\overline{Clear}$ and $\overline{Preset}$ in its commercial or library implementations. Find below its logical symbol and function table.

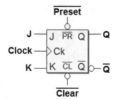

\overline{Clear}	\overline{Preset}	J	K	clock	Q	\overline{Q}	JK-PET Flip-Flop (with $\overline{Clear}$ and $\overline{Preset}$)
0	1	–	–	–	0	1	Action of Clear
1	0	–	–	–	1	0	Action of Preset
0	0	–	–	–	1	1	(invalid)
1	1	0	0	⌐⌐	Q_p	$\overline{Q_p}$	Previous value
1	1	1	0	⌐⌐	1	0	SET command
1	1	0	1	⌐⌐	0	1	$RESET$ command
1	1	1	1	⌐⌐	$\overline{Q_p}$	Q_p	Outputs reversed

5.6.4 T-PET Flip-Flop

If the two inputs of a JK flip-flop are connected, we get a type T (*Toggle*) flip-flop. The figure below shows the network based on the JK-PET and the corresponding logical symbol.

The function table of the T-PET is derived from the JK-PET table eliminating the rows where J and K are different.

T	Clock	Q	$\overline{Q}$	The T-PET Flip-Flop
0	⌐⌐	Q_p	$\overline{Q_p}$	Previous state
1	⌐⌐	$\overline{Q_p}$	Q_p	Toggle

The T flip-flop can also be obtained from the D type through the same procedure as above, deriving the JK from the D. By doing this, we get:

$$D = (J\,\overline{Q} + \overline{K}\,Q)$$

given that $T = J = K$, the expression is reduced to:

$$D = T\,\overline{Q} + \overline{T}\,Q = T \oplus Q$$

From this, we get the network below. In the following, we will come back to this a few times.

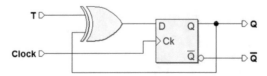

5.6.5 Synchronous Initialization of Flip-Flops

Networks at a certain level of complexity use a *synchronous initialization* structure where the *Clock* synchronizes the actions of $\overline{Clear}$ and $\overline{Preset}$. The figure below provides an example:

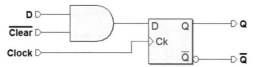

It shows a possible structure of the synchronous $\overline{Clear}$ for a D-PET-type flip-flop. When input $\overline{Clear}$ is activated (at 0), it prevails over input D and forces the output to 0 on the active edge of the *Clock*.

5.7 Timing Parameters of Flip-Flops

Flip-flops, like the combinational components they are made of, are subject to *propagation delays*. As we saw with combinational networks, propagation times are measured by taking 50% of the signal edges as a reference. Propagation times are published on the producer data sheets on statistical bases, as their *minimal*, *typical*, or *maximum* values are declared.

Aside from *propagation times*, there are other timing parameters that are statistically quantified such as *safety margins*, which must be observed to guarantee the flip-flop works as expected. The table below gives a succinct definition of the main timing parameters.

t_{PLH}	*Propagation time* measured from the activation of the *Clock* to the output's transition from low to high (L-H)
t_{PHL}	*Propagation time* measured from the activation of the *Clock* to the output's transition from high to low (L-H)
t_s	*Setup Time*: the time interval the value of a synchronous input must remain stable *before* the active edge of the *Clock*
t_h	*Hold time*: the interval when the value of a synchronous input must remain stable *after* the active edge of the *Clock*
t_w	*Minimum width* of an input signal
T_{min}	Minimum *Clock* period
F_{max}	Maximum *Clock* frequency

The timing diagram in the figure below shows the timing parameters of a D-PET flip-flop in the acquisition sequence of a 1 followed by a 0. The **tw** in the figure refers to input D and in this example, it is given by the sum of the **ts** and **th** times.

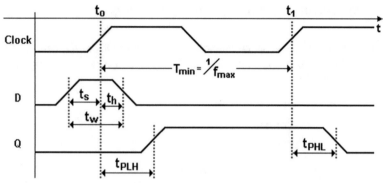

If the *safety margins* are not observed, unpredictable behavior can result. This is called *metastability*, and it will be dealt with in the next section.

5.7.1   Relationship Between Propagation and Hold Times

Let's use the example from the figure below to study the relationship between propagation times and hold times.

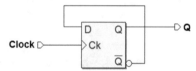

The circuit is composed of a D-PET flip-flop whose input D is connected to its negated output $\overline{Q}$. At each active edge of the *Clock*, it will toggle its value, as it would in a T-PET-type flip-flop with an input of $T = 1$.

If the value of input D is to be acquired correctly, it must remain stable for at least a time **th** after the rising edge of *Clock*.

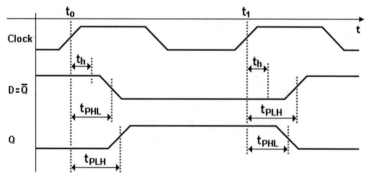

The situation is exemplified in the timing diagram above where we see that **th** must be shorter than both the two propagation times (**t**$_{PHL}$ and **t**$_{PLH}$). This condition is at the base of all sequential networks and all the flip-flops are designed to satisfy it.

5.7.2 Maximum Clock Frequency of a Network with Flip-Flops

A flip-flop's timing parameters enter into play when evaluating the *maximum clock frequency* of the network that uses it. We have previously seen the implementation of the T-PET flip-flop based on the D-PET. In the network in the figure below, we continually force T to 1, imposing a configuration in which the output *changes value* at every rising edge of the *Clock*.

The purpose of this example is to take into account the propagation delay of a combinational network along the feedback path.

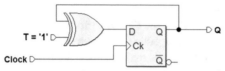

Input D is driven by output Q through the XOR gate which works like a NOT gate since it has an input at 1. For the network to function correctly, the signal at D must be stable at least from time **ts**, at the moment when the *Clock* sees the rising edge until time **th** after the transition.

Notice that the time scale in the following three timing diagrams corresponds to **1 nS** (between two notches on the time axis). The values: **ts = 2 nS, th = 1 nS, t** PHL(FF) = **7 nS**, **t** PLH(XOR) = **3 nS** have been taken.

What the three situations have in common is that signal D is stable after time **t** PHL(FF) **+ t** PLH(XOR) regardless of the *Clock* period. In the first diagram below, the sequence is characterized by *Clock* with period **T = 20 nS** (corresponding to a frequency of **50 MHz**):

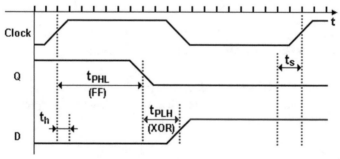

The next rising edge samples data **D**, which was stable for a time longer than **ts**, so the network functions correctly. Also, the condition of **th** is maintained since **th** is shorter than the propagation time of the flip-flop (in the next two figures, the representation of **th** is omitted).

In the figure below, we see the limit case where D is stable for exactly a time **ts** before the edge with a *Clock* period of **12 nS**. The network still functions correctly.

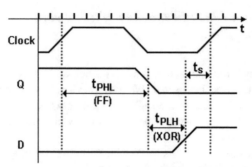

The period **12 nS** corresponds to a frequency of about **83 MHz**, the *maximum clock frequency* of the network.

   In the figure below, however, the *Clock* period is **11 nS** (corresponding to about **91 MHz**). The following *Clock* edge occurs when *D* is already stable but the **ts** is not observed. In this case, there is no guarantee the network will function; we are over the network's *maximum working frequency*.

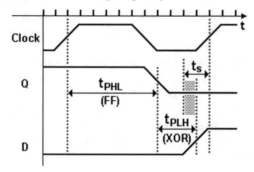

5.8   Flip-Flops: Graphic Symbols and Tables

At the beginning of this chapter, we described flip-flops by their *logical type* and their *command type*. This section summarizes the characteristics and graphic symbols of the most commonly used flip-flops.

5.8.1   *Logical Types*

The following *function tables* recap the behavior of the *logical types* of flip-flop that were introduced in this chapter (SR, JK, D, E, and T). They do not describe the command structure but only the *logical inputs*, assuming they are *active-high*.

SR			
SET	*RESET*	*Q*	
0	0	Q_p	Previous state
1	0	1	*SET* Command
0	1	0	*RESET* Command
1	1	1	Invalid

JK			
J	*K*	*Q*	
0	0	Q_p	Previous state
1	0	1	*SET* Command
0	1	0	*RESET* Command
1	1	Q_p	Toggle

D		
D	*Q*	
0	0	Memorizes 0
1	1	Memorizes 1

E			
E	*D*	*Q*	
0	–	Q_p	Previous value
1	0	0	Memorizes 0
1	1	1	Memorizes 1

T		
T	*Q*	
0	Q_p	Previous state
1	Q_p	Toggle

5.8.2   Command Types

This summary includes only the most commonly used flip-flops. Let's recap the command configurations that flip-flops use to acquire the inputs and update the outputs.

- **Direct command**:
  A command is *direct* when the action of the *logical inputs* does not depend on any other input. They directly control the flip-flop's behavior, which is *asynchronous* in this case.

- **Level-enabled commands**:
  A command is *level-enabled* when there is an added input that enables/disables the inputs according to its logical level. Another term to identify this type is *Latch commands*.

- **Edge-triggered command**:

  We have an *edge-triggered* command when there is an added synchronization input, usually called *Clock*, which enables the inputs to function in response to their rising or falling edges. These flip-flops are *synchronous*.

Direct Commands

The direct command structure is currently used only with the *Set-Reset* logical type. The figure below left shows the symbol without initialization inputs. The one on the right shows a symbol that includes $\overline{Clear}$ and $\overline{Preset}$.

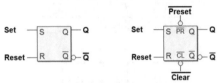

$\overline{Clear}$ and $\overline{Preset}$ might seem superfluous since they do the same thing as the logical inputs, but it is often useful in a project to separate the initialization network from the rest.

Level-Enabled Commands

Level-enabled command structures are mainly used in D-Latch type-flip-flops and often in large structures that employ hundreds of thousands of them, as the *read/write memories* (not covered in this book). The level-enabled Set-Reset is used as internal block of more complex flip-flops. The considerations about initialization inputs are valid here as well.

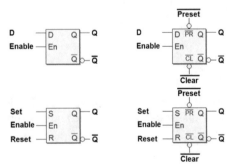

Edge-Triggered Commands

PET or NET *edge-triggered command* structures are the most commonly used especially in the D and E logical-type flip-flops. The use of the E type is spreading in complex networks while use of the JK type is in decline. The figure below shows both PET and NET flip-flop symbols. They generally have the initialization inputs $\overline{Clear}$ and $\overline{Preset}$.

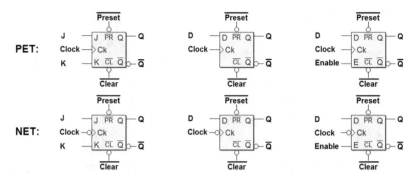

5.8.3 Excitation Tables

Excitation tables give us another way to describe the behavior of a flip-flop. Function tables, which we have seen before, indicate the assumed value of an output in relation to the inputs. Excitation tables provide values to assign to the logical inputs of a flip-flop in relation to the desired *output transition*.

Below left is the function table of the *Set-Reset* flip-flop. On the right is the corresponding excitation table where each of the four possible output transitions is shown next to the input configuration needed to obtain it.

Set-Reset Function Table				Set-Reset Excitation table		
Set	*Reset*	Q		$Q_p \rightarrow Q$	*Set*	*Reset*
0	0	Q_p	Previous state	$0 \rightarrow 0$	0	–
1	0	1	SET command	$0 \rightarrow 1$	1	0
0	1	0	RESET command	$1 \rightarrow 0$	0	1
1	1	1	Invalid	$1 \rightarrow 1$	–	0

The excitation table is easily derived by the function table. For example, the output transition of Q from 0 to 0 (written as: $0 \rightarrow 0$) can be obtained by keeping the previous state of the flip-flop (*Set* = 0, *Reset* = 0, row 1 of the function table), or with a Reset command (*Set* = 0, *Reset* = 1, row 3).

This is translated in the express terms by the top row of the excitation table. *Set* must be at 0, while the value of *Reset* is *don't-care*. The same holds for the transition $1 \rightarrow 1$, while the other two transitions have no *don't-cares*.

The function and excitation tables of the JK flip-flop are shown below.

JK Function table				JK Excitation table		
J	K	Q		$Q_p \to Q$	J	K
0	0	Q_p	Previous state	$0 \to 0$	0	–
1	0	1	SET command	$0 \to 1$	1	–
0	1	0	RESET command	$1 \to 0$	–	1
1	1	$\overline{Q_p}$	Toggle	$1 \to 1$	–	0

This excitation table shows a *don't-care* for each of the four transitions. For example, the transition $0 \to 1$ is obtained through a Set command ($J = 1$, $K = 0$), or a *toggle* command ($J = 1$, $K = 1$). In short, about the transition $0 \to 1$ the value of K is *don't-care* while J must be at 1.

To be thorough, we have put the tables for the D flip-flop below even though the excitation table is immediately derived from the function table. The table for the E flip-flop, when it is enabled by input E, is identical.

D Function table			D Excitation table	
D	Q		$Q_p \to Q$	D
0	0	Memorizes 0	$0 \to 0$	0
1	1	Memorizes 1	$0 \to 1$	1
			$1 \to 0$	0
			$1 \to 1$	1

5.9   Exercises

For each network below, complete the timing diagram (all the diagrams to be completed are also available in *PDF* in the *digital contents* of the book, on the *Deeds* Web site).

We suggest to draw first the diagrams without using the simulator and then check your answers with it (the network files are on the Web site, too).

1. Set-Reset flip-flop (*direct command*)

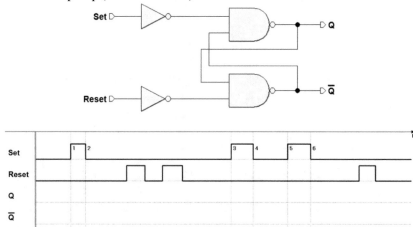

2. Set-Reset flip-flop (*direct command, NAND base cell*)

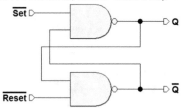

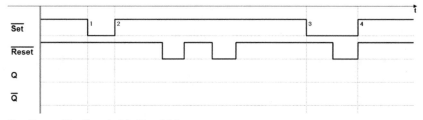

3. Set-Reset flip-flop (with *Enable*)

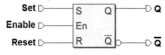

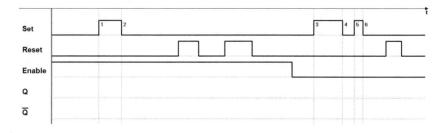

4. D-PET flip-flop (with $\overline{Preset}$ and $\overline{Clear}$)

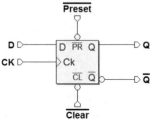

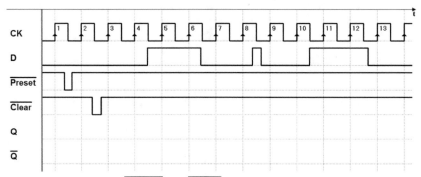

5. JK-PET flip-flop (with $\overline{Preset}$ and $\overline{Clear}$)

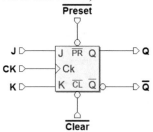

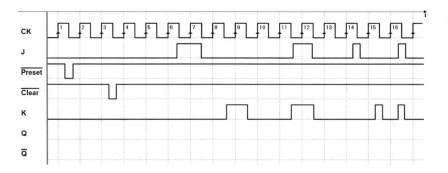

6. E-PET flip-flop (with $\overline{Preset}$ and $\overline{Clear}$)

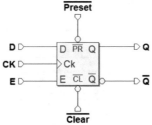

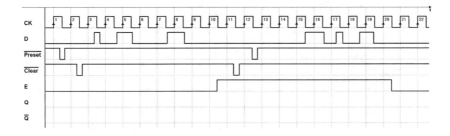

5.10 Solutions

The timing diagrams below were obtained using the *Deeds* timing simulation. The files of all the networks are available on the *digital contents* of the book so that the answers can be checked through the simulator.

1. Set-Reset flip-flop (*direct command*)

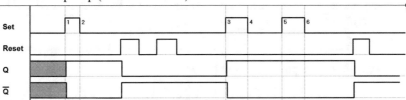

2. Set-Reset flip-flop (*direct command, NAND base cell*)

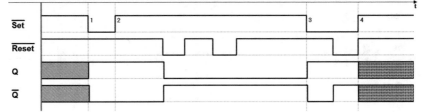

3. Set-Reset flip-flop (with *Enable*)

4. D-PET flip-flop (with $\overline{Preset}$ and $\overline{Clear}$)

5. JK-PET flip-flop (with $\overline{Preset}$ and $\overline{Clear}$)

6. E-PET flip-flop (with $\overline{Preset}$ and $\overline{Clear}$)

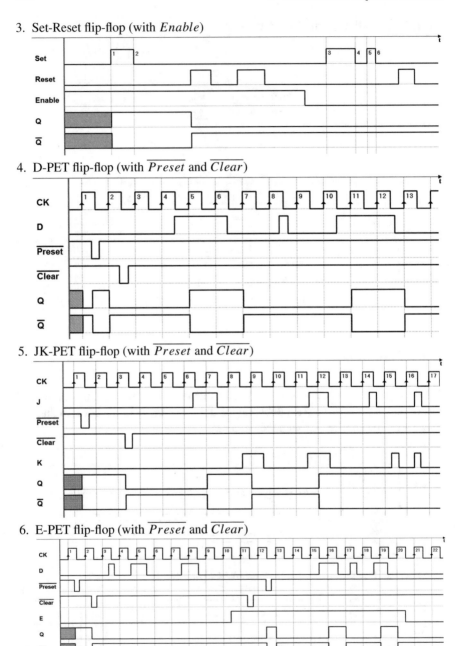

Chapter 6
Flip-Flop-Based Synchronous Networks

Abstract The flip-flops are the building blocks of all sequential networks. A regular structure made of flip-flop and combinational networks can implement any sequential circuit. In this chapter, the structures are not designed but either assembled in an intuitive fashion or taken from standard building blocks. The presentation of counters and registers introduces progressively the real full-featured components that are available for design. Sequential network analysis in the time domain is the important skill that is developed at the end of the chapter.

In this chapter, we will examine the most commonly used flip-flop-based networks that can be used as *functional standard blocks* to create more complex digital networks. We will analyze these networks *intuitively*, whereas in later chapters will deal with their systematic design. We will represent the networks through logical schematics that is in the form of a set of components, such as logical gates and flip-flops with their connections.

In Chap. 5, we examined simple sequential networks, which are generally formed by combinational networks with feedback. This was to gain an understanding of the structure and functionality of various types of flip-flops, the elementary memory cells. Theoretically, a more complex sequential network could be designed in the same way, as a set of combinational networks with feedback. This approach would, however, pose a number of problems especially on the design and testability level.

It is preferable to design a complex sequential network in a more structured way, by using flip-flops as the base *sequential elements*, connected by *purely combinational* networks. This makes it possible to clearly divide the memorization function, the typical quality of a sequential network (entrusted to the flip-flops), from the function that determines its logical behavior and evolution over time (entrusted to the combinational networks).

As seen in Chap. 5, the output of a sequential network is not only a function of the inputs at that moment but also of the values that they took on in the past.

In a sequential network made up of flip-flops and combinational networks, the history of the network only leaves a trace on the values taken on by the flip-flop, since combinational networks, as such, have no memory.

The *set of values* memorized by the flip-flops is called the *state* of the networks. The outputs of a sequential network will thus be a function of the *inputs* and the *state*.

© Springer International Publishing AG, part of Springer Nature 2019
G. Donzellini et al., *Introduction to Digital Systems Design*,
https://doi.org/10.1007/978-3-319-92804-3_6

A *flip-flop-based synchronous network* is a network in which the flip-flops *share the same clock*, and the flip-flop asynchronous inputs (i.e., $\overline{Clear}$ and $\overline{Preset}$) are used *only for their initialization*. If these two conditions are not met, the network falls into the category of *asynchronous networks*.

Due to their regularity, synchronous networks have great advantages from the design and testability point of view, and it is possible to design the simplest structures without relying on formal synthesis methods. The networks examined in this chapter use D, E, and JK logical-type flip-flops with PET behavior.

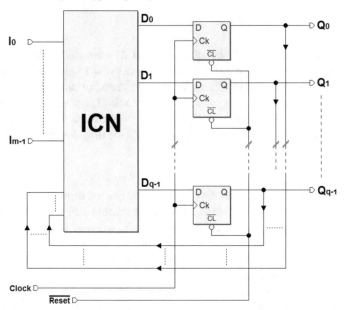

The figure above shows the basic schematic of a *flip-flop-based synchronous network* in a simplified form. The flip-flops share the clock signal. This guarantees one of the properties of synchronous sequential networks, the simultaneity[1] of the change in the flip-flop outputs. The active-low $\overline{Reset}$ makes it possible to initialize all the flip-flops.

The D inputs of the flip-flop are generated by the *Input Combinational Network* (ICN), which processes inputs $I_0..I_{m-1}$ coming from outside and the outputs $Q_0..Q_{q-1}$ of the flip-flops themselves.

At every *positive edge* of the *Clock*, the values of $Q_0..Q_{q-1}$ are replaced by those the ICN provides. It is therefore easy to understand how the combinational network shapes the sequential network's *behavior* through the *external inputs* and the *values memorized* in the flip-flops (the *state* of the network). This is why the ICN is called also the *next state combinational network*.

---

[1] Within the limits of the propagation delay dispersion of real components.

The general structure of a synchronous network is shown in the figure below:

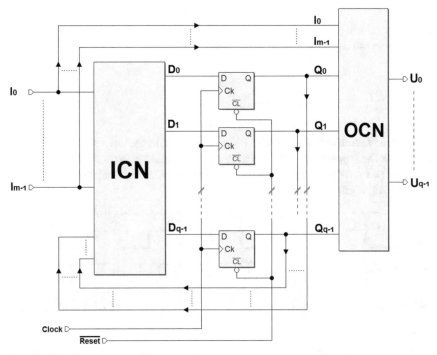

This structure is different from the previous simplified one because of addition of the *Output Combinational Network* (OCN), which allows for greater flexibility in generating outputs. Each of the outputs $U_0..U_{p-1}$ can be generated as a combinational function of the state memorized in flip-flops $Q_0..Q_{q-1}$ and of the external inputs $I_0..I_{m-1}$.

6.1 Synchronous and Asynchronous Signals

Since flip-flops share the same clock in a synchronous network, their outputs change in response to the edges of the clock; that is, they are *synchronous* with the clock. All the *network's outputs* and *internal* signals obtained through the flip-flop outputs' combinational networks maintain a fixed temporal relation with the clock even in presence of propagation delays.

In principle, *input signals* have no relationship with the network's clock since they are generated by external systems. In general, an input signal is *asynchronous* unless it comes from another *synchronous network* that uses *the same clock*. Before using an *asynchronous signal*, it makes sense to *synchronize it* using an appropriate synchronization network.

6.1.1 Synchronizer

Let's now consider the network below, which is made up of a simple D-PET flip-flop
that receives an *asynchronous* signal in the input.

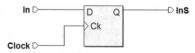

We want to examine this network as the *synchronizer* of the input signal. Output *InS*
represents the *synchronized* version of the signal applied to input *In*. The timing diagram below shows the evolution of the input. Note that the fact that it is *asynchronous*
has been highlighted.

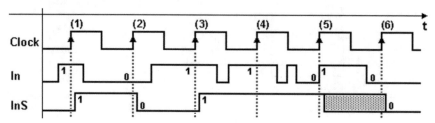

On the rising edges of the *Clock*, from (1) to (4), output *InS* takes on the value of
input *In* and keeps it until the next active edge. Any *intermediate variation* of the
input is not detected, however, as we can see in the time intervals between edges
(3) and (5).

On edges (1) to (4), we assume the *setup times* (**ts**) and *hold times* (**th**) of the
flip-flop have been respected. On edge (5), however, they have been violated. The
input makes an upward transition that is *simultaneous* to that of the *Clock*. Because
of that, we cannot predict the output of the flip-flop, which is shown as *indeterminate*
from the edge (5) and (6).

Under these anomalous conditions, a physical circuit may generate *invalid logical
levels* or *oscillations* for a brief period.

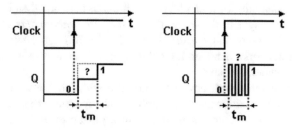

Two examples of this phenomenon that is called *"metastability"* are reported in the
figure above. On the left, output *Q* of the flip-flop is brought to an invalid logical
level for time **tm** before stabilizing at value 1.

On the right, the output goes through a few oscillation cycles between the two levels before settling.

The logical network that reads these types of signal can produce an *error*. This is a *random* behavior in terms of the probability of it happening and for how long time **tm** lasts. These errors depend on the physical characteristics of the flip-flop and the frequency of the signals in play. The closer we get to the speed limits of the specific component in use, the greater the probability of metastable behavior becoming significant. The relationship is exponential.

In any case, errors due to metastability are very rare in well-designed flip-flops in non-critical conditions. The average time between two consecutive errors could be on the order of hundreds of years. Since the probability of an error in the digital circuit due to other causes (a circuit failure, electromagnetic disturbance, for example) is much higher, we can say that signal synchronization through this technique is generally reliable for many applications.

6.1.2 Multistage Synchronization

Under critical conditions or when higher security margins are requested, it is necessary to use a configuration that is more complex than the previous one and uses multiple synchronizer flip-flops connected in ripple fashion.

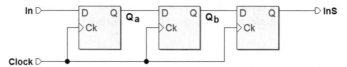

By extending the synchronization procedure this way, we can reduce the probability of error due to metastability to acceptably low limits. This means that any metastable behavior by the first flip-flop is filtered by the second and so forth. Aside from the question of metastability, it is interesting to consider the temporal relation between input *In* and output *InS* as shown in the figure further on.

At every rising edge of the *Clock*, the first flip-flop transfers the value of input *In* onto output Q_a, as with the previous synchronizer. Q_a thus reproduces the evolution of input signal *In* in clean, synchronized form.

Seeing that the first flip-flop generates the very output Q_a after the edge of the *Clock*, the second reads the new value of Q_a on the edge of the next *Clock*. Thus the timing path of Q_b is the same as that for Q_a, but delayed by one *Clock* period.

The next flip-flop echoes the process so output *InS* is still identical to Q_a, but delayed by another *Clock* cycle. On the next figure, we have highlighted the fact that the value generated by a flip-flop on edge (n) is read by the next flip-flop at edge (n + 1).

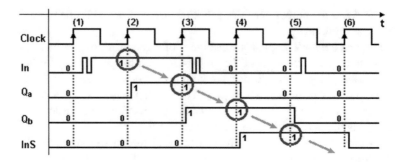

Further on, we will often revisit the concept of *reading the value at the next Clock cycle* in the information exchange between two synchronous networks.

6.2 Registers

Registers are important logical structures used to memorize data. It is possible to *"write"* binary data on a register, keep it for a period of time, and *"read it"* as many times as necessary. From this perspective, the D flip-flop is also a *register*, since it can carry out the operations described on *one single bit*.

A register is usually made up of a number of flip-flops equal to the *number of bits* of the data that needs to be memorized. Registers allow for two ways to store and retrieve data: the *parallel* mode and the *serial* mode, as we will see in the next few sections.

6.2.1 Parallel Registers

The networks in the figure below are two examples of *parallel registers* that can memorize four bits of information.

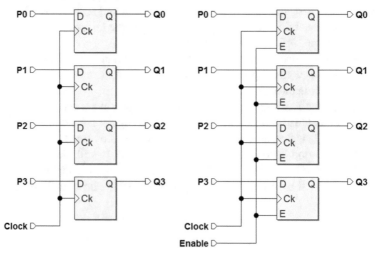

The register at the left uses D-type flip-flops while the one on the right uses the E type. These are *synchronous* networks since the flip-flops receive the same *Clock*. An older term for this type of register is "PIPO" (*Parallel Input–Parallel Output*). Normally, registers have an initialization input, which for simplicity's sake is not shown in this figure.

In the D-type version, input data *P3..P0* is memorized *in parallel* in the flip-flops at a rising edge of the *Clock*. The data is maintained on outputs *Q3..Q0* and remains available until the next writing. If there is no new writing, they remain in the register indefinitely as long as the network is in operation (i.e., its power supply is on).

In the E-type variant, we have the added opportunity to enable/disable the writing under the control of the *Enable* input. This is the most commonly used version in complex systems where, for example, there can be many registers and the one that will store the information needs to be *selected* each time.

Below there are two examples of 8-bit synchronous parallel registers taken from the *Deeds* simulator library. These are two versions of the same component that differ by the way their connections are represented.

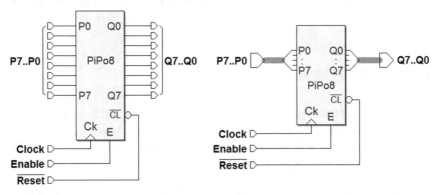

In the version on the left, the terminations are represented individually: the eight data inputs *P7...P0*, the eight outputs *Q7...Q0*, and the *Clock* inputs, the *E* enable, and the asynchronous initialization input $\overline{CL}$ (*Clear*). On the right, inputs *P7...P0* and the eight outputs *Q7...Q0* are shown in the form of *multi-wire connections*, in short, *bus type*.[2]

The figure below shows the typical operation sequence of an eight-bit register with *bus-type* connections and enabling. In the timing diagram, the hypothesis is that the register will initially contain 0 in all the flip-flops. As we can see, the graphic representation of the values is *"cumulative"*; that is, it represents all the bits in the register on one single track, a bar formed by two parallel lines that cross at the transitions of the signals. The value taken by the signals between one transition and another (in this case, represented in hexadecimals) is written inside the bar.

---

[2]Representing multi-wire connections as *bus* allows us to simplify the logical schematic and make it more readable, especially if complex.

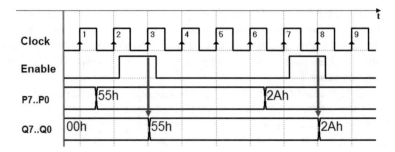

In our example, lines $P7..P0$ are set over time to values 01010101 ($=$ 55h) and then 00101010 ($=$ 2Ah). *Enable* is only driven to load the new data onto outputs $Q7..Q0$ in response to edges 3 and 8 of the *Clock*.

Parallel registers are widely used in digital systems where data is generally organized into "words" made of multiple bits and stored in registers. Note that in the structure of the *generic synchronous sequential network*, featured at the beginning of this chapter, the *state of the network* is memorized in a register.

6.2.2   Shift Registers

The other mode of data input in registers is *serial*, where the bits that make up a word to memorize are presented to the input one at a time in succession. One register that allows for this is called *shift register* (SHR). It can be made of D, E, or JK logical-type flip-flops. We have already been introduced to this structure, with the D type, used as a synchronizer.

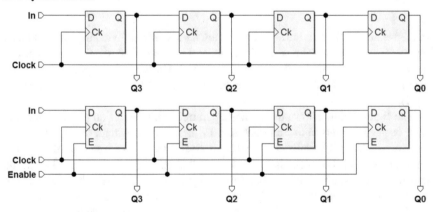

In the figure here above, we see two examples of four-bit shift registers: the first with a D-type flip-flop and the second, with the E type. We can see that the output of each flip-flop is connected to the input of the next while the *Clock* is shared by all, since this is a synchronous network. For simplicity's sake, the initialization network is omitted. The *"serial"* input of the register is *In*. The outputs in both cases are $Q3..Q0$; the *Enable* input is available only in the second version.

The timing diagram below is an example of how the type E register functions. The hypothesis is that outputs $Q3...Q0$ will initially be at 0 and that input *In* will be activated in correspondence with edge 2 of the *Clock*. Also, *Enable* is activated for four *Clock* cycles, from 2 to 5.

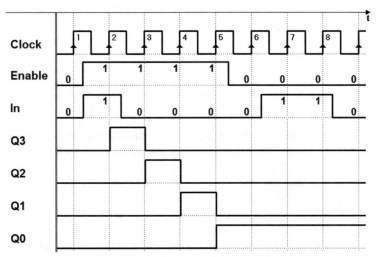

As with the synchronizer, the flip-flop farthest to the left transfers the value of input *In* onto output $Q3$ at every rising edge of the *Clock*. This means that output $Q3$ reproduces the evolution of input signal *In*, synchronized to the *Clock*. The second flip-flop reads the changes in $Q3$ on the next rising edge so $Q2$ is identical to $Q3$, but *delayed* by one *Clock* cycle. The same goes for all the other flip-flops.

From edge 6 on, *Enable* is read at 0. With no enabling, the register's content remains unchanged from edge 5 on, so the new activation of *In* is ignored.

Let's now look at another use of the shift register. Here, it is used as a base element of a *serial sequence receiver*. In the next figure, it is assumed that input *In* receives a sequence of bits with a pre-established format (protocol). In this case, the format requires the bits to be put into groups of five with a *"start bit"* at 1, three *"information"* bits (D0, D1, D2) and a *"stop bit"* at 0. Each of these bits has the same duration as a *Clock* cycle, as shown.

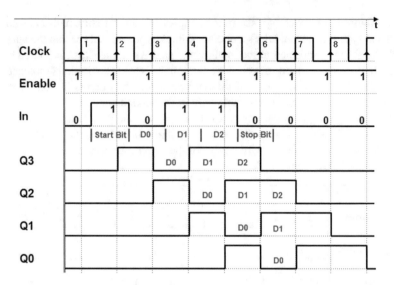

Assume that when line *IN* is in the *"idle"* state, that it is normally defined at 0, so the *start bit* at 1 signals the beginning of the sequence. The *stop bit* separates two consecutive groups with a 0.

In our example, the *start bit* is read by the first flip-flop on edge 2, so the information bits (D0 = 0, D1 = D2 = 1) are read by edges 3, 4, and 5, and the *stop bit*, by edge 6. When receiving the sequence, the register memorizes and shifts the bits received one by one. After edge 6, information bits D0, D1, and D2 are made available *in parallel* on outputs *Q*0, *Q*1, and *Q*2. In this simplified example, the shifting continues with each new *Clock* cycle, and so data is progressively lost.

Serial/parallel conversion is widely employed in digital telecommunication systems where serial communication is mainly used to cover great distances. It is also used in data processing systems when it is necessary to reduce the number of wires connecting the system's modules, to reduce cost or improve usability and practicality. USB (*Universal Serial Bus*) connections are an example of serial format communication. The parallel format provides quicker and more efficient data processing and is generally used in computational systems.

Below are two examples of 8-bit shift registers from the library of *Deeds*:

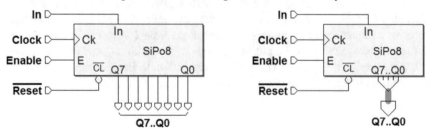

The same device is shown with different connections. The one on the right uses the *bus* type for outputs $Q7..Q0$. On the component, we see the acronym "SIPO," which stands for *Serial Input–Parallel Output*.

There are also "SISO" (*Serial Input - Serial Output*) registers whose internal structure is identical to that of the SIPO type. It is clear that, in principle, one can simply choose any of the outputs Q to obtain a serial output. SISO registers as such (those with *one single serial output*) are normally made with many, sometimes thousands of flip-flops.

They are used to obtain a delayed signal of as many clock cycles. Due to the large number of connections required, it is impractical to make the intermediate outputs externally available.

6.2.3 Shift Registers with Parallel Load

We have seen that registers with serial input and parallel outputs directly carry out serial–parallel conversion. To obtain the opposite conversion, from the *parallel* format to the *serial* format we must first load the data onto the register *in parallel*, and then make it shift *serially*.

We can do this by combining the structures we have already seen in the parallel and shift registers, with the help of multiplexers. The figure below shows the schematic of a 4-bit shift register with parallel load (we have omitted the initialization network).

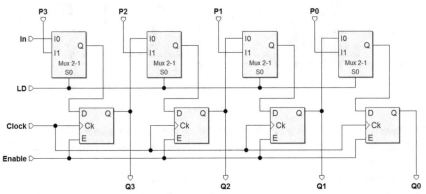

We see the four E-type flip-flops with their outputs $Q3..Q0$, as in the previous register types. Here, however, we have the serial input *In* and the parallel inputs $P3..P0$.

The new input *LD* (*"Load"*) controls the parallel load through the four multiplexers that allow to choose the data to send as input to the flip-flops.

As we can see in the next figure, when $LD = 1$, the selectors route the parallel inputs $P3..P0$ to the flip-flop. If *Enable* $= 1$, the parallel data in the input is loaded onto the flip-flops at the next rising edge of the *Clock*. The data appears on outputs $Q3..Q0$.

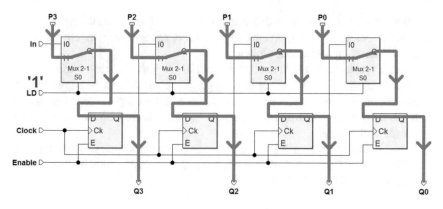

If $LD = 0$, the data is shifted, as we can see in the figure below. The serial input *In* is routed to the first flip-flop; its output is brought to the second and so on. Obviously, the shifting is carried out on the next rising edge of the *Clock*. Note that parallel load and serial shifting are mutually exclusive operations.

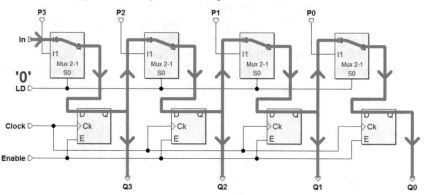

We have previously seen an example of *serial sequence receivers*. A shift register with parallel load, like the one we examined above, is perfectly suited to *transmit a serial sequence*.

In the timing diagram below, let's assume we want to transmit a bit sequence to output $Q0$ according to the serial receiver format defined in the example. Remember that the format requires a group of five bits: the *start bit* at 1, then three *information bits* D0, D1, and D2 followed by the *stop bit* at 0. Each bit has the same length as a *Clock* cycle.

We choose D0 = 1, D1 = 0, and D2 = 1. As seen in the next figure, we set these values on the register's parallel inputs: $P1 = D0$, $P2 = D1$, and $P3 = D2$.

Notice that *P0* has been forced to 1 since it is the *start bit* to be transmitted first, while the serial input *In* is forced to 0, because we need to transmit the *stop bit* last.

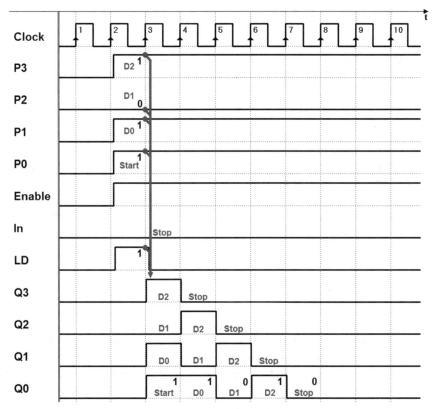

Together with the data, we enable the register by activating the *Enable* input. We bring *LD* to 1 for one *Clock* cycle so that the parallel load can be carried out on edge 3 of the *Clock*. The *start bit* is brought to output *Q0* and is maintained for one *Clock* cycle.

Given that $LD = 0$, the register shifts right at edge 4 bringing the value of D0 to output *Q0*, which is maintained for one *Clock* cycle.

Meanwhile, note that serial input *In* inserts a 0 from the left, so the register's content progressively returns to zero. In the end, the *stop bit* is transmitted at 0. In this simplified example, the shifting continues on the following *Clock* cycles and the register continues to send 0 on line *Q0*.

In the next page are two examples of 8-bit shift registers ("PiSo8") with parallel load taken from the library of *Deeds*.

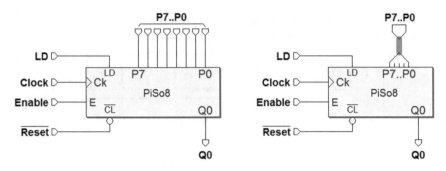

The only difference between these two components is in the connections. The one on the right uses the *bus type* for inputs *P7..P0*. These components belong to the "PISO" (*Parallel Input–Serial Output*) classification.

Note, however, that their only output is line *Q0* and there are also a few small differences in the enabling logic. Here, input *E* controls the enabling of the shifting only. To do a parallel load of the data, one needs to activate *LD*.

6.2.4   Universal Shift Register

The registers shown so far make it possible to load and read data in serial or parallel format but shifting is only done in one direction (to the right, $Q3 \rightarrow Q0$). Other registers allow shifting in both directions.

The *universal register* allows *bidirectional shifting* and can be loaded and read in *serial* and *parallel* format.

Below, the schematic of a 4-bit universal register (the initialization network has been omitted also in this case).

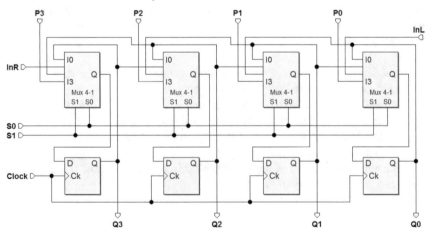

For a summary of this register's terminations and their functions, let's consider the component "Univ4" shown below (from the *Deeds* library).

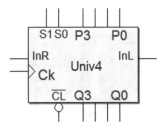

As in parallel registers, here there are inputs $P3..P0$ and outputs $Q3..Q0$. Obviously, clock (Ck) and clear ($\overline{CL}$) inputs are also present.

InR and InL are the serial inputs for *right* and *left shifting*, respectively.

Inputs $S1$ and $S0$ select component's operation according to the table below.

$S1$	$S0$	Function
0	0	Maintaining information
0	1	"Right" shifting
1	0	"Left" shifting
1	1	Parallel loading

The configuration ($S1 = 0$, $S0 = 0$) maintains the information because connects each input D with the output Q of the same flip-flop (see the figure).

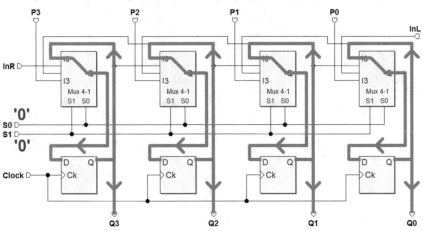

On the rising edge of the *Clock*, the register reloads the previous values on the flip-flops.

The combination ($S1 = 0$, $S0 = 1$) configures the register for *right shifting* (see the figure below).

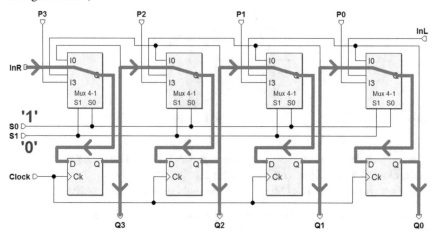

On the rising edge of the *Clock*, the data at $Q1$ is loaded onto the farthest right flip-flop and appears on $Q0$. The one on $Q2$ is transferred to $Q1$, and $Q3$ to $Q2$. *InR* is copied onto the first flip-flop and appears on $Q3$.

The next figure shows the *left shifting* operation. It is obtained through the combination ($S1 = 1$, $S0 = 0$).

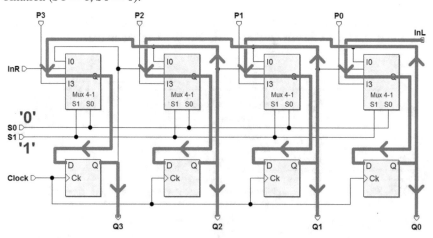

Despite the seemingly complex paths in the figure, the routing here is analogous to that of the previous figure, only in the opposite direction. In this mode, the serial input is *InL*. On the rising edge of the *Clock*, the value of *InL* is loaded onto $Q0$, while the other outputs shift one stage to the left.

Finally, the configuration ($S1 = 1$, $S0 = 1$) routes the value of parallel load input P onto each corresponding input D. On the rising edge of the *Clock*, the register will carry out a parallel load as shown in the figure below.

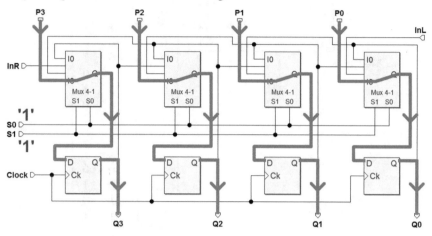

6.3 Counters

Another commonly used type of sequential network is the *counter*. This term indicates a network that generates a numerical sequence in a particular code (think, for example, of an increasing sequence made up of binary numbers represented by a certain number of bits). The network's *active edge* of the clock input causes the passage from one element of the sequence to the next. The counter is *synchronous* when the flip-flop network that creates it is *synchronous*.

6.3.1 Binary Counters

The following figure depicts an example of a *natural binary 4-bit counter*. The table on the right shows the outputs' *16 combinations*. This is an *increasing sequence*, so it is an *"up counter"*.

A *counting cycle* is made up of a sequence of 16 different configurations that can be generated: it is the case of a *"module 16"* counting. When it gets to the highest number "1111", the count continues cyclically at "0000". The rising edge of the *Clock* advances the count. Input $\overline{CL}$ (*Reset*) makes it possible to initialize the count at the value "0000".

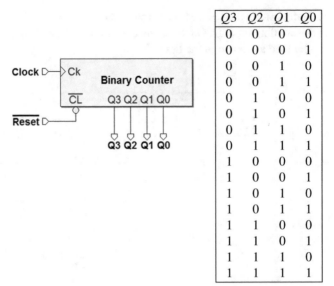

Q3	Q2	Q1	Q0
0	0	0	0
0	0	0	1
0	0	1	0
0	0	1	1
0	1	0	0
0	1	0	1
0	1	1	0
0	1	1	1
1	0	0	0
1	0	0	1
1	0	1	0
1	0	1	1
1	1	0	0
1	1	0	1
1	1	1	0
1	1	1	1

The internal structure of a counter is a synchronous sequential network like the one described at the start of this chapter. It is made up of a D-PET flip-flop parallel register and a combinational network that controls its behavior. In this counter, the function required to the combinational network is to increase the binary number on the outputs by 1, as shown in the figure below.

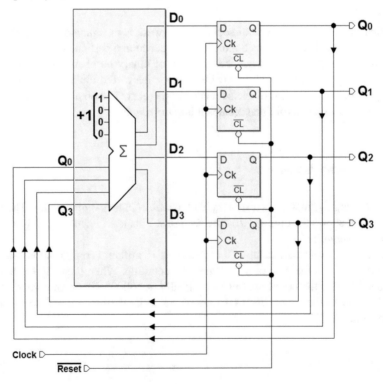

Intuitively, the count could be done by a full adder that adds the constant $+1$ to the number on the flip-flops' outputs at that time. The result is submitted to the flip-flops' inputs that will load the new number at the next rising edge of the *Clock*. Note that the carry from the fourth bit of the sum is ignored, since there are only four bits. This is how we get module 16 counting.

Let us now proceed more systematically and describe, in this truth table, the behavior that the combinational network should have.

$Q3$	$Q2$	$Q1$	$Q0$	$D3$	$D2$	$D1$	$D0$
0	0	0	0	0	0	0	1
0	0	0	1	0	0	1	0
0	0	1	0	0	0	1	1
0	0	1	1	0	1	0	0
0	1	0	0	0	1	0	1
0	1	0	1	0	1	1	0
0	1	1	0	0	1	1	1
0	1	1	1	1	0	0	0
1	0	0	0	1	0	0	1
1	0	0	1	1	0	1	0
1	0	1	0	1	0	1	1
1	0	1	1	1	1	0	0
1	1	0	0	1	1	0	1
1	1	0	1	1	1	1	0
1	1	1	0	1	1	1	1
1	1	1	1	0	0	0	0

The left side of the table shows the 16 possible combinations of flip-flop outputs $Q3..Q0$ and, on the right, the corresponding values of inputs $D3..D0$ that the combinational network must produce.

Since the rising edge of the *Clock* loads the value processed by the combinational network onto the flip-flops, the table actually links the *current state* of the network with the *next state*.

Remember that a D-PET-type flip-flop connected to an XOR through feedback reproduces the functioning of the T type, as we have seen before.

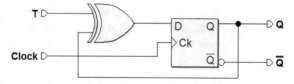

If we put $T = 0$ at the input of the XOR the flip-flop keeps the previously memorized value. If $T = 1$, then outputs are inverted.

With this in mind, we synthesize the combinational network described in the table above. Omitting all the intermediate steps, we derive the schematic of the adder based on D-PET flip-flops connected to form a T-type flip-flop.

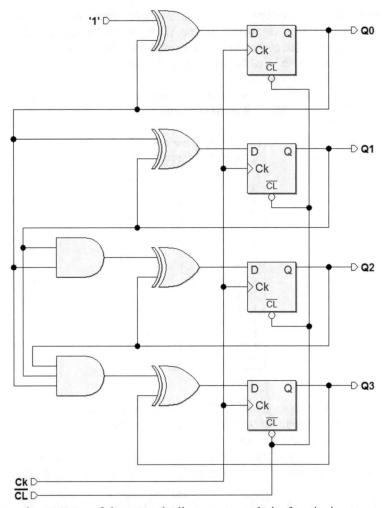

The regular structure of the network allows us to study its functioning even on an intuitive level, with the help of the timing simulation below.

Output $Q0$ changes values at each active edge of the *Clock* since the corresponding flip-flop always receives the inverse of $Q0$ in the input. With $Q1$, however, the inversion condition is only true when $Q0$ is 1 at the XOR, whereas if $Q0 = 0$, $Q1$ keeps its previous value.

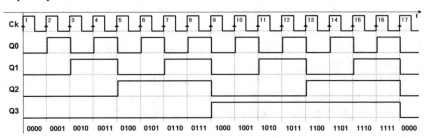

Likewise, $Q2$ changes when outputs $Q0$ and $Q1$ are both at 1, due to the two-input AND gate. Finally, $Q3$ toggles only when $Q0$, $Q1$, and $Q2$ are all at 1, due to the three-input AND.

The simulation shows that an output changes when all the lesser significant outputs are high. This observation allows us to intuitively extend the binary counter to any number of bits.

Let us now look at the inversion function, which can be obtained through the JK-PET flip-flop (this is implicit in this logical type and it is enough to connect the inputs J and K together):

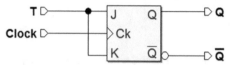

By substituting the D flip-flops with as many JKs, we simplify the previous schematic as seen in the following figure:

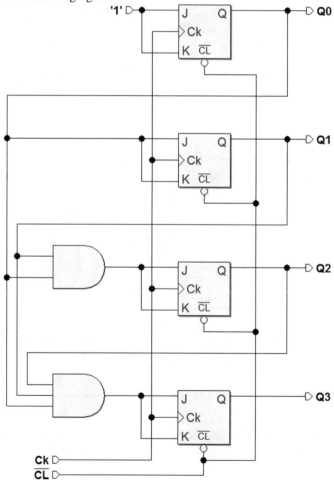

We can also draw the same network placing the flip-flops horizontally, as in the figure below. The advantage is a more intuitive view of the flip-flop inputs' *"cascade"* driving.

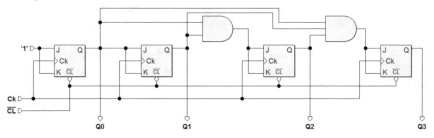

The network simplifies even more if we take AND's associative property into consideration. It allows us to use simple 2-input ANDs to arrive at the structure shown in the figure below:

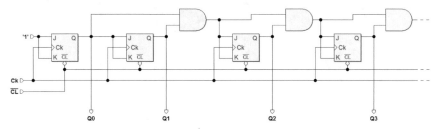

Nevertheless, when the number of bits increases in this type of structure, the maximum operating frequency declines since the number of levels in the combinational network rises linearly along with the number of flip-flops.

Let's go back to the binary counter's timing simulation. An interval of this is seen below:

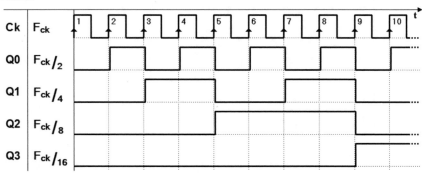

If we consider the relation between the signals' timing evolution (the so-called *waveforms*), we see that the *period* of $Q0$ is twice that of the clock Ck. Likewise, the period of $Q1$ is twice that of $Q0$ and four times the clock.

If *Fck* is the frequency of the clock signal, then *Fck/2* is the frequency of the *Q0* signal, *Fck/4* the frequency of *Q1* and so on. In our example, the waveforms of the outputs are also *symmetrical*; that is, the lengths of their high and low intervals are identical.

A counter therefore can be used as a *"frequency divider"*, a network that provides periodic signals derived from the clock with frequencies equal to that of the clock divided by a power of 2.

The following figure shows the "UCnt4," an example of a *synchronous, binary, 4-bit up counter* taken from the *Deeds* simulator library. This counter is functionally identical to the one described above with an added *TC* (*"Terminal Count"*) output.

TC activates when the number generated by the counter reaches the highest value, according to the simple *combinational function TC = Q3 · Q2 · Q1 · Q0*.

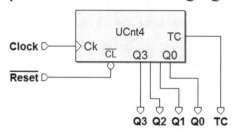

The figure below shows an example of the component's timing simulation that highlights the activation of *TC*.

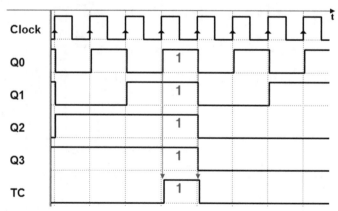

6.3.2 *Counters with Enabling*

The figure in the next page shows a counter that is similar to the previous one. It has an extra *enable* input *En* that controls the counting function.

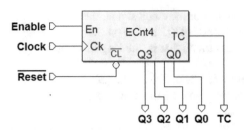

The count is *enabled* only if *En* is at 1. When the count is disabled, the counter's outputs do not change despite the edges of the clock. The counter here is the "ECnt4" (*"up counter with enable 4 bits"*), taken from the library of *Deeds*.

Let's examine how enabling functions in principle. The following figure describes the function of a binary up counter 4 bits *with enabling*:

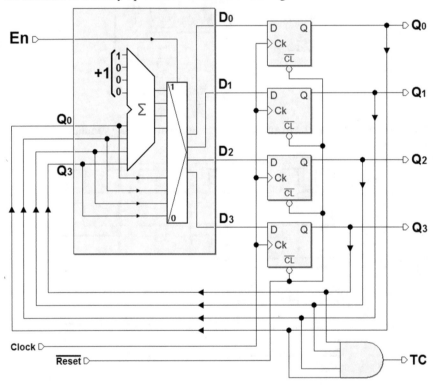

Compared to the last structure we saw, this one has a multiplexer in front of the inputs of the flip-flop that is controlled by *En*. When *En* is high, the multiplexer connects the output of the adder with the flip-flops, which brings us back to the normal count as seen previously.

When *En* is low, the multiplexer feeds back the outputs into the inputs of the flip-flop. Thus, at each edge of the clock, the previous values are confirmed (and the count halts). The *TC* is generated in function of the value taken on by the flip-flops. If the value "1111" is on the outputs $Q3..Q0$, it activates.

The timing diagram depicted in the next figure shows an example of the functionality of a counter with enabling. The record starts with *En* set low and the count halted (say, at the value of "1101"). The count only proceeds on the rising edges of the clock if *En* = 1. In this simulation, it is activated three times for the duration of one clock cycle.

At the first (1) activation of *En*, the count gets to the value "1110" and stops. The value will increment ("1111") at the next (2) activation of *En*. Note that the entire time the counter has this combination of outputs, the output *TC* is activated, an indication that the counter is at its "*terminal*" value.

Then, the counter moves up one increment at the last (3) activation of *En*: its outputs go from "1111" to "0000", because the count is cyclical and uses only 4 bits.

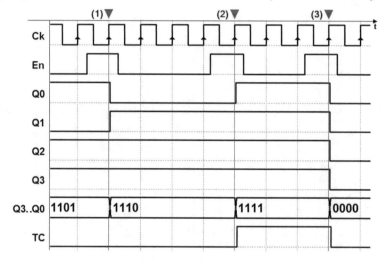

Note that *TC* = 1 signals also that the count on the next active edge of the clock will go from the maximum value to zero. In the next section, we will see that this will be useful in connecting multiple "*cascading*" counters in order to get a count with a higher number of bits.

In sum, counters with enabling make it possible to *count the number of times* input *En* is activated in response to the active edge of the clock, all while keeping the operations rigorously synchronous.

6.3.3 Up/Down Counters

Up/down counters allow to count *up* or *down*. For example, take the "DCnt4" ("*up/down counter with enable 4 bits*") from the *Deeds* library:

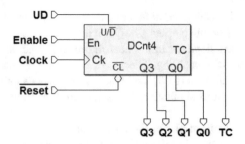

Here, the *up/down counter* has the added input $U/\overline{D}$ that sets the count direction.

The up/down counter in the figure below repeats the structure of the counters seen previously but with one main difference: the content of the register can be increased or decreased.

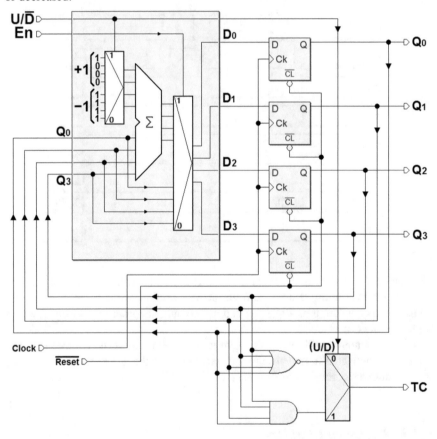

We can do this by presenting the constants $+1$ or -1 (represented in *two's comple-ment code*) to the adder through the multiplexer seen in the upper left-hand corner of the figure. The choice of $+1$ or -1 is based on the value of the input direction control $U/\overline{D}$.

$U/\overline{D}$ controls also the multiplexer in the lower right-hand side of the figure that generates the *Terminal Count* coherently with the count direction. When counting up, TC is activated when the highest number "1111" is reached. When $U/\overline{D} = 0$, the counting is down, *TC* is set to 1 if we have reached the lowest number "0000".

This block description finds a possible circuital synthesis in the following figure. The network is very similar in structure to that of the *up counter without enable*. Keeping the similarities in mind, let's try to interpret the elements of this new network from an intuitive perspective.

The D-PET flip-flops are connected as T type, but four XORs have been added in the feedback loop.

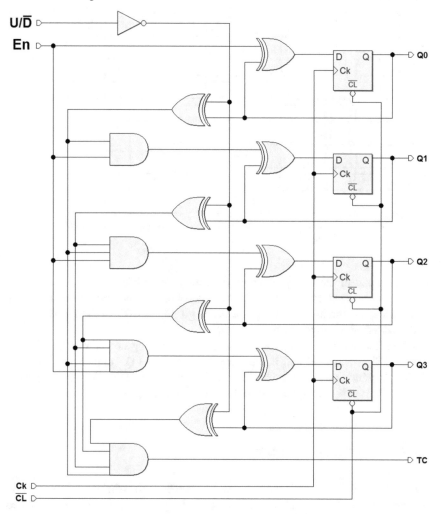

With the input $U/\overline{D}$, we can invert, or not, the value taken from the outputs of the flip-flops. If $U/\overline{D} = 1$, these XORs *do not invert*, and the network works like an up counter. If $U/\overline{D} = 0$, the network works like a down counter.

Input *En* controls all the flip-flops. If *En* = 0, all of them are forced to recharge their own value (at each active edge of the clock), so the outputs do not change, and the count does not change. If *En* = 1, however, everything works as if that input were not there and the count is enabled.

Independently of *En*, output *TC* is generated by an AND gate. AND's four inputs come from the outputs of the (direct or negated) flip-flops in function of input $U/\overline{D}$. *TC* will be activated when the outputs of the flip-flops get to "1111" if the count is *up* (or to "0000", if it is *down*).

The figure below shows an example of the timing simulation of the *up/down counter with enable*:

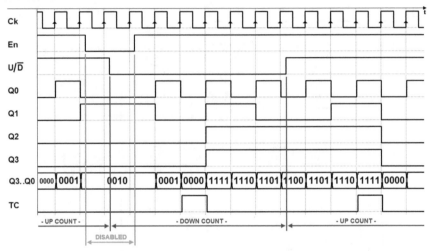

In the first part of the timing diagram, the counter is enabled (*En* = 1) and counts up ($U/\overline{D}$ = 1). Then for two clock cycles, the counter is disabled (*En* = 0) and the count stops, staying at the last number it had got to, "0010". In the meantime, the counter is asked to count down ($U/\overline{D}$ = 0).

When it is enabled again, (*En* = 1), it starts counting down. When it reaches "0000", *TC* is activated and it starts counting from "1111" until it gets down to "1100", at which point we order it to count up again ($U/\overline{D}$ = 1). When it gets to "1111", *TC* is activated again and starts from "0000" (and so on).

6.3.4 *"Universal" Counters*

The most complete counter is the *"universal"* counter. It adds the possibility to *preset* the number the counter contains, like in parallel registers. The structure presented improves enabling and *TC* output, too.

In the figure next page, we see an example of a universal counter, the "Cnt4" (*"counter 4 bits"*), from the *Deeds* library.

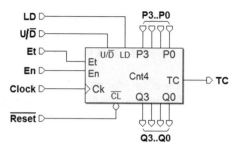

The figure below shows a universal counter block by block. The counter has the *preset* inputs $P3..P0$, in the same number as the outputs $Q3..Q0$, and the *load* command input LD.

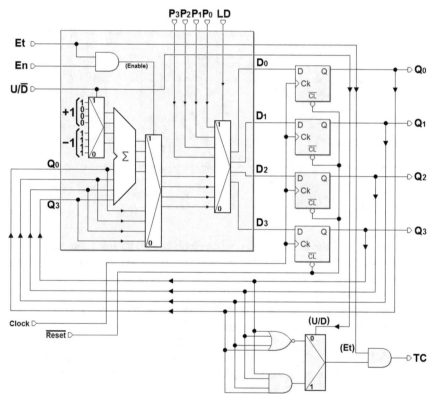

Here, a multiplexer controlled by input LD has been added to the structure of the counters seen previously. If LD is at 1, the number set on the inputs $P3..P0$ is routed to the D of the flip-flops, and it will be loaded into the counter on the next rising edge of the clock. If $LD = 0$, however, the counter works like the previous types. For example, the functionality of the direction input $U/\overline{D}$ is the same.

There are now two enable inputs: *En* (*"Enable"*) and *Et* (*"Enable Terminal Count"*). Both *En* and *Et* must be active to enable the count. Since *Et* enables the generation of output *TC*, it is used separately from *En* with multiple cascading counters, as we will see further on.

Remember that in all the counters seen previously, *TC* function could not be disabled and its value depended only on the direction of the count and the output values.

Counter Extension

The synchronous structure of a universal counter lends itself to the *extension of the number of bits* by using multiple interconnected *"cascading"* devices. For example, we see in the figure below that a 12-bit counter has been obtained through three 4-bit counters.

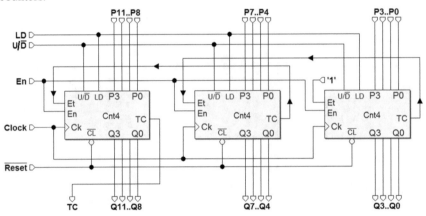

As we can see, all three devices share the clock signal, so the entire structure is *synchronous*. Inputs $\overline{Reset}$ and $U/\overline{D}$ are also shared, so the three counters will be initialized together and will always count in the same direction.

The enable input *En* is shared as well as the load command *LD*. The counter on the far right-hand side of the figure is used for the less significant bits ($Q3..Q0$) and the farthest left, for the most significant ($Q11..Q8$).

Now let's assume that the *En* command is set to 1 and the direction is up. The counter that generates $Q3..Q0$ will be enabled since its input *Et* is at 1. So that the middle counter, which produces $Q7..Q4$, counts only when it should, we connect its input *Et* to the *TC* of the counter that generates $Q3..Q0$.

This way, *TC* is used like a *"carry"*: when the farthest right counter gets to the maximum number, it instructs the middle one to count up one unit by enabling it with $Et = 1$.

There is an analogous connection between the *TC* of the middle counter and the farthest left *Et* input. The *TC*, which is produced by the farthest left counter, will only be active when *all three* counters reach "1111".

This structure can be extended to more bits, but we must remember that every counter we add increases the overall propagation delay of the combinational network, which propagates the *Et* enable signals through the *TC* of the various components.

In this book, we have used 4-bit counters for simplicity's sake. Obviously, in CAD system libraries, we find components of any size, like the 8- and 16-bit counters in the figures below.

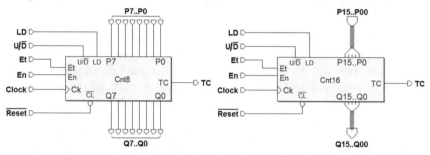

The 8-bit counter on the left is shown in its normal version. The 16-bit counter on the right is shown in the *bus-type* connections' version.

6.3.5 Asynchronous Counters

A counter is *asynchronous* when the flip-flops it is made of *do not all share* the same clock. The network in the figure below represents an asynchronous, binary up counter where all JK flip-flops but the first use the $\overline{Q}$ of the one before it as a clock signal.

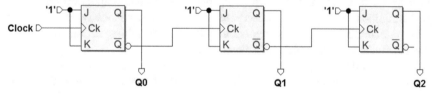

Every flip-flop in this network (active on the rising edge of its *own Ck* input) changes state when the output *Q* of the previous flip-flop switches *from one to zero*. Because of this behavior, this is called a *"ripple counter"*, a term that recalls a wave-like propagation.

The evolution of the outputs is described in the figure below where the propagation times of the flip-flops have been highlighted (in approximate terms). Note that the delay between the *Clock* at the input of the counter and any given output grows *proportionally* to the position (weight) of that output.

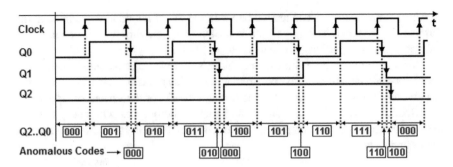

If we ignore the delays, this counter follows an *up binary sequence*, but the asynchronous commutation of the outputs generates anomalous codes that alter this sequence, as shown in the figure. These codes have a short duration (the same as the propagation time of the flip-flop), but they can create problems in a network that processes its outputs, for example a decoding network. This is why asynchronous counters are used only in special cases.

As we know, a counter can be considered a *frequency divider*, when we consider the timing relation between the waveform of the *Clock* and that of any of the outputs. An asynchronous counter can be used to good advantage to this purpose. As we see in the figure below, the frequency of outputs $Q0$ is $1/2$ of that of the *Clock* and outputs $Q1$ and $Q2$ provide a frequency signal of $1/4$ and $1/8$, respectively.

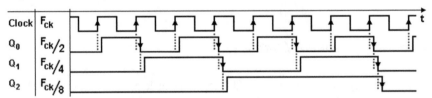

In counters used as frequency dividers, the asynchronous outputs do not pose a problem since the signals generated are often used independently of each other. Also, the simplicity of the asynchronous counter versus the synchronous one is a great advantage for very high *"division ratios"*.

Frequency dividers are used in telecommunications devices where they generate signals whose frequency is a submultiple of that of a *Clock* generator, which works at a higher frequency.

To be thorough, below, we show an asynchronous *down* counter. In this network, the input *Ck* of each flip-flop except the first is connected to the output *Q* of the one before it.

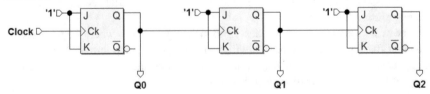

In the timing diagram below, we see that the flip-flop's change in outputs in the backward count occurs when the one before it makes a transition *from zero to one*.

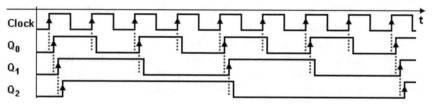

6.4 Network Analysis Examples

One essential step in beginning to design digital systems is to understand the behavior of a given sequential network. The timing analysis of a network provides familiarity with some general aspects of the interaction between the combinational and sequential components of a logical circuit. The familiarity with the *low-level* behavior of sequential networks is an important step for the designer who is mindful of the workflow from the project specifications through to the final product.

Next, we will show a series of examples of analyses through the *timing diagrams* of simple sequential networks that have a given logical schematic associated with suitable input signal sequences. We will carry out a functional network test; that is, we will study the evolution on time of the outputs as a function of the inputs.

6.4.1 Example 1

In this section, our goal is to analyze the function of the network of flip-flops depicted in the figure below. A simple observation of the schematic gives useful indications on how to analyze it. This is a synchronous network made of D-PET flip-flops with an asynchronous initialization input $\overline{Reset}$ (which acts on the inputs $\overline{Clear}$ of the flip-flops). The network generates the three outputs $Q2$, $Q1$, and $Q0$.

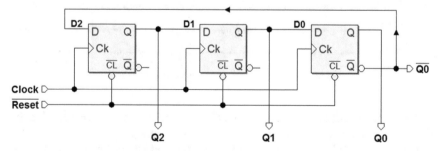

The structure of the network is also easy to identify: a shift register where the serial input $D2$ is connected to the negated output $\overline{Q0}$ of the last flip-flop. It is not necessary to identify the specific structure to analyze it. The procedures shown here work for any synchronous network of flip-flops.

To start, we must first have a *timing diagram* where we can sketch the evolution of *Clock*, and the inputs and outputs of the network. In the case at hand, we will insert signal $D2$ ($= \overline{Q0}$) in the diagram for ease of examination.

The next figure shows the track of the diagram to construct. Here, the *Clock* and the initialization signal $\overline{Reset}$ have been defined. We suppose $\overline{Reset}$ active at the beginning of the diagram and then deactivated in the interval between the edges (1) and (2) of the *Clock*.

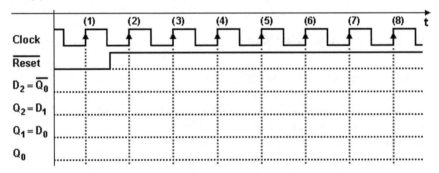

As long as input $\overline{Reset}$ is kept active (low), outputs Q of the flip-flops are *forced* to zero and edge (1) of the *Clock* cannot provoke changes. Note that we must draw $D2$ with the value of 1 in this initialization phase since it is connected to the negated output $Q0$.

Remember that deactivating $\overline{Reset}$ does not change the state of the network and the signals remain unchanged until the next active edge of the *Clock* (2).

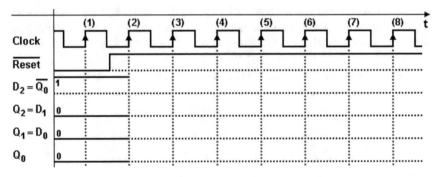

At every active edge of the *Clock*, the D-PET flip-flops transfer the logical value that is on their own input D at that moment onto their output Q.

On edge (2) of the figure above, inputs $D2$, $D1$, and $D0$ of the flip-flop are 1, 0, and 0, respectively. Therefore, let's draw the outputs of the flip-flops after edge (2) the figure in the next page.

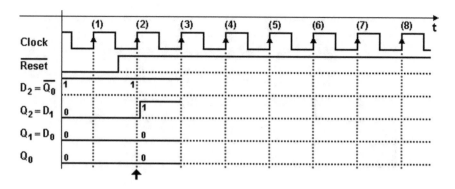

In the previous figure, we have chosen to highlight the propagation delay between the edge of the *Clock* and the change of output $Q2$. Note that the new value taken by $Q2$ (and thus by $D1$) will be acquired by flip-flop $Q1$ on the next edge (3).

Up until edge (3), the situation remains the same, given that flip-flops change their state only on the edge of the *Clock*. On edge (3), the values on inputs D are transferred to outputs Q, in the same way as on edge (2). In the figure below, we see the situation after edge (3). Note that there is a delay in the activation of $Q1$.

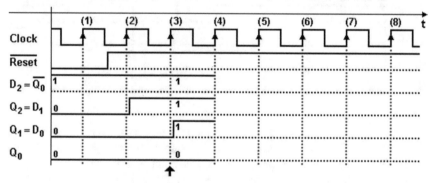

In the two following diagrams, we continue drawing the diagram in relation to edge (4) and (5) by applying the same criteria.

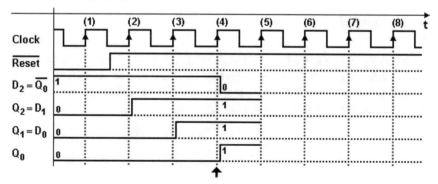

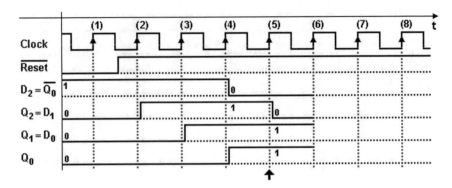

Finally, in the figure below, we see the complete timing diagram. The figure shows the typical behavior of a *shift register*.

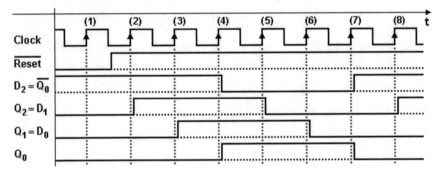

The network we have examined is called a 3-bit *"Johnson Counter"*. It can be made with a shift register with any number of bits by connecting the negated output of the last flip-flop to the input of the first. It has the advantage of a simple structure and the corresponding disadvantage of a counting code which is different from pure binary (but can be easily decoded).

6.4.2   Example 2

The network in the following figure is made up of three D-PET flip-flops with shared *Clock* and $\overline{Reset}$ signals. It is easy to find the base structure of the shift register but the input of the farthest left flip-flop, $D2$, is obtained by an XOR tree that processes the outputs $Q1$ and $Q0$, and the input *Seed*.

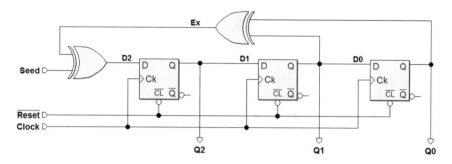

We know that, at every active edge of the *Clock*, the flip-flops transfer the logical value on input *D* at that moment onto output *Q*. The network analysis consists in evaluating inputs *D2*, *D1* and *D0* in relation to those active edges. Based on the figure above, we can write the Boolean expressions. Note that only the first one describes a network with logical gates, while the others are simple connections.

$$D2 = Seed \oplus Q1 \oplus Q0; \qquad D1 = Q2; \qquad D0 = Q1$$

These three expressions provide the values of inputs *D2*, *D1* and *D0* as function of the values of input Seed and flip-flop outputs *Q2*, *Q1*, and *Q0* (the current state of the network). Once these are loaded on the flip-flop, they will constitute the *next state* of the network. Here too, we trace the behavior of the network on a timing diagram seen in the figure below. We prepare traces for all the network's inputs and outputs, the flip-flops' inputs and even the intermediate signal *Ex* for ease of analysis.

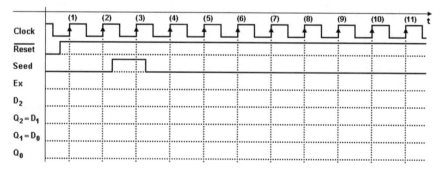

The next figure shows the timing analysis up until edge (4). The $\overline{Reset}$ signal, which is active before edge (1) of the *Clock*, sets the flip-flop's outputs to 0. It is easy to understand why the *Seed* activation is necessary to make the network evolve. In its absence, the network would remain in the situation set by $\overline{Reset}$ indefinitely.

Inputs *D1* and *D0* are also at 0, since they are connected to outputs *Q2* and *Q1*. *D2* is at 0, as we can see from the Boolean expression since the external input *Seed* is set to 0. Therefore, edge (1) of the *Clock* does not change its outputs, which remain at 0. The same goes for edge (2).

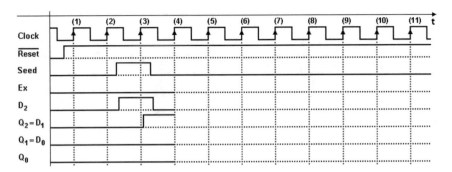

In the initial setting of the diagram, we assumed that the input *Seed* is activated for the duration of the *Clock* cycle between edges (2) and (3). The immediate result is that $D2$ is activated: on edge (3), the output $Q2$ switches to 1, while the other flip-flops do not change their values ($D1 = D0 = 0$).

After edge (3), the input *Seed* returns to 0 and remains to this value until the end of the diagram, so the evaluation of $D2$ is simplified, since it now depends only on the variations of outputs $Q1$ and $Q0$.

To continue the analysis, after *Seed* is brought to zero, the networks that generate $D2, D1$, and $D0$ produce the values 0, 1, and 0, respectively. These are transferred onto the flip-flop outputs at edge (4). With the same method, let's continue the analysis until we complete the diagram shown here:

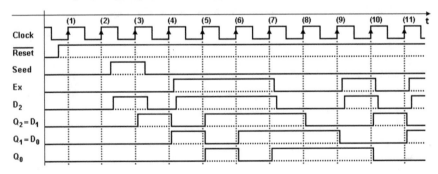

We can make some general comments about the complete diagram. All the flip-flop outputs commute at the active edges of the *Clock* with a delay that is equal to their propagation time. Signals Ex and $D2$ show an added delay due to the combinational network that generates them. As we can see in the diagram, signal $D2$ can change asynchronously with respect to the *Clock* between edges (2) and (4) because it is dependent on external input *Seed*.

The network we have analyzed is a simplified example of a *pseudo-random number generator*. The number generated in this case is made up of outputs $Q2, Q1$, and $Q0$. A pseudo-random number generator is normally created with a high number of flip-flops (ex. 32) because the sequence generated is only *apparently* random; it actually repeats cyclically.

6.4.3 Example 3

The network in this example uses two D-PET flip-flops with shared *Clock* and $\overline{Reset}$ signals and without command inputs. Outputs $U0$ and $U1$ are taken directly from outputs $Q0$ and $Q1$ of the flip-flops while TC is obtained through a logical gate.

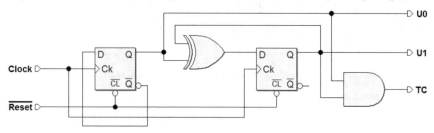

Let's apply the same analytical process as before. We evaluate $D0$ and $D1$ at the active edge of the *Clock* since at that time they will be loaded onto the flip-flops. It may be useful to "*separate*" the combinational networks that produce $D0$, $D1$ and the outputs $U0$, $U2$, and TC from the overall schematic. Let's re-draw them apart, in the form of a circuit and as Boolean expressions.

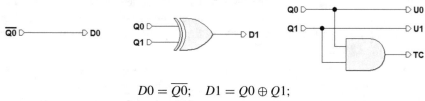

$$D0 = \overline{Q0}; \quad D1 = Q0 \oplus Q1;$$

$$U0 = Q0; \quad U1 = Q1; \quad TC = Q0 \cdot Q1$$

By using the schematics or expressions just described, we can trace the timing diagram. Aside from the *Clock*, let's trace the signal $\overline{Reset}$ in the diagram, as active from the beginning and deactivated just before edge (1) of the *Clock*.

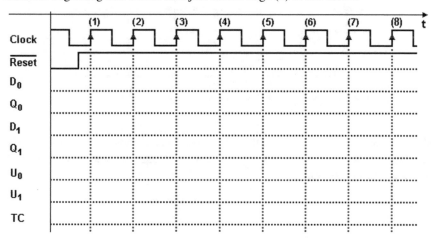

The first step is to consider $\overline{Reset}$, which forces the flip-flops to 0 at the beginning and then is removed. The state of the network changes at *Clock* edge (1) when the two flip-flops take on the values of *D*0 and *D*1.

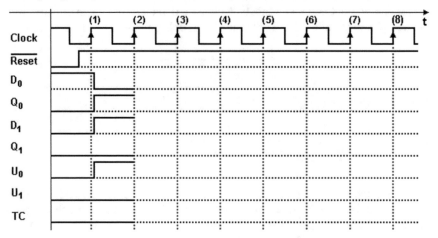

Now, we suggest that for practice the readers continue the analysis on their own, following the criteria suggested so far. The timing diagram will look like the following figure, where we see the *cyclical* quality of the sequence, which repeats every four edges of the *Clock*.

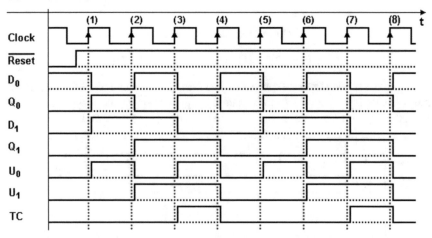

This device behaves as a *counter*; the values taken on by outputs $U1$ (MSB) and $U0$ (LSB) follow a *module 4 binary natural up count*. *TC* signals when the outputs have reached their highest value.

6.4.4 Example 4

The network in the figure below is very similar to the previous one. It uses two D-PET flip-flops and has the same outputs $U0$, $U1$, and TC, but it has an added input $\overline{SyncRes}$ ("Synchronous Reset") and the logic associated with it. The *Clock* and $\overline{Reset}$ network are identical to the previous case.

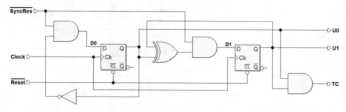

A couple of intuitive points can help the analysis of this network.

If the input $\overline{SyncRes}$ is 1, the two ANDs conditioned by this signal transmit the value of their other input to their respective outputs and the network is functionally identical to the one in the previous exercise.

If $\overline{SyncRes}$ equals 0, the outputs of both the ANDs are at 0, setting $D0$ and $D1$ to 0 and so the two flip-flops are brought to zero *synchronously*.

Let's derive the combinational networks that produce $D0$ and $D1$, and/or the expressions. The $U0$, $U1$, and TC network is identical to that of the previous exercise.

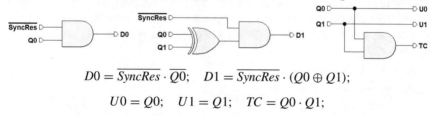

$$D0 = \overline{SyncRes} \cdot \overline{Q0}; \quad D1 = \overline{SyncRes} \cdot (Q0 \oplus Q1);$$

$$U0 = Q0; \quad U1 = Q1; \quad TC = Q0 \cdot Q1;$$

Let's set up the timing diagram as in the figure below, with the *Clock* and $\overline{Reset}$ signals set as in the previous exercise. Let's assume that the $\overline{SyncRes}$ command is activated for two Clock cycles as drawn here:

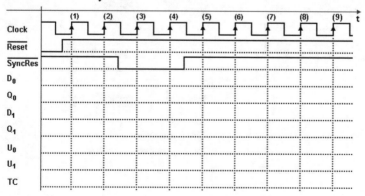

Now, it is the reader's job to complete the timing layout, which will turn out to be like the figure below. Notice that signals $D0$ and $D1$ respond immediately (except for propagation times) to $\overline{SyncRes}$ command variations.

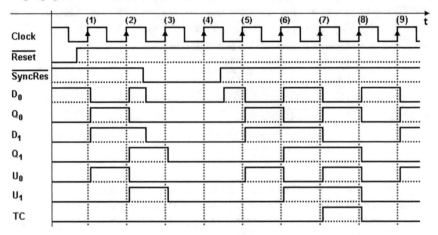

6.4.5   Example 5

The figure below shows a network that uses two D-PET flip-flops that share the same *Clock*. Both flip-flops are connected to the asynchronous initialization input $\overline{Reset}$. The two inputs are *EN* and *DIR*. The outputs generated by the networks are $U0$, $U1$, $U2$, $U3$, and *MAX*.

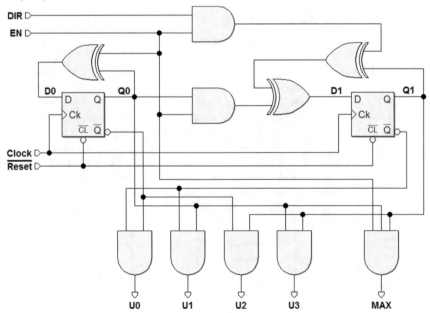

As before, the fundamental step to analyze the network is to evaluate $D0$ and $D1$ at the active edges of the *Clock*. Let's separate the combinational networks that produce $D0$ and $D1$ from the full schematics and describe them in terms of Boolean expression as well.

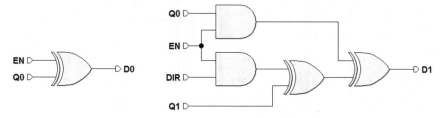

$$D0 = EN \oplus Q0; \qquad D1 = (EN \cdot Q0) \oplus (Q1 \oplus (EN \cdot DIR))$$

These networks combine the values of inputs EN and DIR with the values of outputs $Q0$ and $Q1$ of the flip-flops (the *"state"* of the network), and they produce a new value for $D0$ and $D1$. On the active edge of the *Clock*, these values will be loaded and will constitute the *"next state"* of the network.

The network's outputs $U0$, $U1$, $U2$, $U3$, and MAX are combinational functions of the flip-flops' outputs and of the input EN, as shown below, both as a network schematic and in terms of Boolean expressions.

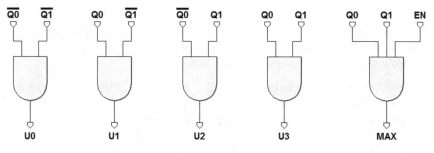

$$U0 = \overline{Q0} \cdot \overline{Q1}; \quad U1 = Q0 \cdot \overline{Q1}; \quad U2 = \overline{Q0} \cdot Q1; \quad U3 = Q0 \cdot Q1;$$

$$MAX = Q0 \cdot Q1 \cdot EN$$

For the timing analysis, inputs EN and DIR should be set in a way to avoid an unrepresentative timing diagram. We have chosen to include the flip-flops' inputs and outputs in the diagram to make the analysis easier.

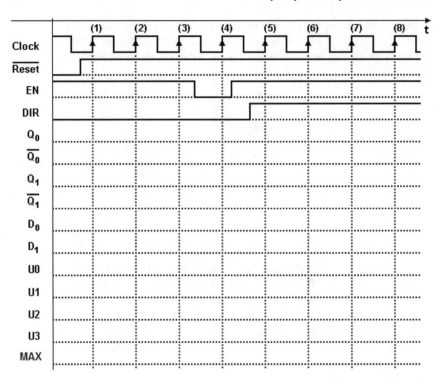

The same criteria of analysis we have seen before are appropriate to analyze this network. We suggest that the reader carries out the analyses personally.

Below we provide some advice on how to proceed:

1. We should focus initially on the evolution of the state of the network and then afterwards on the generation of outputs. So, we should first trace the evolution of signals $Q0$ and $Q1$ (direct and negated), and $D0$ and $D1$.
2. Even if we assume, as usual, that the propagation delays are short with respect to the *Clock* period, it is very useful to represent them in the diagram, albeit in a qualitative way.
3. Note that after the activation of $\overline{Reset}$, the network starts to evolve with edge (1) of the *Clock*.
4. After each active edge of the *Clock*, we must recalculate the values of $D0$ and $D1$, which will be loaded onto the flip-flops at the next active edge.
5. Between edges (3) and (4), *EN* changes. As a result, $D0$ and $D1$ immediately follow this change and the values of $D0$ and $D1$, which are transferred onto $Q0$ and $Q1$ are sampled by the flip-flops at edge (4).
6. Likewise, between edges (4) and (5), both *EN* and *DIR* change.
7. When the analysis of the next state is finished, we trace the outputs $U0$, $U1$, $U2$, and $U3$. These outputs can only change on the active edges of the *Clock*. This does not apply to *MAX*, which depends also on an input.

The figure below reports the results of the analysis. We see in the timing diagram that the flip-flops' outputs $Q0$ and $Q1$ follow a binary up counting sequence in the first few cycles after $\overline{Reset}$ is activated.

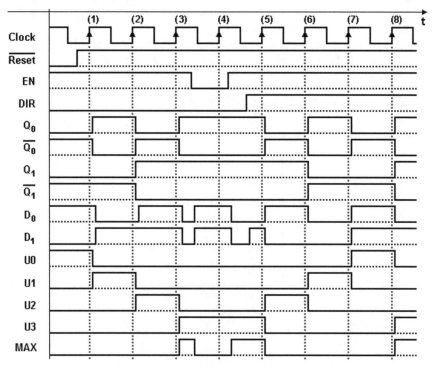

When input EN is at 0, the network loads the preexisting values onto the two flip-flops. On edge (4), $Q0$ and $Q1$ do not change. As on edge (5), the count changes direction since the value on input DIR changes.

The circuit, evaluated on outputs $Q0$ and $Q1$, behaves like an *up/down binary counter*. Input EN enables the count (if $EN = 1$), while DIR sets the direction (*down* if $DIR = 1$).

As we see in the timing diagram, outputs $U0$, $U1$, $U2$, and $U3$ are activated by combinations 00, 01, 10, and 11 of $Q1$ and $Q0$, respectively. Output MAX decodes the same combination of $U3$, but it is enabled only if $EN = 1$. Thus, the output is brought to 0 asynchronously following the evolution of EN between edges (3) and (5).

Finally, note that $D0$ and $D1$ evolve asynchronously between edges (3) and (5), since they follow the changes of inputs EN and DIR. What matters, however, is that their values should be stable at the moment they are read (i.e., on the rising edge of the *Clock*).

6.5   Exercises

Analyze the following synchronous sequential networks by completing the timing diagrams on the side page. The templates are also available on the simulator Web site, as *PDF* files, on the *digital contents* pages.

It is advisable to complete the layouts on paper *without* the help of *Deeds*, using it only to check the solutions. The Web site also provides the Deeds files of the networks proposed.

1. Exercise 1—(timing diagrams in the next page)

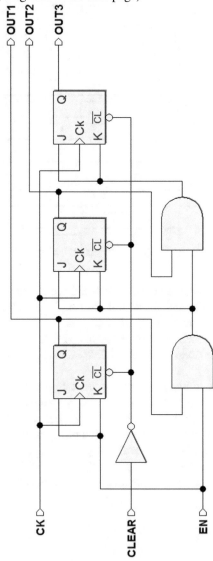

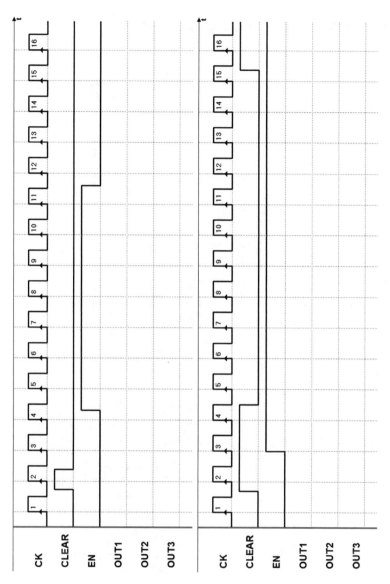

2. Exercise 2—(timing diagrams in the next page)

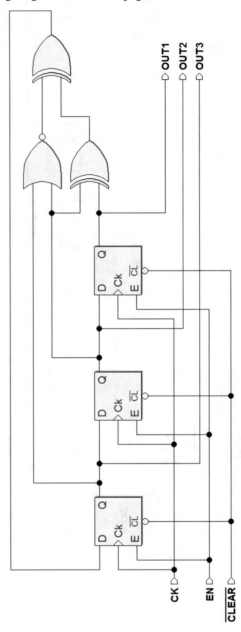

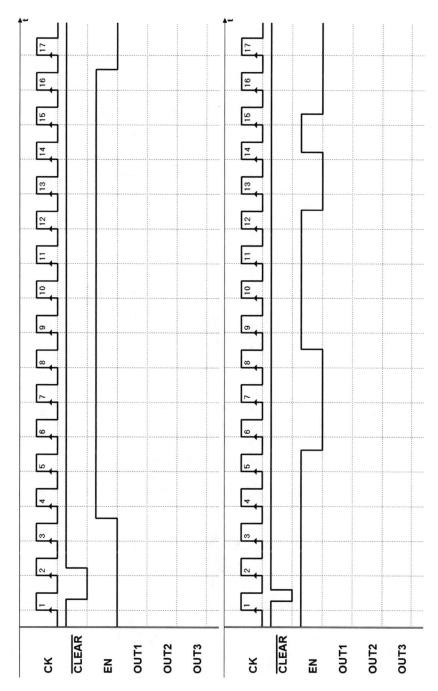

3. Exercise 3—(timing diagrams in the next page)

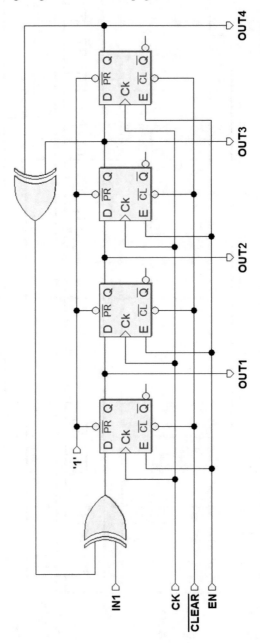

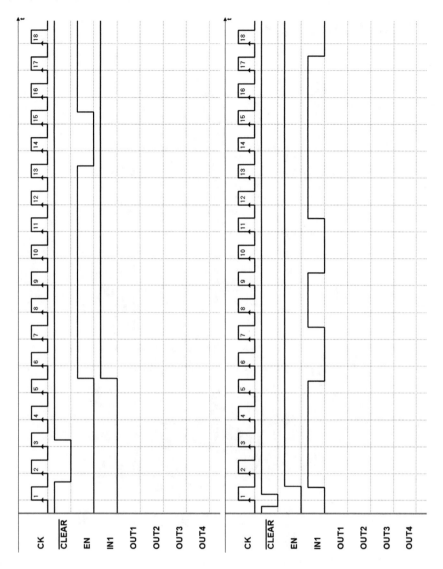

4. Exercise 4—(timing diagrams in the next page)

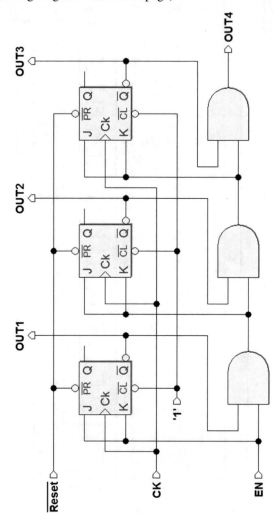

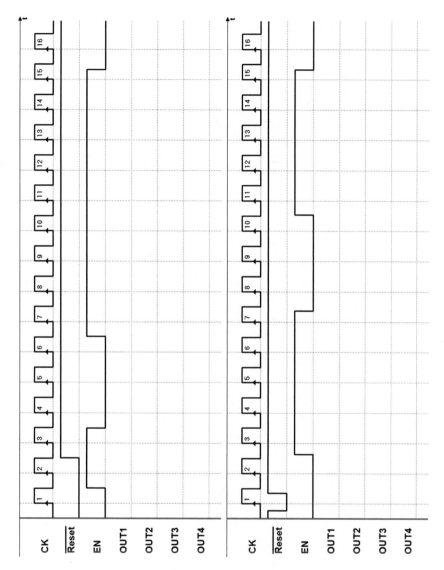

5. Exercise 5—(timing diagrams in the next page)

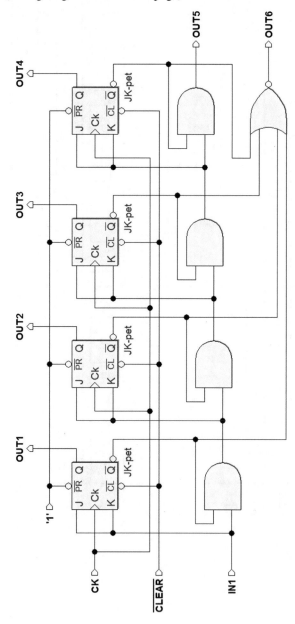

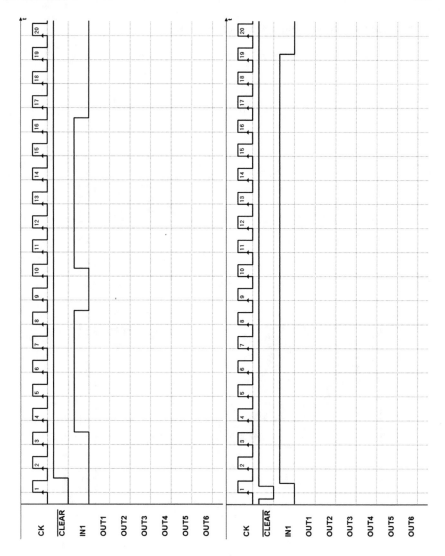

6. Exercise 6—(timing diagrams in the next page)

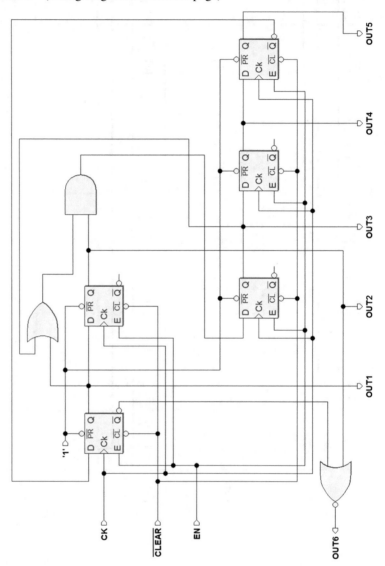

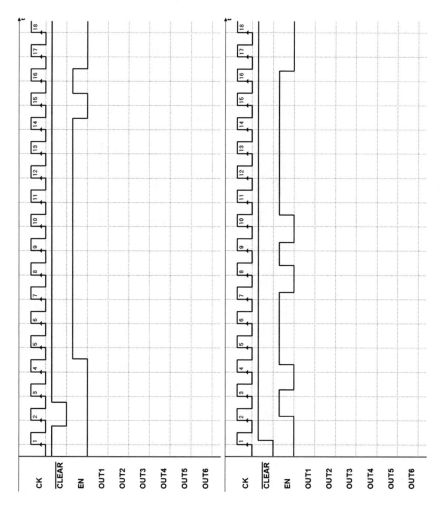

7. Exercise 7—(timing diagrams in the next page)

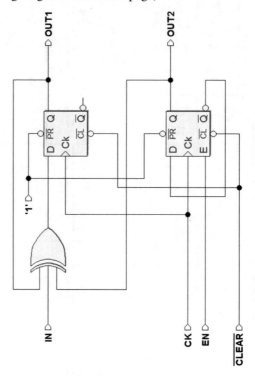

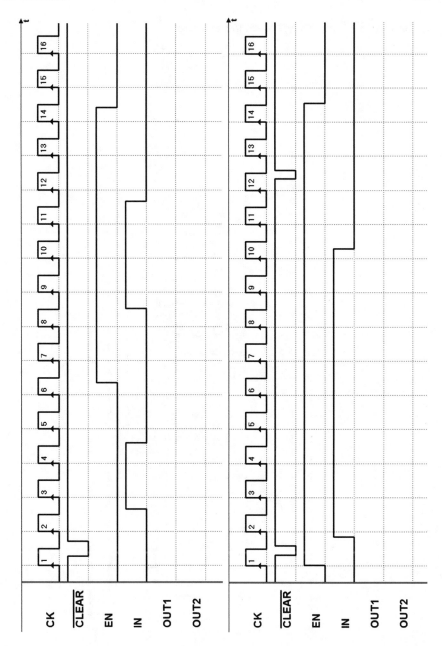

8. Exercise 8—(timing diagrams in the next page)

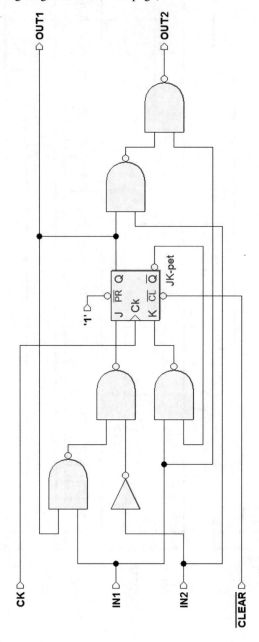

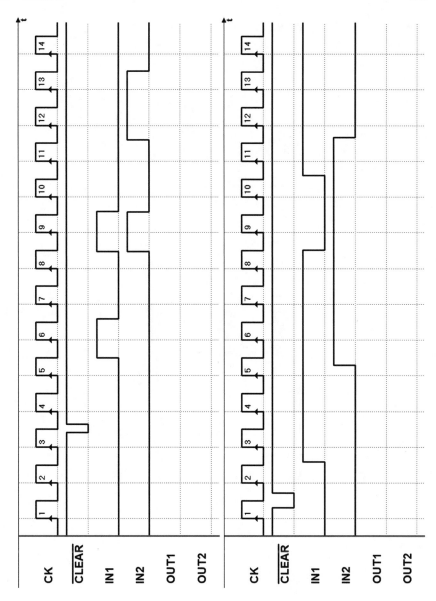

9. Exercise 9—(timing diagrams in the next page)

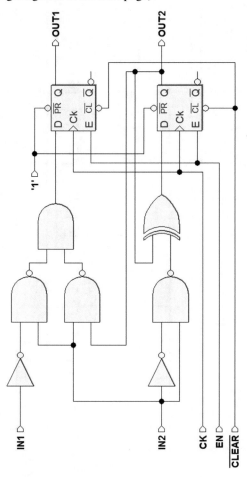

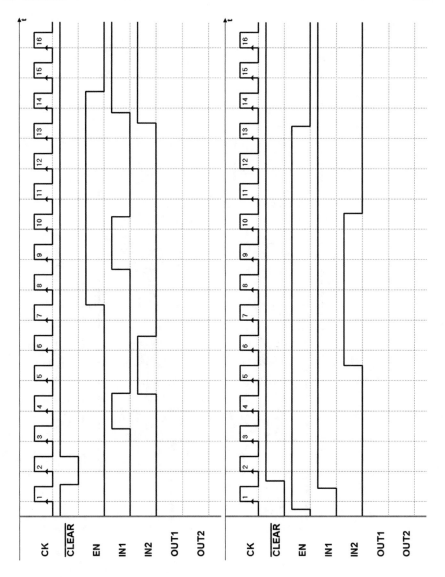

10. Exercises 10—(timing diagrams in the next page)

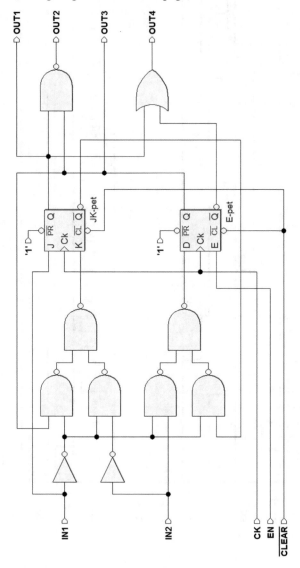

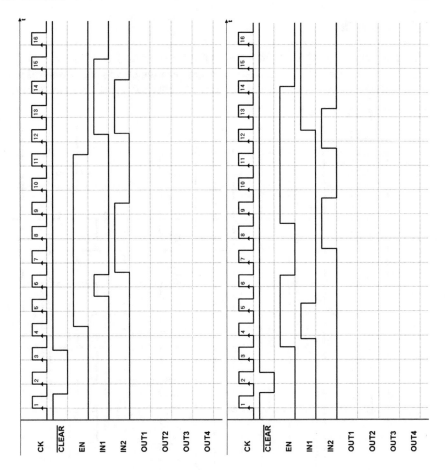

6.6   Solutions

The timing diagrams reported here were obtained through the *Deeds* timing simulator. The files of the networks assigned here are available on the *Deeds* site and on the *digital contents* pages of the book, so the solutions can be checked on the simulator as well.

1. Exercise 1 (solutions)

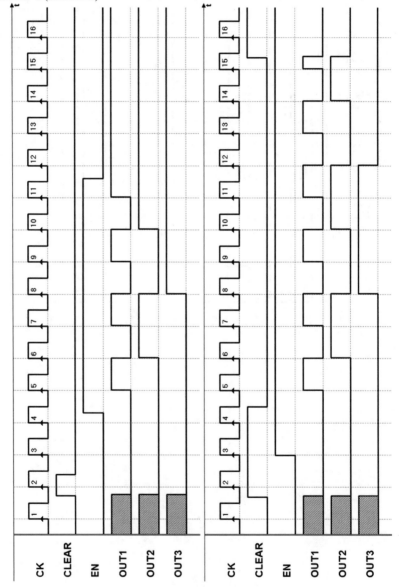

2. Exercise 2 (solutions)

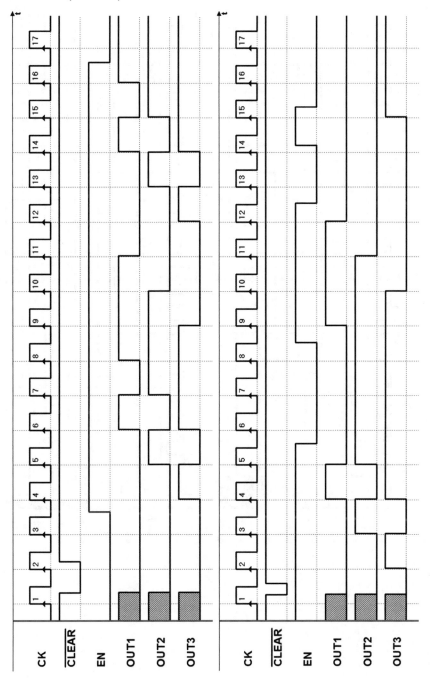

3. Exercise 3 (solutions)

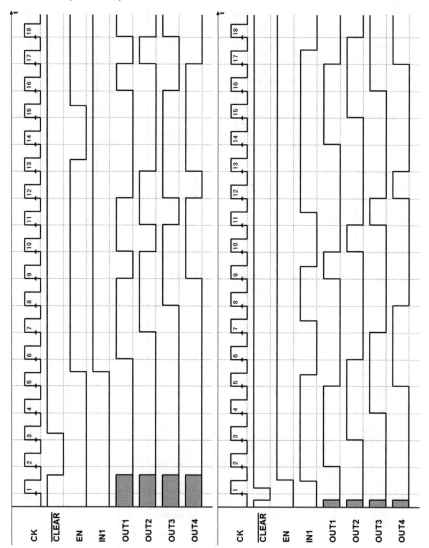

4. Exercise 4 (solutions)

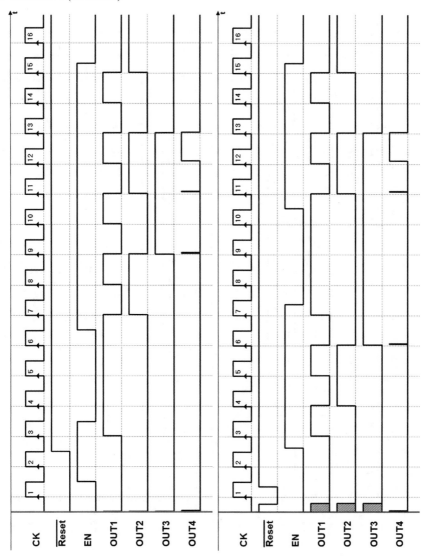

5. Exercise 5 (solutions)

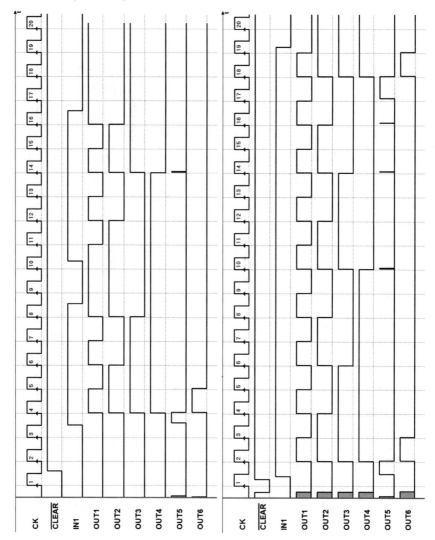

6. Exercise 6 (solutions)

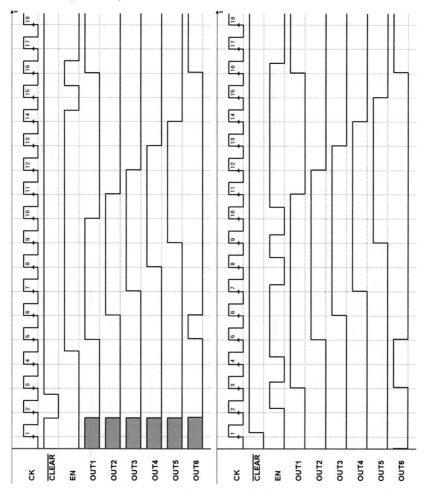

7. Exercise 7 (solutions)

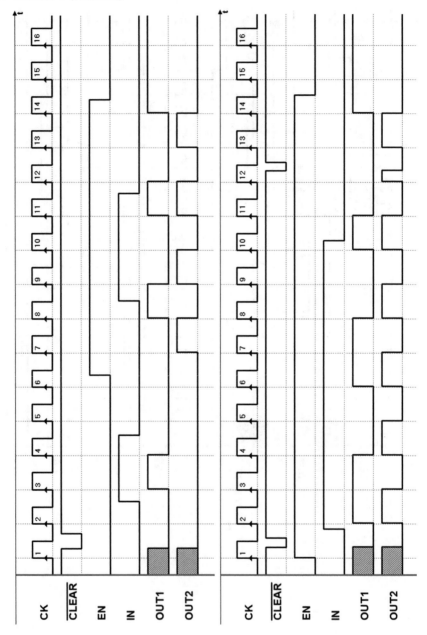

8. Exercise 8 (solutions)

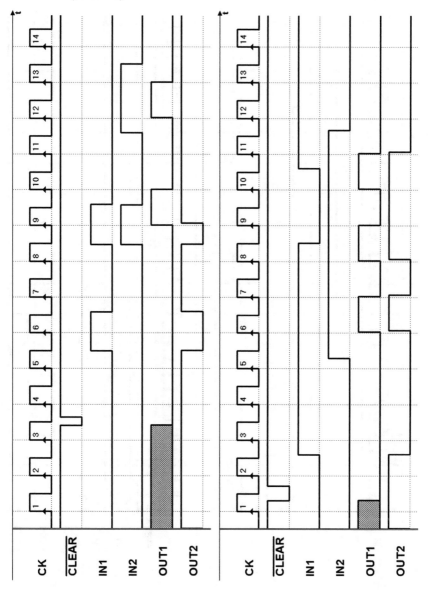

9. Exercise 9 (solutions)

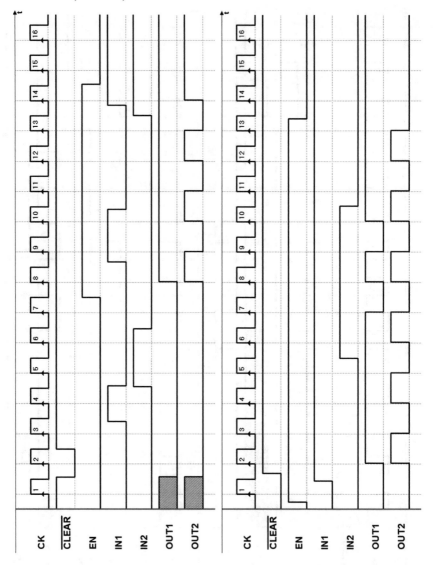

10. Exercise 10 (solutions)

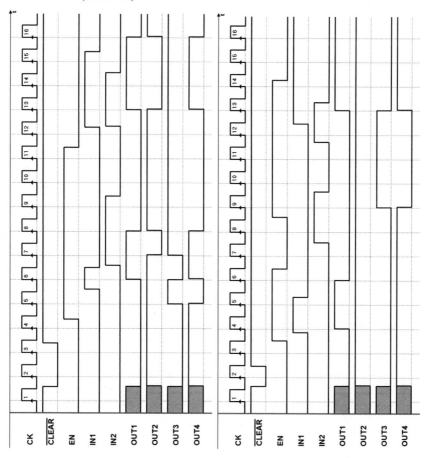

Chapter 7
Sequential Networks as Finite State Machines

Abstract The term Finite State Machine indicates a regular and standard structure that is able to describe and synthesize sequential networks in a general fashion. Algorithmic State Machine (ASM) is the tool adopted in the book and developed through the chapter. The attention is focused on synchronous stand-alone machines, their properties, and timing behavior.

In past chapters, we have seen different examples of sequential networks whose functions we analyzed through timing diagrams. All the synchronous networks we examined can be described by the general structure we see here:

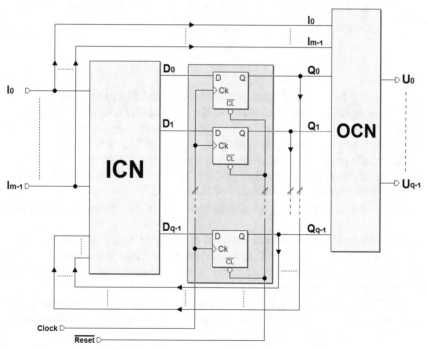

© Springer International Publishing AG, part of Springer Nature 2019
G. Donzellini et al., *Introduction to Digital Systems Design*,
https://doi.org/10.1007/978-3-319-92804-3_7

We called *state* the set of values memorized by the flip-flops, which we grouped into a *parallel register*, the *state register* of the network (highlighted in the center of the figure). Our purpose in this chapter is to learn to design synchronous sequential networks with a general *method*. To do this, we work with the standard model of a *Finite State Machine* (FSM), that we can use to create sequential networks that can perform any logical algorithm that requires a finite number of operations.

7.1   Finite State Machines: Standard Model

The standard model described below is valid for both synchronous and asynchronous sequential networks. A general sequential network can be depicted as having *three blocks*, as shown in the figure below.

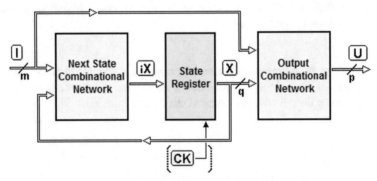

Let *I*, *X* and *U* be the set of *input*, *state* and *output* variables, respectively. A set of variables is called a *vector*. For example, above, $I = \{I0, I1, I2..Im - 1\}$, where *m* is the number of input variables. In the figure, the lines carrying the variables *I*, *X* and *U* are shown with a *slash* indicating the number of lines they group.

The middle block is the *state register*, that memorizes the *state vector X*, defined by the set of *q* variables of the state $\{X0, X1, X2.. Xq - 1\}$. Each of the 2^q variable combinations identifies a specific *state* of the network.

The clock *CK* is shown *in parentheses* because it is only present in synchronous networks.

The *Next State Combinational Network*, previously referred to as ICN, receives the vector of inputs *I* and state *X*, and based on this information, it generates vector *iX*, which is brought to the input of the register. As we have seen, vector *X* will take on the value of *iX* at the appropriate time.

The *Output Combinational Network*, the OCN from Chap. 6, has in its input a vector of state *X* and a vector of input *I* and generates the *output vector* of the network $U = \{U0, U1, U2... Up - 1\}$.

7.1.1 Synchronous and Asynchronous Machines

The network's timing evolution is conditional upon the specific structure of the state register. If the state register is made with *synchronous flip-flops* timed to the same clock, the sequential network is described by the model of the *synchronous FSM*. The clock scans *the evolution over time* of the sequence of states taken by the machine (see the following figure left).

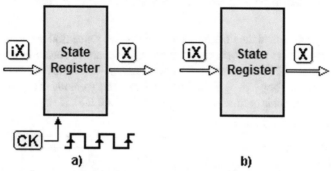

a)   b)

If the state register uses *asynchronous flip-flops*, it *does not* employ a clock input (upper right-hand figure) and is described by the *asynchronous FSM* model. Here, the network's timing evolution depends on inputs changes and the internal delays of the network.

7.1.2 Moore and Mealy Machines

There are two architectural variants for the modality of generating outputs. The standard model shown above where the outputs are a *function of the state and of the inputs* is normally called a *Mealy machine*. In the *Moore machine*, the outputs are *a function of the state only* since they do not directly depend on the inputs. The figure below summarizes the four possible variants:

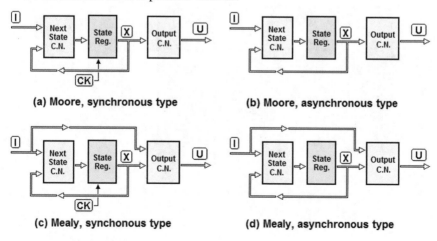

(a) Moore, synchronous type   (b) Moore, asynchronous type

(c) Mealy, synchonous type   (d) Mealy, asynchronous type

The models on the left (a) and (c) are *synchronous*. The synchronization of the clock makes the state change on the active edges of the clock. The models on the right (b) and (d) are *asynchronous*. In the *Moore* machines, (a) and (b), there is no connection between the inputs and the output combinational network. By contrast, in the *Mealy* machines, (c) and (d), there is.

7.1.3   An Example of Synchronous Finite State Machine

In Chap. 6, we analyzed the function of a synchronous sequential network that could count up (module 4) and synchronously bring the count to zero. Let's re-examine it now in terms of FSM. As we see in the schematic below, the network uses two D-PET flip-flops. They can be considered to be memory elements that make up the *state register*, and their outputs as variables of the state ($X0$, $X1$):

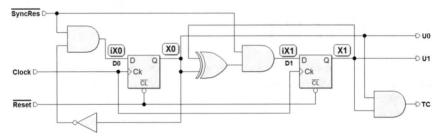

In the figure, we see a set of logical gates that calculate the values of the flip-flops' inputs D as functions of the external input $\overline{SyncRes}$ and the variables of state $X0$ and $X1$. This part of the circuit can be considered the *network of the next state*. Likewise, we see that outputs $U0$, $U1$, and TC are combinational functions of the variables of state $X0$ and $X1$. In light of these observations, let's re-draw the network and organize the schematic according to the standard model of *Moore's synchronous* FSM:

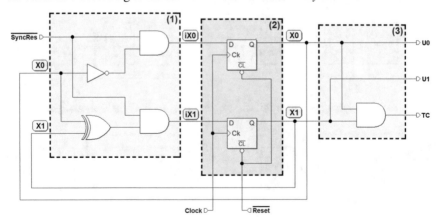

Block (1) generates the next state by offering it to the input of flip-flops ($iX0$ and $iX1$), based on input $\overline{SyncRes}$ and the current state of $X0$ and $X1$. Let's derive their Boolean expressions:

$$iX0 = (\overline{SyncRes} \cdot \overline{X0})$$
$$iX1 = (\overline{SyncRes} \cdot (X0 \oplus X1))$$

Block (2), the state register, memorizes the values of $iX0$ and $iX1$ at each active edge of the *Clock*, thus updating the state represented by $X0$ and $X1$. The $\overline{Reset}$ signal is used for the asynchronous initialization of the flip-flops.

Block (3) generates outputs $U0$, $U1$, and TC of the network, that depend on the variables of state $X0$ and $X1$ in the following way.

$$U0 = X0; \qquad U1 = X1; \qquad TC = (X0 \cdot X1).$$

7.1.4 General Equations of the Next State and the Outputs

In a *synchronous* FSM, the clock determines the instants $n - 1$, n, $n + 1$... when the FSM can change state. Cyclically, at every clock edge, vector iX is transferred onto the state register, thus updating the value of X. Time is reduced to a succession of numerable events.

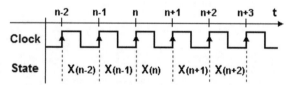

Let $X(n)$ be the *state* of the FSM in interval $[n, n + 1]$. In response to $n - 1$, n, $n + 1$, ..., the state assumes values $X(n - 1)$, $X(n)$, $X(n + 1)$... (in the figure, these events occur at the rising edges of the clock).

Let's now turn our attention to *Moore's synchronous* FSM shown here:

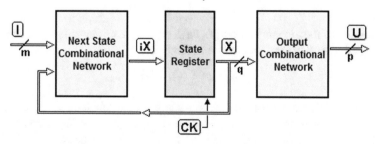

Let's express the state $X(n)$ in the interval $[n, n + 1]$ as function of the previous state $X(n - 1)$ and of inputs $I(n - 1)$ in interval $[n-1, n]$:

$$X(n) = f(X(n-1), I(n-1))$$

This expression is called the *"equation of the next state."* Keeping in mind that X is a vector made up of q variables, we will have q scalar equations, one for each variable of the state.

Briefly, the equation above can also be expressed thus:

$$X \leftarrow f(X, I)$$

The arrow "$\leftarrow$" reminds us that X is a *"delayed function"* of the inputs and of the previous state. The function $U(n)$, which represents the system's output in interval $[n, n+1]$, is:

$$U(n) = g(X(n))$$

The above expression is the *equation of the outputs* for *Moore's* FSM. It represents the outputs' dependency on the *state* it takes on in interval $[n, n+1]$. Given that state $X(n)$ remains constant for *the whole interval*, the outputs will as well.

Let's also keep in mind that if p is the number of variables that make up the vector of the outputs, we will have p scalar functions, one for each output.

For *Mealy's synchronous* FSM, however, the function $U(n)$ is different since there is a connection between the FSM's inputs and the combinational network of the outputs (see the figure below).

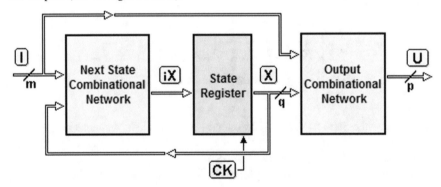

The *equation of the outputs* for *Mealy's* FSM describes the dependency between the outputs and state $X(n)$ and inputs $I(n)$ of the FSM.

$$U(n) = g(X(n), I(n))$$

Note that while state $X(n)$ remains constant for the whole interval $[n, n+1]$, the same cannot be said for inputs $I(n)$, which will condition the value of outputs $U(n)$ during the interval since the inputs generally can vary at any instant.

The outputs that partially depend on input values are called *"conditional outputs."*

Hereinafter, we will use *interval n* to refer to the timing interval $[n, n+1]$, and *state n* to refer to state $X(n)$.

7.2   ASM Diagrams

When designing a sequential network, it is wise to adopt a *behavioral* rather than a circuital approach. This means starting by defining the algorithm that the FSM must execute and working *top-down*. The algorithm will only be implemented in circuital form afterward, through a process called *synthesis*.

There are several methods for describing and designing a state machine's algorithm. The *ASM diagrams (Algorithmic State Machine diagrams)*, used through the book, represent a simple and intuitive method to design and synthesize a state machine.

Another graphical method, still in use in books and documents, is the *State Diagram*. Appendix B provides the simple rules for converting state machine representation from one method to the other one.

ASM diagrams (or *charts*) seem similar to the *flow diagrams* used in programming, but they are structured so that they describe *the evolution of the states* of a network. They were introduced in the 1970s by Christopher R. Clare in his book *"Designing Logic Systems Using State Machines"* (McGraw-Hill, 1973, out of print). ASM diagrams can describe both *synchronous* and *asynchronous* FSMs. Below, ASM diagrams are applied to *synchronous machines*.

7.2.1   Description of States

Here is the first example of an ASM diagram which has four *state blocks* (the rectangles), interconnected by lines that represent the states' *logical flow*.

Let's try to interpret the figure at the right. The diagram describes the *algorithm* i.e. the logical behavior of a device that can take *four different states*, represented by blocks (a), (b), (c) and (d).

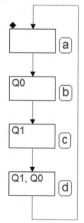

The letters on the right of the blocks in the ovals represent the *symbolic names* assigned to the states. These four states happen *one after the other*, in the order indicated by the *arrows*.

Outputs can be associated to the states. The diagram shows only the *active* outputs, state by state inside of the rectangles. For example, in state (c), output $Q1$ is active, while in state (d) outputs $Q1$ and $Q0$ are active. Convention dictates that *only the active outputs* are indicated inside state blocks, so *unreported* outputs are understood as *inactive*.

In our example, $Q1$ and $Q0$ are the device's only two outputs since the diagram does not mention others. x *No inputs appear* in the diagram. Indeed, this device has no inputs that have an effect on *states' flow* (the next examples, however, do have them).

In the ASM description of the network, the initialization and synchronization inputs (*Reset* and *Clock*) are never represented since they do not affect its behavioral description.

Furthermore, an ASM diagram does not explain the *timing modalities* with which the machine *goes from state to state*, which depend on the structure of the state register (whether the network is *synchronous* or *asynchronous*).

When the designer describes the FSM in algorithmic terms, (s)he has already decided if the network will be *synchronous* or *asynchronous*. In our example, the device was planned to be *synchronous*, so the *Clock* will govern *the progress from state to state*. Hereinafter, let's assume all the devices we will analyze and design to be *synchronous*.

The figure below left shows a <u>block</u> schematic that describes the device. It has *Clock* inputs, initialization inputs $\overline{Reset}$ as well as two outputs $Q1$ and $Q0$.

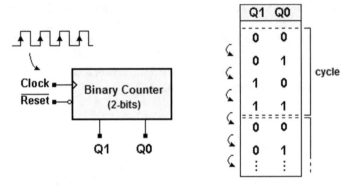

The figure on the right shows the sequence of outputs, from state to state as indicated in the ASM diagram. This sequence shows clearly that our device behaves like a 2-bit *synchronous binary counter*. It counts from 0 to 3 cyclically (module 4).

Activating Reset

When the initialization input $\overline{Reset}$ is activated, the flip-flops of the state register are forced to a known value. This means that the network is brought to a pre-defined state that, at the design level, we call the *"reset state"*.

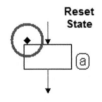

As per convention, the *reset state* is shown in the diagram with a *small rhombus* (an ace of spades!) at the upper left-hand corner of the rectangle.

The figure below highlights a *sequence* where we can examine the behavior of the network upon initialization. As we can see in the timing path, the choice was made to begin the simulation with $\overline{Reset}$ inactive.

Given that we assume nothing about the state of the network *before* the activation of $\overline{Reset}$, we represent the state as initially *unknown* (shown by the question marks), and the outputs as *indefinite* (see figure in the next page).

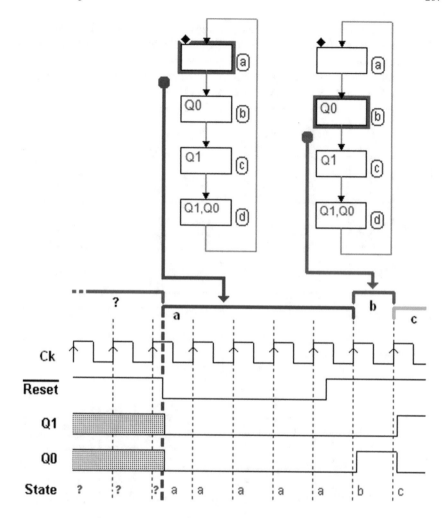

When the input $\overline{Reset}$ is *activated*, the network is brought to state (a), the *reset state* specified by the ASM diagram. Remember that $\overline{Reset}$, being an external signal applied to the FSM, can change at any time and acts *asynchronously* (i.e., independently of the *clock*). The network remains forcedly in state (a) until $\overline{Reset}$ is *deactivated*.

After $\overline{Reset}$ is *deactivated*, the network is brought to state (b) at the *first rising edge* of the *clock* and continues on the sequence laid down by the ASM diagram, as seen in the figure above.

The Relation Between the ASM Diagram and Timing Diagram

The following figure highlights the relation between the *state flow*, *output timing diagram* and the *clock*. The machine is synchronous; it remains in each state for the duration of one *clock* cycle.

Assuming that the FSM is in state (a) at a given time, the ASM diagram tells us that the next state will be (b). The timing diagram shows the passage from state (a) to state (b), which occurs at the *rising edge* of the *clock*.

With the passage to state (b), output $Q0$ activates. At the next *rising edge*, the FSM goes to state (c), where it activates output $Q1$, and so on.

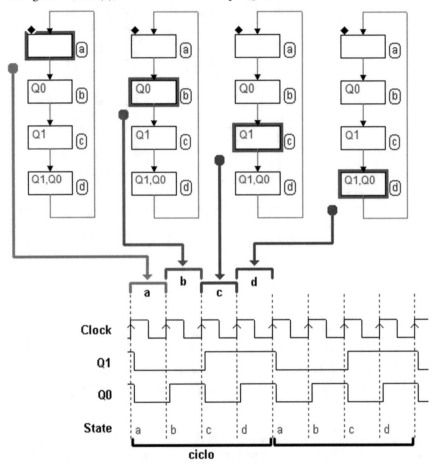

The short delay highlighted in the timing diagram represents the *physical delays* of the logical network. The change of outputs occurs *"after"* the rising edge of the *clock*. The timing diagram also shows the *cyclical nature* of the network well; there, we see the repetition over time of the generated sequence.

7.2.2 *Inputs*

The figure below displays another, broader example of an ASM diagram. This one represents a *variable module counter*, which can count in *pure binary* (module 16) or in *BCD 8421 (Binary Coded Decimal)*.

There are 16 states in the diagram, and one *decision block*. The decision block, which we are meeting for the first time, takes account of an input called *MOD*.

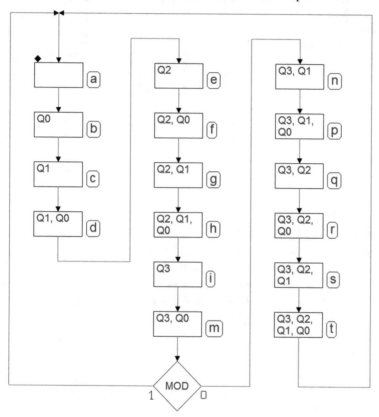

If its value is 0, the machine goes through all 16 states in the order shown in the diagram. If its value is 1, after state (m), the machine returns to (a), thus going through a sequence of 10 states.

If the value is 0, the sequence of states produces $Q3$, $Q2$, $Q1$, and $Q0$ at the outputs, a *binary sequence from 0 to 15*. If the value is 1, the outputs take on the values of the BCD codes corresponding to the decimal numbers 0 a 9.

The figure below left shows the block schematic of the variable module counter. The block schematic shows inputs $\overline{Reset}$ and *Clock*, as well as the input *MOD* and the four outputs $Q3$, $Q2$, $Q1$, and $Q0$. The table shows the *two sequences generated* by the counter as the input setting varies.

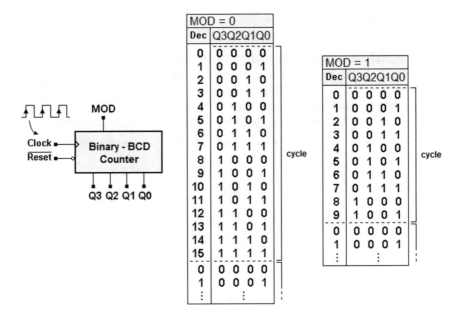

Using Decision Blocks

In ASM diagrams, the network's inputs have an influence on the *sequence of the states* through the *decision blocks*. Argument of a decision block is normally an *input variable* or a *Boolean function of multiple input variables*.

A decision block allows to choose between *two logical paths* according to the Boolean value of its argument. The figure below shows two examples taken from standard ASM diagrams.

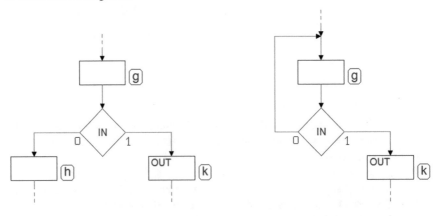

On the left, the decision block allows input *IN* to define what state will follow state (g). The answer is (h) if *IN* = 0, or (k) if *IN* = 1 (this case is analogous to the one examined previously).

On the right, however, state (g) has two possible next states: *itself* or state (k). If $IN = 0$, the machine remains in state (g); the next edge of the clock will produce no change in state. In $IN = 1$, at the active edge, the machine will go to state (k).

The machine stays in state (g) *waiting* for input *IN* to go to 1 before continuing. In (g), the network is said to be in a *waiting state*.

Timing Aspects of State and Decision Blocks

The *state blocks* determines the timing of the FSM. The passing of time is related to the states; to remain in a certain state corresponds to a certain *length of time* and a sequence of states *evolves over time*. In synchronous machines, one state follows another *to the rhythm of the clock*, so a state will last *one clock cycle* or a *multiple* depending on the algorithm.

The *decision blocks* represent just a *combinational* function, which is independent from time. Looking back at the example of the variable module counter, when the counter is in state (m), input *MOD* decides which state to switch to at the next active edge of the clock. The decision block performs a logical operation: *determining the path* toward the next state, either (a) or (n). The effects of the decision appear *upon transition* in the next state.

Example: The Up/Down Binary Counter

We have seen the FSM that describes a simple *2-bit binary counter*. The counter was designed to generate an up binary sequence. Now, let's describe the behavior of a *up/down* counter. We start with the circuit block seen previously and add a command input that we call *DIR*. See the figure below:

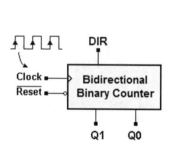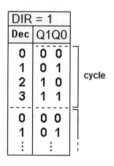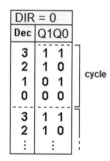

The command input will set the count *up* if it is at 1, or *down* if it is at 0. The figure shows the sequences called for as input *DIR* varies. As a further specification, we want the count to be able to *change direction* at any state the device is in.

Let's proceed stepwise and change the previous counter. Its ASM diagram is shown in the figure below left. The four-state sequence corresponds to the *up* count. To also allow for a *down* count, we insert a decision block after every state block to condition the sequence to the value of input *DIR*.

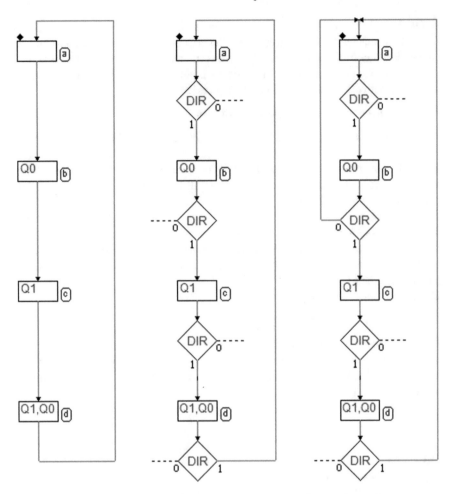

In the central figure, we can see the four conditional blocks that were added, but the diagram is still incomplete. The logical paths related to the value 0 of *DIR* is 1 are missing and the sequence of states *counts up*.

Let's add the missing logical paths making sure that if $DIR = 0$, the count *decreases* one unit at every clock cycle. In other words, we want the sequence of states to go backward.

For example, the right-hand side of the figure shows that if $DIR = 0$, the logical path brings the machine from state (b) to state (a).

The following figure adds the *down count paths* one by one: at left, from (c) to (b), and center, from (d) to (c).

To complete the sequence, we must now identify the path for $DIR = 0$, in state (a): if, in the *up count* we go from (d) to (a), in the *down count* we must, obviously, go from (a) to (d).

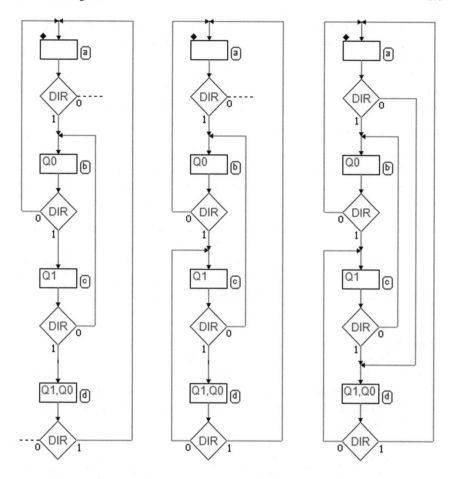

The complete ASM diagram is shown on the right-hand side of the figure. Note that the machine's behavior in state (a) is analogous from a topological perspective to that of all the other states. This is in spite of the fact that it could seem different from a graphic perspective.

The figure here below shows the timing simulation of the counter.

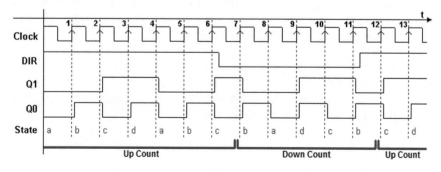

We see the *Clock* and *DIR* inputs and outputs $Q1$ and $Q0$, as well as an indication of which *State* the FSM is in. The active edges of the clock are numbered for ease of reading.

On the first active edges of the clock (numbered 1–6), input *DIR* is high so the count sequence is *forward*. From edge (7) to edge (11), input *DIR* is low so the counter goes *backward*, and then goes forward again on edge (12).

Sample Project: Edge Detector

An *Edge Detector*, as its name states, can signal that its input has changed value from 0 to 1 or vice versa. Its behavior is shown in the figure below with a timing diagram. Every time input *IN* has a (rising or falling) *edge*, the detector reports the event by generating a pulse on output OUT:

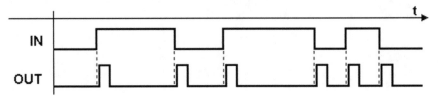

The path represented here is only approximate. The precise timing evolution of the output will depend on the way we choose to build it.

If we take *Moore's* machine as a model, the output OUT will be synchronous with the clock and so the pulse generated will have a minimum duration equal to a clock cycle. The figure below shows the timing diagram with adjustments made in view of these choices.

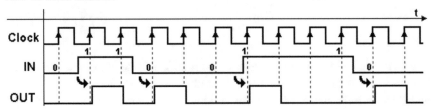

At every transition of input *IN* ($0 \rightarrow 1$ or $1 \rightarrow 0$), a pulse is generated with a duration of *one clock cycle*. Note that because of asynchronicity between the input and the clock, we must accept the inevitable and random delay between the input transition and the output pulse generation.

The delay will never be longer than one clock cycle. Based on these specifications, let's now derive the ASM diagram of the edge detector.

To begin, we must assume that the FSM is *"in a certain state"* and that the input has *"a specific value"* consistent with the design.

To this end, let's assume that, at a certain moment, input *IN* is 0 and the machine
is in state (a) where the output is inactive (see the figure below left).

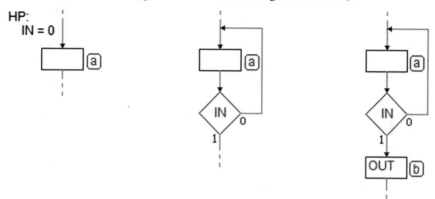

In this case, the machine must wait for the only possible event, a transition of *IN*
from low to high. Therefore, we insert after the state a decision block that controls
the input (at the center of the figure), so that the machine stays in state (a) in the
absence of input changes.

When the input changes to 1, the machine must report the transition and activate
the *OUT* output. To achieve this, we must add a state (b) that will generate *OUT* for
one clock cycle (on the right-hand side of the figure).

It is wise at this point to take a few moments to
think about the evolution over time of the signals
involved here in relation to the ASM diagram we
are building. The figure on the right shows the timing sequence of the signals in the transition from
state (a) to (b). The image highlights that the FSM
stays in state (a) while *IN* = 0 in the initial clock
cycles.

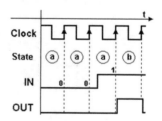

Then we see the transition to (b) at the rising edge of the clock, after *IN* has taken
the value of 1. The output *OUT* is activated when the machine is in state (b).

As we get back to drawing the ASM diagram, we must keep in mind that the
output *OUT* must be no longer than one clock cycle so state (b) must necessarily
be followed by a state with no output *OUT*. Also, since input *IN* is now at 1, the
machine must wait for it to go to 0.

Let's add a state (c) where the machine will wait for the input to go back low (see
the figure in the next page).

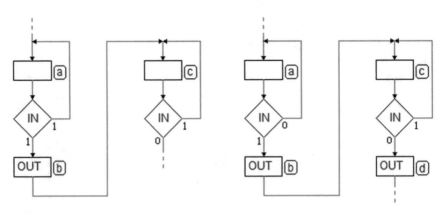

When the input goes to 0, the machine must produce the output *OUT* again. As before, to activate the output, we add a state (d) to follow state (c) when *IN* is 0 (the right-hand side of the figure).

Note the very close analogy between the pair of states (a) and (b), and the pair (c) and (d). The behavior of the machine is similar between them but the difference in the input value dictates *two different sequences* of states.

After *OUT* is generated in (d), the machine must wait for *IN* to go back high again. This is the very behavior of state (a), which was introduced with the assumption that the input would be 0.

So, let's close (d) over (a), completing the diagram as shown in the figure below. There are no more *paths to close*, and the initial assumption about state (a) is satisfied.

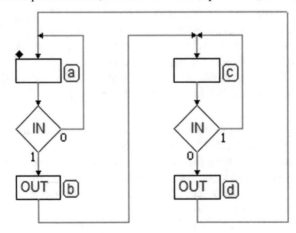

We must still decide which *reset state* the machine will go to on initialization. Intuitively, states (a) and (c) should be the most suitable for this since the FSM does not activate the output there, but simply waits for the input.

States where the machine waits for events without generating outputs are called *"idle states."* In the absence of further specifications, we will choose state (a), shown in the figure with a small rhombus on the upper left-hand side.

Here below is the timing diagram of the edge detector with the input sequence from before, plus the sequence of states taken by the FSM.

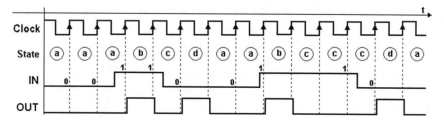

Looking closely at the timing path and the ASM diagram, we see that the evolution of the input signal must be *subject to some restrictions* so that the FSM works correctly. In short, the input changes must be far enough apart.

In the first part of the path above, where the input signal is high for two consecutive cycles, the machine still has time to detect the falling edge, and the evolution of the output *OUT* is what we expect.

This is a limit situation for this FSM. If the input signal changes more frequently, the machine can no longer follow and will produce a incorrect output sequence (it would not satisfy the given specifications), as in the figure below:

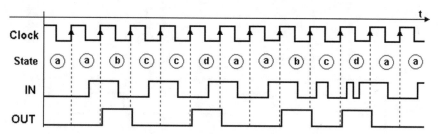

Here we see that the output *OUT* is no longer generated at each edge of the input signal. At the end of the path, where we see an input signal that changes multiple times within the *same clock cycle*, the transitions are *totally invisible*.

This type of problem, normal in digital design, is linked to the type of operation to carry out, the algorithm used but, mostly, the clock frequency. Here, the edge detector would work if the clock frequency were raised appropriately.

7.2.3 Conditional Outputs

We have seen that the state block and the decision block suffice to define *Moore's* machines. To describe *Mealy* FSM ASM diagrams also use the *conditional output*. As we have seen, in *Mealy's* machines the outputs can depend not only on the state, as with *Moore's* machines, but also *directly on the inputs* in terms of combinational logic.

Conditional output blocks make it possible to describe this type of dependence on an ASM diagram. The machine described in this ASM diagram on the right has only two states: (a) and (b). The conditional output block is represented by a rectangle with rounded corners, with inside the name of the output (*UC* in this case). The decision block that evaluates input *C* here is not involved in changing states. If we observe the paths the block can take, we see that whatever the input value, the next state of (a) will always be (b) just as the next state of (b) is (a).

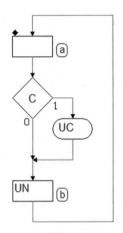

The decision block, however, serves the purpose of the conditional output block. Output *UC* is generated when the machine is in state (a) if and only if $C = 1$. The timing diagram shown in the figure below demonstrates this behavior.

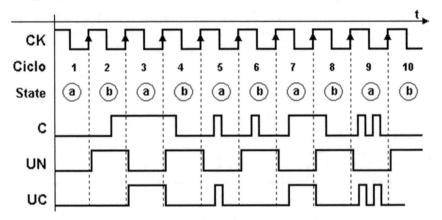

In line with the ASM diagram, the path here shows that output *UC* is not generated in clock cycles 2, 4, 6, 8, and 10 since the machine is in state (b) in those cycles. In the other cycles, the machine is in state (a) so *UC* activation depends on the input value:

1. In all of cycle 1, input *C* is low, so *UC* is inactive.
2. By contrast, in 3, *C* is high for the whole cycle, so *UC* is active.
3. In cycle 5, input *C* is active only for a brief interval and output *UC* is consequently activated (as we can see, the evolution over time of *UC* copies that of input *C*).
4. In cycle 7, the behavior is similar to cycle 5 with one exception. *UC* is active when input *C* goes high and then goes down to zero along with the transition to state (b) since there is no provision for output *UC* in (b).
5. Finally, in cycle 9, the behavior is the same as in cycle 5.

For comparison, the path also features the (unconditional) output of state UN, which is activated every time the machine is in state (b). Since over time, the two states (a) and (b) alternate at every clock cycle, UN evolves periodically.

Example: A New Edge Detector

We have already designed an *edge detector* as a *Moore's* machine with four states. Let's re-think this device in light of the possibilities offered by *Mealy's* machines, i.e., using *conditional outputs*. We will see the advantages and disadvantages of this approach.

In the previous device, the output activation in synchronicity with the clock produced a delay between the actual arrival of the signal edge and the generation of the output. This delay is equal or less than the duration of one clock cycle.

With *Mealy's* machine, however, we can control the outputs directly through the inputs. We want to take advantage of this to obtain a device that can generate the output pulse as soon as the input changes without waiting for the active edge of the clock.

In the figure below left, we start setting up the solution, assuming a state (a) where we start with input IN at 0, and wait until the transition to 1. At the center of the figure, we add another state (b), which is symmetrical to (a), to manage the opposite case, leaving the paths between them sketched for the moment.

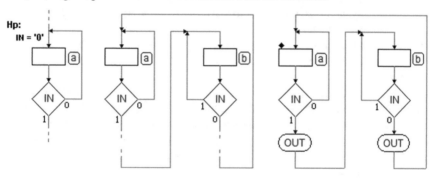

Set up this way, the machine still has no outputs but it can already do the important job of *following* the input signal. The machine remains in one of two different states, each corresponding to a specific input value. It goes from one to the other synchronously as it changes.

Let's add the conditional output OUT (on the right side of the previous figure) to state (a). In state (a) when the machine waits for IN to go high, no output is generated. As soon as IN goes to 1, the output OUT is activated. Let's complete the ASM diagram by introducing the conditional output OUT into (b). It will be activated when $IN = 0$.

It is important to note that decision blocks express both state transition conditions and the logical function that links OUT to input IN. In fact, an ASM diagram that contains conditional output blocks describes *two different* aspects of the MSF's behavior with a single drawing.

The first is the *state transition*, which is not influenced by conditional output blocks. The second is the aspect of the *conditional outputs*, which are expressed in the form of combinational functions of the inputs and of the state where they are instantiated.

For comparison, the timing diagram below gives an example of the behavior of the conditional output OUT in function of the same input sequence used for the test of the edge detector with state outputs.

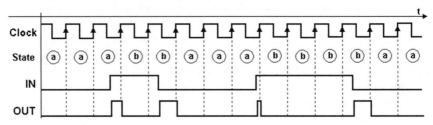

By the time of the first transition, the machine is in state (a). The input's change in value makes the output OUT activate immediately. It remains active until going to state (b). In (b), the output OUT is conditioned by $IN = 0$, so when it is in that state it activates only when the input goes to zero. The other input transitions in the figure show analogous behavior.

Nonetheless, in the timing sequence we have used here, input IN is *asynchronous* with the machine's clock. The *duration* of output OUT is *unverifiable*, as shown in the diagram.

Consider the third input transition, that produces a very brief pulse. If the transition occurs too late respect to the edge of the clock, the pulse might not be generated due to the physical delays of the components, or be too short to be readable.

It is clear that this machine cannot be used with *asynchronous* inputs. A synchronous input, however, makes it possible to avoid limit cases and obtain an output signal of the proper duration. In the timing diagram below, the IN signal is synchronous with the machine's clock and so the duration of the output is constant.

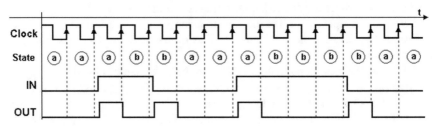

Finally, note that the machine uses only two states. Generally, a *Mealy* machine needs *fewer states* than *Moore's* machine.

Example: Version 3 of the Edge Detector

Now let's discuss the advantages and disadvantages mentioned earlier to obtain another version of the detector that maintains its conditioned outputs' *response readiness* but generates the output for *at least one clock cycle*. Let's begin with the ASM diagram of the four-state version, dealt with previously, and add two conditional outputs as in the figure below:

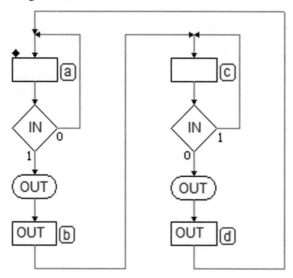

In state (a): waiting for *IN* to go high, the conditional output *OUT* is inactive. As soon as *IN* goes to 1, *OUT* is immediately brought to 1 while we are still in (a). At the next edge of the clock, when the FSM goes to (b), it will continue to generate *OUT* since it is defined as a state output.

In the timing diagram below, the arrows indicate the instants the output *OUT* is activated. They come *before* they would have in *Moore*'s machine, where we would have waited for the edge of the clock to activate the output.

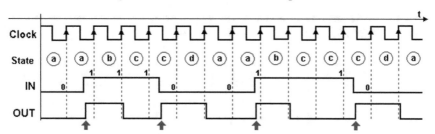

This new structure allows us to generate a readable output with a minimum guaranteed duration, even though input *IN* behaves *asynchronously*.

7.3   Examples of ASM Diagram Construction

Here, we show a collection of synchronous FSM design examples to clarify the basic concepts regarding the construction of ASM diagrams subject to assigned specifications. The examples in this chapter will be on an FSM that directly manages the inputs and outputs of the system we want to design.

7.3.1   Introductory Examples

For now, let's deal with simple synchronous FSMs whose specifications are defined verbally and through *timing diagrams*. This will clarify the relationship between the input/output signals' *evolution over time* and *the evolution of the states*.

Example 1

Using ASM diagrams, design an FSM that generates a periodic OUT signal which is synchronous with the clock CK, whose value is high for one cycle and low for the next, as in the figure.

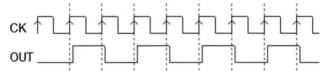

Solution

In a standard synchronous FSM *with no inputs*, the output changes only if the state changes (*Moore's* model).

In our case, to make the values 0 and 1 alternate at each clock cycle, we must change state cyclically. See the figure at the right:

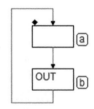

Example 2

Draw the ASM diagram of a synchronous counter that generates continously the sequence (ABC) 000, 001, 010, 011, 100, 000, etc.

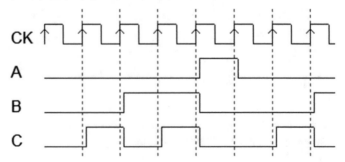

Solution

As in the previous example, the outputs will only be a function of the state (there are no inputs) and at every cycle the combinations requested are different.

Thus, we should introduce a sequence of states, each with a different output combination into the ASM diagram. The sequence will repeat cyclically.

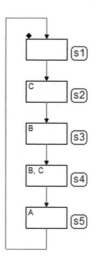

Example 3

Using the ASM diagrams, design an FSM with input *IN* (synchronous), that generates a synchronous *OUT* signal for the whole time input *IN* is at 1, as described in the figure below:

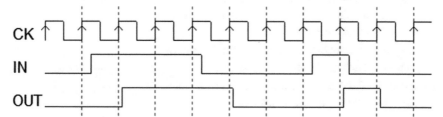

Solution

So that the output *OUT* is synchronous with the clock, it should be a *state output*, so we will use *Moore's* model. This means the input will be evaluated by the FSM at the rising edge of the clock. The solution is shown in the figure on the right.

In state (a) we wait for the input to go to 1 without activating the output *OUT*.

When the input changes, the state changes, generates the output *OUT* and waits for the input to return to 0.

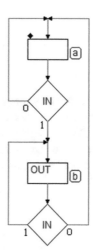

Example 4

Using the AMS diagrams, design an FSM that generates the following cyclical signal sequence: *A*, *B*, *AB* at the low-high transition of input *IN* (where *A* and *B* are the two outputs of the machine):

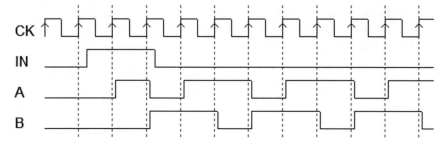

Solution

The signal sequence generated by a synchronous FSM will obviously be synchronous with the clock, as we see in the timing diagram. So, let's use *Moore's* model.

The input is evaluated in the waiting state (a). As soon as the input goes to 1, we enter into the cyclical sequence of states (b), (c) and (d).

In each state, we activate a different combination of outputs as defined by the specifications.

Note that the FSM will return to state (a) only upon the activation of the asynchronous reset.

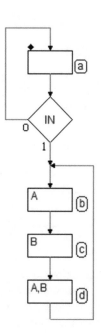

Example 5

Using the ASM diagrams, design an FSM with one input *STR* (*Start*) that is synchronous with the clock and three outputs: *ONE*, *TWO*, and *TRE* (abbreviation for *Three*). When a pulse with the duration of one clock cycle hits *STR*, we want the machine to activate the *TRE*, *TWO*, and *ONE* outputs in sequence. The count cannot be interrupted and ends with all the outputs at zero.

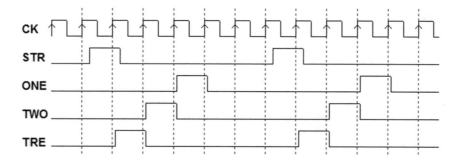

Solution

The solution is similar to that of Example 4, except that the initial state (a) is included in the cyclical sequence of the states.

See the figure at the right. The input, which is impulsive, is evaluated in state (a). Note that if input *STR* evolved not impulsively but *continually*, the machine would still work, but it would continue to generate a cyclical sequence for all the time *STR* was high and then stop in state (a) after the command was removed.

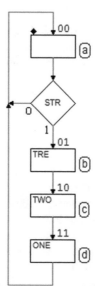

Example 6

Design an FSM with the same functionality as that of example 5 but with a different specification for activating a sequence by input *STR*, represented in the figure below. The sequence should start when input *STR* goes from low to high. The count cannot be interrupted and should end with all the outputs at zero.

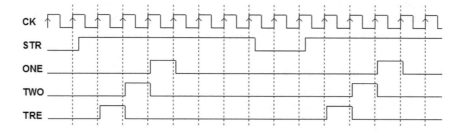

Solution

The solution for the previous example might
at first seem to work here but that is actually
not the case. Here, the event that can activate
the sequence is the input's *transition from low
to high.*

To be sure to detect this transition, we must
first verify that *STR* is low (or has returned to
that level).

So, to the diagram of the previous example,
we add an extra state (rs), *before* state (a).

In state (rs) we wait for *STR* to get to 0 (or
verify that it is already there) before moving
on to the check for 1 in state (a).

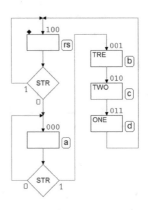

Example 7

Using the ASM diagrams, design a synchronous FSM with inputs *A* and *B* that are
synchronous with the clock. The FSM must generate an output *ERR* every time both
inputs are high. Output *ERR* must go to zero when the inputs have different values.

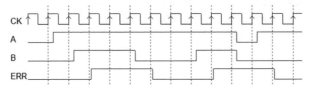

Solution

We've chosen to design a *Moore* model synchronous FSM. In a synchronous machine,
two inputs are determined to be *contempo-
raneous* when they are checked *in the same
clock cycle.*

In the waiting state (a) we *contemporaneously* check inputs *A* and *B*, and move to
state (b) if they are both 1. In state (b) we
generate output *ERR* and re-check inputs *A*
and *B* together.

By the way the decision block paths are
connected, the FSM only stays in (b) if *A*
and *B* are equal; otherwise it returns to state
(a).

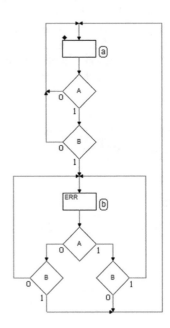

Example 8

Using the ASM diagrams, design an FSM that checks a synchronous input *IN* and generates an *ERR* signal lasting one clock period if *IN* stays high for more than one clock cycle (see the timing diagram below).

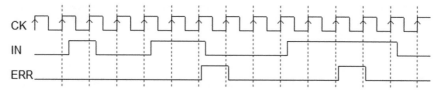

Solution

We've chosen to design a *Moore* model synchronous FSM (see the figure to the right). In waiting state (a) we check input *IN* and move to state (b) when it is at 1.

In state (b) we perform the same check. If the input is at 1 in state (b) as well, we generate output *ERR* and move to state (c) to report that the input is high for more than one clock cycle. Otherwise we go back and wait for the next 1.

Output *ERR* must be active for no more than one clock cycle so we generate it only in state (c).

In (c) we check input *IN*. If it has gone to 0 in the mean time, we return to state (a).

If, however, *IN* were still at 1, we would have to wait for it to finally go back to 0 before returning to state (a).

If we were to go directly, we would actually be checking the same signal again.

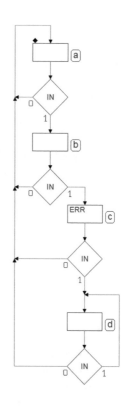

Example 9

Design an FSM that generates an output *OUT* that is high when input *IN* is high but stays high for one clock cycle (see figure):

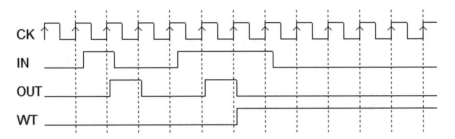

If input *IN* lasts longer, the FSM stably activates the output *WT* (*Wait*), entering into an infinite cycle that it can get out of only when reset is applied.

Solution

We have chosen to design a *Moore* model synchronous FSM. As we see in the figure at the right, we check input *IN* in waiting state (a) and we move to state (b) if it is high.

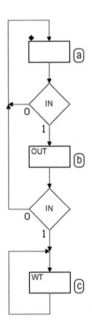

In state (b) we generate output *OUT* and if the input has changed again, the FSM goes back to state (a) to wait for the next transition.

If input *IN* stayed low, the FSM would move definitely to state (c) generating output *WT* in an infinite cycle, from which it could only exit upon activation of the Reset command.

Example 10

Design an FSM with two signals in the input, *ENB* (*Enable*) and *SEL* (*Select*), which are synchronous with clock *CK*, and two outputs, *GO* and *HLT* (*Halt*). The FSM waits for *ENB* to take the value 1. If *ENB* is at 1, the machine chooses whether to activate *GO* or *HLT* depending on *SEL* being at 1 or 0, respectively, for all the time *ENB* stays at 1. When *ENB* goes back to 0, the machine goes back to the waiting state and does not generate active outputs.

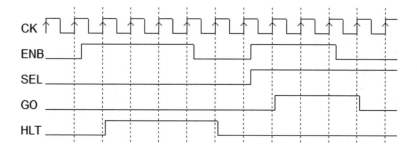

Solution

We have created a *Moore* model synchronous FSM. In the waiting state (a) we check input *ENB* and, if it is at 1, we move to state (b) if *SEL* is at 1, otherwise we move to (c).

In state (b) we generate output *GO*, waiting for *ENB* to return to 0. Likewise in state (c), we activate *HLT* and wait for *ENB* to change value.

Finally, when *ENB* goes back to 0 the machine returns to state (a).

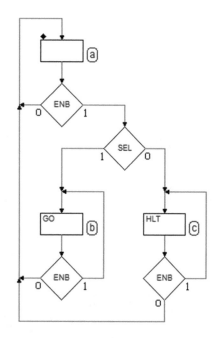

Example 11

Here we have the same specifications as in Example 10 with one variant: if *ENB* is at 1, the value of input *SEL* will let us choose which signal is activated, *GO* or *HLT*, while they are generated.

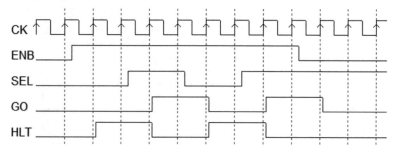

Solution

We have created an initial version using *Moore's* model (the figure on the right). In practice, the machine is identical to the one in Exemple 10 but in states (b) and (c) input *SEL* is checked.

The cycle will occur in state (b) or (c) depending on the value of *SEL*.

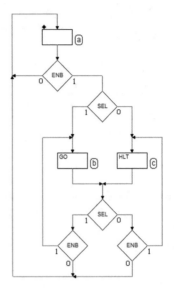

The figure on the left shows another solution using *Mealy's* model. The machine has only two states and while it is in state (b), it activates *GO* or *HLT* depending on the value of *SEL*.

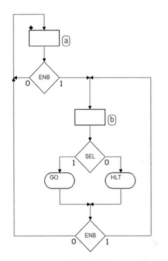

This will work a bit differently from a timing point of view (see the figure below). To obtain the same sequence as in the diagram above, the *SEL* check must be postponed by one clock cycle.

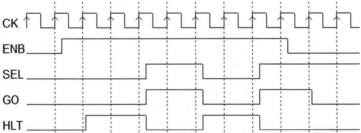

Example 12

Using the ASM diagrams, design an FSM that has an input *IN* that is *synchronous* with clock *CK* and an output *OUT* that activates at the *rising edge* of *IN* and stays active until the *next rising edge* of *CK*.

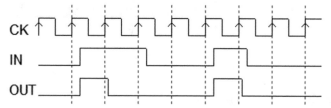

Solution

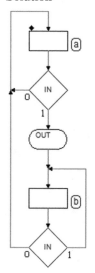

Since the output must activate at the edge of *IN*, we must use the *Mealy* model (see the figure at the left).

In state (a) we wait for *IN* to go from 0 to 1; the conditional output allows us to generate *OUT* as soon as *IN* is high, while we are still in state (a).

At the next edge of *CK*, the machine goes to state (b) where no output is instantiated so *OUT* is deactivated as soon as the machine leaves state (a).

From state (b) we go back to (a) at the edge of *CK* that follows *IN*, returning to 0.

7.3.2   Pulse Generators with Adjustable Duty Cycle

We want to design a synchronous FSM that generates a continuous succession of pulses with an *adjustable duty cycle* on output *PWM* (*"Pulse Width Modulation"*). See the figure at the right.

The output period, *five times* that of the clock *CK*, is fixed. The duty cycle is regulated through inputs *P*1 and *P*0.

The figure below describes the expected evolution of the output depending on inputs *P*1 and *P*0.

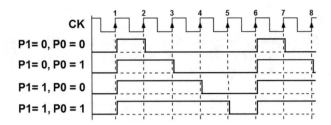

Solution

Let's set up the diagram by introducing state (a) where we activate output *PWM* (see the figure aside).

All the sequences share this state, which corresponds to the clock cycle between edge 1 and 2 in the timing diagram.

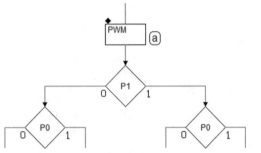

Four logical paths corresponding to the four combinations of $P1$ and $P0$ branch off from state (a). Let's complete the part of the diagram that relates to the first sequence in the timing diagram ($P1 = 0$ e $P0 = 0$):

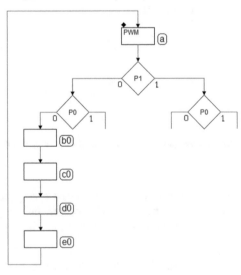

We have added four states with no active outputs that will take four clock cycles before returning to state (a) so that the period is five clock cycles in total, as specified. If inputs $P1$ and $P0$ are both equal to 0, the machine generates a PWM that is at 1 for one clock cycle and then at 0 for four other cycles in states (b0), (c0), (d0), and (e0).

Let's complete the second sequence. It is only different from the first because of state (b1) where we activate PWM.

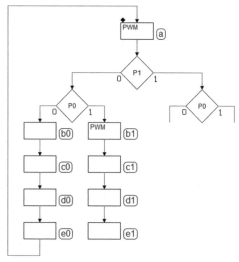

Let's close this sequence and complete the other two similarly, taking care to close all the paths on the state (a) from which all the sequences start.

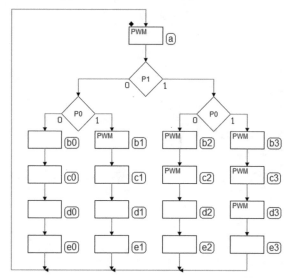

We realize here that there are superfluous state sequences because in some cases they perform the same functions. For example, the state sequence (c1)(d1)(e1) produces the same result as the sequence (c0)(d0)(e0) and both end in state (a) so it would be possible to define the states of the second sequence as in the ASM below:

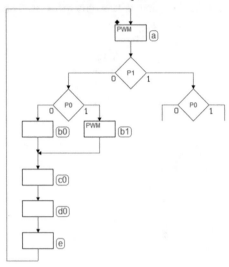

We reused states (c0)(d0)(e0) from sequence 1. Let's continue with the same approach for the third and fourth sequence. We will get the final result of this second version, which is simpler than the first one.

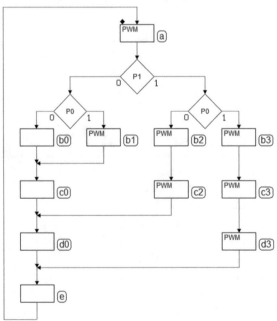

We could stop here. Note that the value of inputs $P1$ and $P0$ is checked onl6pty in state (a). Any change in the inputs during pulse generation will only produce an effect at the generation of the next pulse. This is an appropriate design choice.

Let's assume then that the specifications define that $P1$ and $P0$ can change only in state (a). This means that we can test their value in the following states since they will not change then. From this assumption and an analysis of the four possible sequences to generate, we will be able to further reduce the number of states. In cycle 2 of the sequence, after state (a) (between edges 2 and 3), output PWM does not have to be generated if both inputs $P1$ and $P0$ are 0 (see the figure below left).

Afterward, in the third clock cycle of the sequence (between edges 3 and 4), output PWM must only be generated if $P1$ is at 1, so we continue designing as we see in the figure below, center.

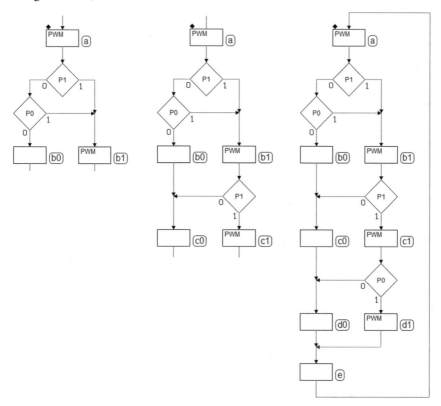

Taking analogous considerations, we complete the diagram (at the right in the figure) testing only $P0$ in the fourth cycle of the sequence (between edges 4 and 5) since the path dictates that $P1$ is at 1.

The final step is to insert the last state (e) corresponding to the last cycle of the sequence (between edges 5 and 6).

7.3.3 Sequence Detector

A synchronous network has an input *LN* and an output *OK*. Input *LN* reads a sequence of zeroes and ones, as shown in the figure below.

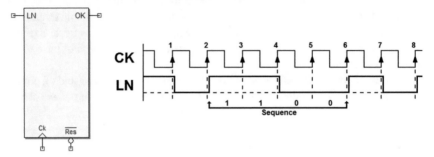

Output *OK* is activated for the duration of *one clock cycle* every time the sequence "1100" is recognized.

Solution

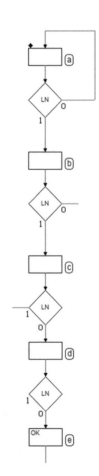

We define a state (a) where the machine waits to read a 1 on input *LN* (see the figure aside). We move to state (b) as soon as *LN* = 1 is read.

Since the value of input *LN* can change at every clock cycle, we must read its value in state (b) and the following ones.

We proceed drafting the ASM diagram, taking into account, for the moment, only the expected sequence "1100", leaving the other paths for the next step.

Thus we get to state (e) after reading the 4-value sequence 1, 1, 0 and 0, one after the other at a distance of one clock cycle (because the FSM is synchronous).

If we look, for example, at the timing diagram shown before, the four values were read in order on edges 3, 4, 5 and 6.

According to the specifications, output *OK* is generated in state (e) to report that the sequence we were looking for has been found.

Now, going in order, let's complete the remaining paths.

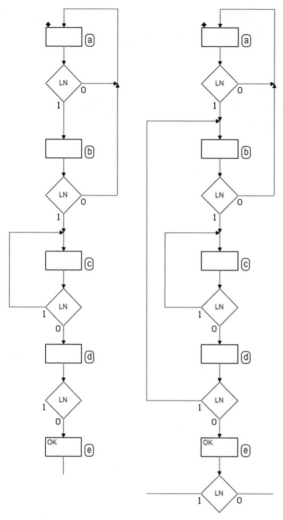

See the figure at the left. If a 0 is read in state (b) after a 1 is read in state (a), we must return to state (a) because we are looking for two consecutive 1s.

We find them there if a 1 is read in state (b). In this case, we move to state (c).

In state (c) we either read a 1 or a 0, which corresponds to the next bit of the sequence we're looking for.

Reading a 1 in state (c) doesn't mean that the sequence does not correspond to the one we want. The specifications do not dictate that there be a 0 before the sequence.

LN could even remain at 1 for a long time. What matters is that after the two last 1s, we read two 0s: we remain therefore in state (c), waiting for a 0.

In sum, we move to state (d) if we have identified a 0 after two 1s. We must decide what to do if the 0 is not read in state (d).

The only assumption we can make about any 1 read in state (d) is that, it is the first 1 of a new sequence, since it follows a 0. This means that we move to state (b) where we would have gone after state (a) for the same reason.

Finally, we must decide how to leave state (e), where the FSM goes after identifying the sequence.

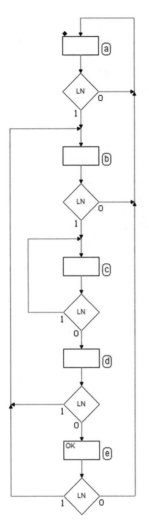

To that end, let's complete the diagram (see figure at the left). Let's assume we have recognized the sequence and moved in state (e).

If we find another 0 we go back to state (a) to wait for a new sequence. Instead, if a 1 is read, it could represent the beginning of a new sequence, so we move to state (b) as we did from state (d).

7.3.4 *Serial Synchronous Transmitters (2 bits)*

Design a 2-bit serial synchronous transmitter. The device reads three synchronous inputs GO, $D0$ and $D1$ and generates the serial sequence on the output LIN.

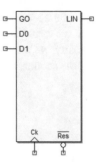

The transmission of the two data bits $D0$ and $D1$ is started by a low-high transition of the command input GO. The following figure describes the format of the transmitted sequence.

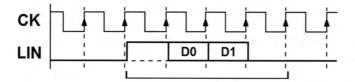

The bit time is the same as the clock cycle *CK*. The sequence begins with a *start bit* at 1, continues with two *data* bits *D0* and *D1* in that order, and finally ends with a *stop bit* at 0.

Solution

Beginning with the assumption that *GO* is initially at 0, we wait in (a) for *GO* to move to 1.

As soon as it does, we make a transition to state (b) where we activate output *LIN* (i.e., we start transmission by generating the *start bit*) for the duration of a clock cycle.

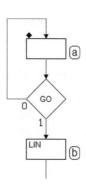

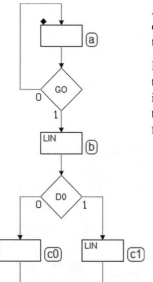

At the next clock cycle, we must transmit the value of *D0*, so we read this input and decide whether to go to state (c0) or (c1).

In (c0), output *LIN* is not active but in (c1) it is. Note that both states (c0) and (c1) are mutually exclusive in terms of state paths, but from a timing perspective, they correspond to the same clock cycle, the one following the cycle where we generated the *start bit*.

With the same criterion (see the figure on the right), we transmit the value of $D1$ in the next clock cycle by setting $LIN = 0$ for state (e0) or $LIN = 1$ for (e1).

The serial sequence must end with a *stop bit* at 0 so we go to state (g) where the output LIN is not activated.

Finally, we must check that input GO is back to 0 before returning to state (a). If we did not check this and GO was still high we would immediately start the transmission of a new sequence when back in (a).

We know that this should only occur when GO has a transition from low to high.

Rather than inserting another state where LIN is not active, let's directly take advantage of the already defined state (g) since it has no outputs, and transform it into a waiting state.

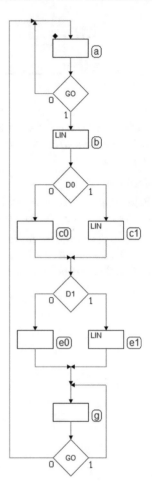

7.3.5 Command Receiver with Serial Synchronous Interface

Let's design a 2-bit command receiver with serial synchronous interface. The device receives the serial sequences on input LN and generates outputs OK and ERR.

The bit time is the same as the clock CK period. The sequences all begin with a *start bit* at 1, continue with two *data* bits $D0$ and $D1$ and finally end with a *stop bit* at 0. The figure below shows the sequence format.

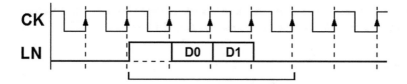

Among the four possible combinations, our receiver must recognize the one where $D0 = 0$ and $D1 = 1$, as in the figure below:

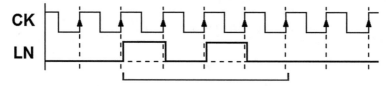

When a sequence has been received, the receiver monitors the *stop bit* to identify any errors. This consists of checking that the serial signal *LN* is at 0 in the bit time corresponding to the *stop bit*. This monitoring is done in all cases, even for combinations that *do not interest us*.

If the *stop bit* is correctly at 0 and the sequence received corresponds to the one we were looking for, the receiver activates output *OK* for the duration of a *CK* cycle, signaling that it *received the expected command*. If the sequence is different, the receiver does not activate any output.

If the *stop bit* is incorrect, the receiver activates output *ERR* (for the duration of one *CK*). After receiving a sequence, the receiver always returns and waits for the next one.

Solution

Let's start designing the ASM diagram setting a waiting loop on state (a), where we wait for a 1 (the *start bit*) on input *LN* (see the figure on the right).

Once the *start bit* is identified, the other bits will be read after it at every clock cycle.

Let's assume the sequence we were looking for has no errors; we continue designing the states without completing the diagram for the other branches yet.

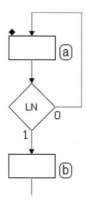

Note that states (b), (c0), and (e0) have been inserted to test *LN* at every clock cycle in accordance with the bits that are received one by one.

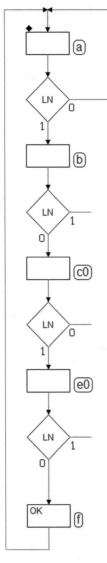

So we get to (f) if we read *D0* = 0, *D1* = 1 and *stop bit* = 0.

The activation of *OK* in (f) concludes the sequence reception, so the machine returns to (a) to wait for the next one.

Now, let's add the remaining paths (see the figure at the right).

If we read 1 in (b) we have a different sequence from the one we *expected*.

We must still complete it because we have to verify that the *stop bit* is correct at the right time.

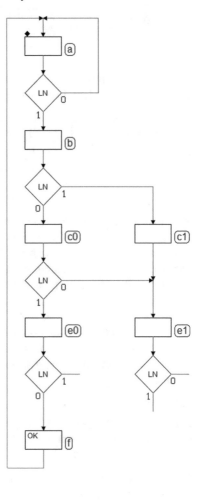

States (c1) and (e1) are added so that the test can be done at the same time as the expected sequence.

Notice that the value of the second bit is not checked on this path. This is because we know that this sequence *is different* from the one expected and we will not activate the output *OK*.

Let's have another look at state (b). If the first bit is read at 0 (what is expected), the machine goes to state (c0) to check the second bit. If it equals 0, the sequence is different from what is expected so we move to state (e1) because we *will not activate* output *OK*.

Finally, let's complete the diagram (see the figure below), noting the valuation of the *stop bit* which occurs in state (e1).

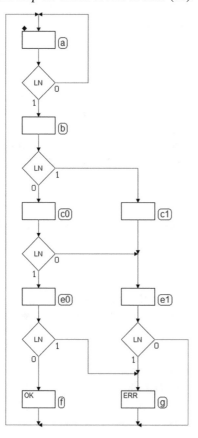

If the bit *is correct*, we go to state (a) to wait for a new sequence. If the bit is *incorrect* we must activate output *ERR* for one clock cycle in state (g) and then return to state (a).

Finally, let's complete the diagram using state (g) for the other path as well; the actual sequence is what we expected but with the wrong *stop bit* so we activate *ERR* rather than *OK*.

7.3.6 Serial Synchronous Receiver (2 bits)

Design a 2-bit *serial synchronous receiver*. The device should receive the serial sequences on line *LIN* and generate outputs $Q0$, $Q1$, *OK* and *ERR*.

The bit time is the same as a clock cycle *CK*; the sequences all begin with a *start bit* at 1, continue with two data bits $D0$ and $D1$ and finally end with a *stop bit* at 0. The following figure shows the format of the sequences received.

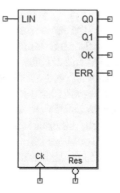

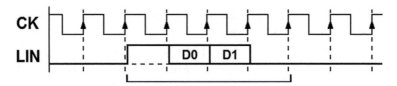

Note that signal *LIN* is synchronous with the clock *CK*, and changes its value a propagation delay time after the rising edge. This is typical when the system that generates *LIN* has the same clock as the device we are designing. Thus, *LIN* is sampled on the edge of the *next* clock that has generated it.

Once it has received the sequence, the receiver checks that the *stop bit* is correct. If the *stop bit* is at zero as it should be, the receiver activates output *OK*, signaling that it extracted the two data bits *D0* and *D1* from the sequence and made them available on outputs *Q0* and *Q1*, respectively. Outputs *Q0*, *Q1*, and *OK* are kept active until another sequence is received.

If the *stop bit* is wrong, the receiver only activates output *ERR* for the duration of one *CK* cycle and then it waits for another sequence.

Solution

We start designing the ASM diagram by setting (a) as a state where we wait for the *start bit* from the input *LIN* (see the figure at the right). Once the *start bit* has been identified, the other bits will be read, one per clock cycle.

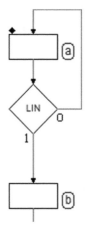

Let's look at state (b): while in this state, the value on input *LIN* corresponds to bit *D0*. We will take different paths depending on the value read on it, as shown in the first figure on the following page.

We move to state (c) if $D0 = 0$ or to state (d) if $D0 = 1$.

We use the same reasoning for *D1*, the second data bit for the sequence, and *separate the paths* afterward based on the value received. Downstream of the reception of bits *D0* and *D1*, the FSM will be in one of the following states: (e), (f), (g), or (h), depending on the four possible values' combinations.

The *paths* have been separated because this is the only way the machine has to *keep track* of the received data *D0* and *D1*. States (e), (f), (g), and (h) *remember* different previous input sequences.

It would not be necessary to activate the outputs *Q0* and *Q1* along the four paths but it could be useful to *monitor* the values received *from outside*.

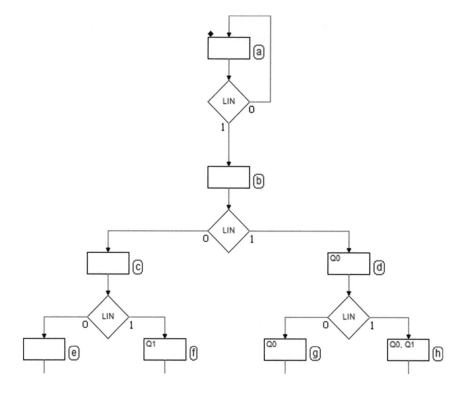

After the data bit, we must test the *stop bit*, which is present on line *LIN* right when the FSM is in (e), (f), (g) or (h).

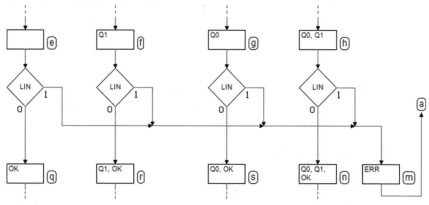

So, as we see in the figure above, we check *LIN*, and if the *stop bit* is wrong, we join together all the paths (we no longer care about the data bits' values), generate *ERR* in state (m) for one clock cycle and then return to (a).

If, however, the *stop bit* is correct for all four paths, we will move to a corresponding state: (q), (r), (s), or (n) where we will activate *OK* and outputs *Q0* and *Q1* depending on the data received.

Let's complete the diagram, as seen in the figure below, with the specification that the values on outputs $Q0$, $Q1$, and OK must be kept until the next sequence. To achieve this, all the states (q), (r), (s), and (n) are enclosed in a loop. The machine leaves the loop when a new *start bit* arrives from line *LIN*.

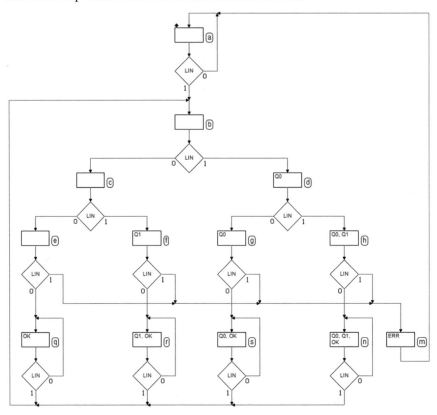

Thus, each of the states (q), (r), (s), and (n) does the same job as (a), waiting for the next sequence. Note that as with state (a), the next state after each of these will be (b).

7.3.7   Push-Button Handling

Design a synchronous sequential network that handles the state of the push-buttons $P1$, $P2$ and $P3$ and generates outputs $L1$, $L2$ and $L3$ and an output *PLS* (*pulse*).

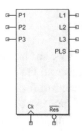

The figure below describes the typical signal evolution at inputs $P1$, $P2$ and $P3$: initially, a push-button is *not activated*, then it is *pressed* and finally *released*.

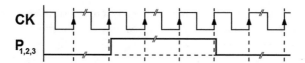

Assume that the switches already have the necessary circuitry to solve the problem of *electromechanical contact bounce* so that the signal they generate is considered bounce-free and synchronous with the clock.

The push-buttons must be managed taking in account their *priority*: P3 first, then P2, last P1. This means that if P3 is pressed, the other push-buttons' states are ignored. If P2 is pressed, P1 is ignored. P1 is considered only if none of the other push-buttons are pressed.

When a switch P_i is pressed, the FSM activates the corresponding output L_i for the time the switch is pressed. When the switch is released L_i is turned off and PLS generates 1, 2, or 3 pulses (depending on which switch is pressed), as shown in the figure below:

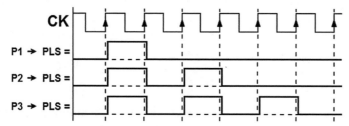

Solution

As we see in the figure to the right, we set (a) as a waiting loop, where we stay if no button is pressed.

Then we add states (b1), (b2) and (b3) to the diagram, where the output corresponding to the pressed button (L1, L2 or L3) is activated.

From the point of view of the algorithm, we move to these states *to remember* that a certain button (P1, P2 or P3) has been pressed.

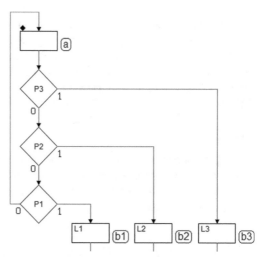

Notice the logic used to test the lines coming from the push-buttons. It reflects the specifications about priority. If P3 is pressed, we do not check the others; if P3 is not pressed, but P2 is, we do not check P1.

Note: the figure could be misleading in that the test of the push-buttons might seem to be carried out *"one after the other,"* in chronological order, but this is not the case. The figure actually describes a *combinational logic* where the inputs are jointly *evaluated at the same time* in state (a). $P2$ and $P1$ are *don't cares* if $P3$ is at 1; $P1$ is a *don't care* if $P3 = 0$ and $P2 = 1$.

The following figure describes the same algorithm in a more redundant fashion, since all the possible combinations of $P1$, $P2$, and $P3$ are considered here.

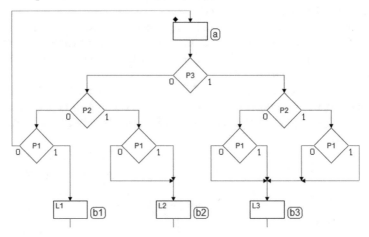

Some of the eight possible paths are collected together according to the rule of priority described above. By eliminating the superfluous decision blocks, making them flow into the same path, we get the figure seen before.

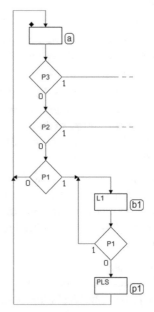

Let's now turn our attention to the function of states (b1), (b2) and (b3). The specifications require that output L_i corresponding to push-button P_i be *kept active* the whole time that it is *pressed*, and that output *PLS* should be activated when P_i is *released*. This means that (b1), (b2) and (b3) must be defined as *wait states*, as shown in the figure aside for the button $P1$.

The figure shows how to generate the required pulse on *PLS* at the release of the button, activating *PLS* for one cycle and then coming back to (a) waiting for a new action.

The figure below shows the complete ASM diagram. After releasing the push-button *P2*, we generate *two pulses* on *PLS* by a sequence of *three* states: (p2) where we activate output *PLS*, (q2) where we deactivate it and finally (p1), already used for *P1*.

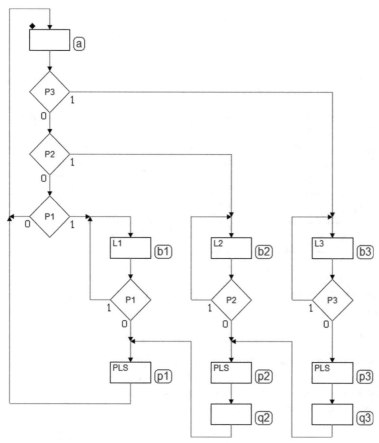

The figure also shows the completed path of *P3*, which reuses states in the same way.

7.3.8 3-Bit Shift Registers

Let's design a 3-bit *shift register* using ASM diagrams. It receives an input *IN* and generates the outputs *QA*, *QB* and *QC*.

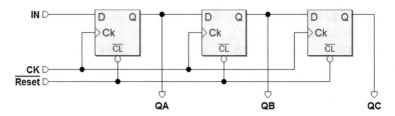

The following figure shows an example of how the register works for a specific sequence applied at input *IN*:

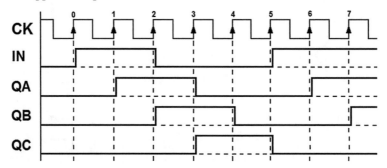

It is not practical to show in the timing diagram all the possible sequences there can be at the input. Rather, an ASM diagram describes the register's behavior thoroughly.

Solution

Let's assume that register's outputs are initially at zero in state (a) (see figure aside). When a 1 appears at input *IN*, the register will load the new value onto output *QA*, on the next active edge of the clock *CK*.

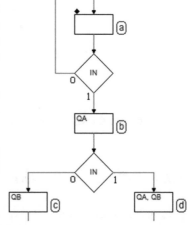

Let's formally describe this new situation by changing states: the FSM goes to (b) where output *QA* is activated.

Note that the input should always be checked in any state because output *QA* will always depend on the value of *IN*.

Now let's check the input in state (b).

On the next active edge of the clock, the FSM changes state. The value memorized in *QA* is moved to *QB*, while *QA* takes the value of *IN*. States (c) and (d) describe the two possible cases depending on the value acquired.

We will use the same reasoning for the next active edge of the clock where the value memorized in QB moves to QC, QA to QB and QA copies IN, giving us the diagram at the top of the next page.

It is not complete yet, but it includes all eight possible states (the combination of the values of outputs QA, QB, and QC). Certain that we do not have to introduce other states since we have exhausted all the possible combinations of QA, QB, and QC, we must determine to what states the FSM will move to depending on the value of IN, from states (e), (f), (g), and (h).

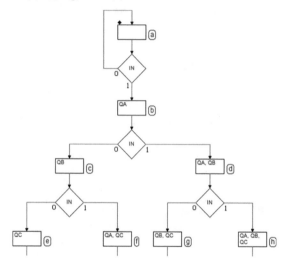

We finish the ASM diagram, obtaining the final version seen below.

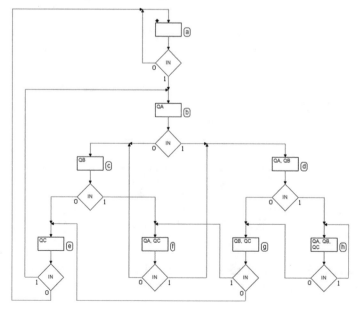

Here are some ideas that helped us complete the diagram from states (e), (f), (g), and (h).

If the FSM is in state (e), after the next active clock edge, QC and QB will be at 0 whatever the value of IN is. This is because they will load the contents of QB and QA, respectively, 0 in this state. The states that already have QB and QA at 0 are (a) and (b). So if IN is at 0, we will go to (a): if it is at 1, we will go to (b).

We will use the same reasoning for state (f) where the next state could only be (c) if IN is at 0, or (d) if IN is at 1.

Following the same criterion for states (g) and (h), we fill in the diagram, closing all the remaining logical paths.

7.3.9   Sequential Networks with Conditional Outputs

Let's describe as an FSM, the functioning of a sequential network based on a *2-bit shift register* with an added output that is a function not only of the state but also of the input (*Mealy* model).

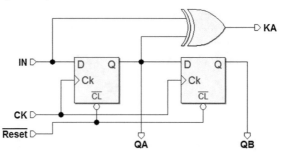

The network reads IN and generates outputs QA and QB. An XOR gate generates output KA, a function of input IN and output QA. The figure below shows how it works for a specific input sequence.

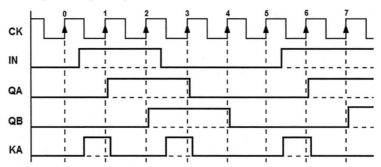

As we can see, KA is generated when IN and QA are different. The *ASM* diagram will describe how the network functions, including the *conditional* output KA.

Some Notes on Conditional Outputs

Even if we assume that the signal at input *IN* is generated by another *synchronous* system with the same clock as the FSM, some factors in play (the length of the connections, possible different technology, etc.) could delay it. Even though the signal *remains synchronous*, it is translated in time (if we look at the timing sequence in the figure above, *IN* has been drawn a bit to the right of the other signals).

The result is that the pulse duration on output *KA*, a combinational function of *IN*, as shown in the same figure, can be much shorter than a clock cycle. In many cases, this could be irrelevant assuming that *KA* is readable on the active edge of the clock.

Solution

To begin, we draw the ASM diagram without output *KA*.

In state (a), *QA* and *QB* are at 0. When a 1 gets to *IN*, output *QA* goes to 1 (on the next active edge of the clock *CK*), moving to state (b) where output *QA* is active (see the figure on the right).

Assuming we will receive a few consecutive 1*s* on input *IN* we continue to draw the diagram. From state (b) we move to state (c) where both outputs *QA* and *QB* are active. We will stay in (c) as long as 1*s* continue to appear at *IN*.

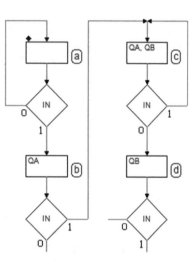

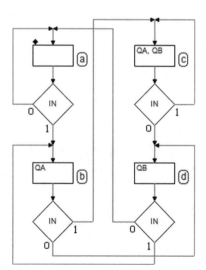

However, if *IN* = 0 we move to (d). The previous *QA* moves to *QB* while *QA* takes the value 0.

Let's now fill in the remaining paths, as in the figure at the left. If we are in (b) and *IN* equals 0, *QA* moves to *QB* and a 0 moves to *QA*, so we will go to (d).

Only *QB* is active in state (d). Since *QA* = 0, whatever the next state may be, *QB* will go to zero. So, depending on the input value we will move to (a) or (b).

Let's add the description of the conditional output. We can see in the network schematic that *KA* activates if *QA* and *IN* are *different*.

Let's focus on state (a) where $QA = 0$. From what we have seen so far, *KA* will activate *while* in (a) if $IN = 1$. So let's add a *conditional output* block along the logical path for $IN = 1$ (see the figure at the right).

$QA = 0$ in state (d) as well. The *conditional output* will be added with the same criterion (for $IN = 1$).

Let's fill in the diagram for states (b) and (c) where $QA = 1$. Output *KA* will activate *while* in these states if $IN = 0$. The block will be placed along the paths for $IN = 0$.

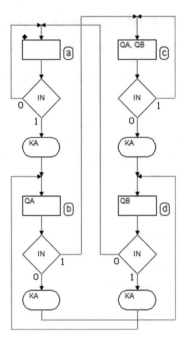

7.3.10 Shift Register with XOR Tree

Let's describe as an FSM the functioning of a sequential network made of a 3-bit *shift register* and an XOR tree. This FSM conducts *parity checks* on outputs $O1$, $O2$, and $O3$ of the flip-flops and compares the results with the value of input *IN*. The product of this comparison is reintroduced into the shift register.

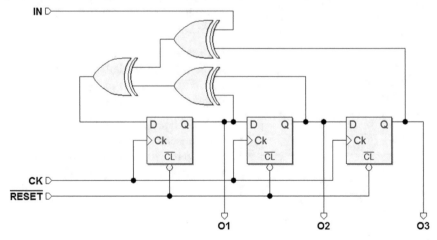

Here is a helpful note about describing the network in FSM terms. If the number of 1s on outputs $O1$, $O2$, and $O3$ is even, the value of input IN is inserted into the first flip-flop on the left. If the number of 1s is odd, its negated value is inserted.

Solution

In the figure below, we see the first steps toward building the ASM diagram. In state (a), outputs $O1$, $O2$, and $O3$ are set to 0.

This means that in state (a), the XOR tree will only copy input IN, so it will go to state (b) and activate output $O1$ at the next active edge of the clock if $IN = 1$.

In state (b) we have an odd number of outputs at 1, so output $O1$ will load the negated value of IN on the active edge of the clock, while $Q2$ will simply copy the current value of $Q1$.

So, let's introduce two new states: (c) and (d). $Q2$ is active in both, while $Q1$ is active only in (c), the state where the FSM will go if $IN = 0$, starting from (b).

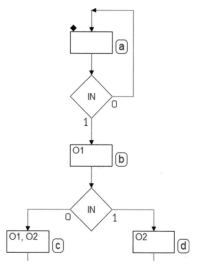

Applying the same criterion to states (c) and (d) (see the figure below), let's introduce four more states: (e), (f), (g), and (h).

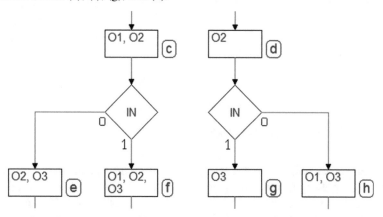

Now the diagram includes eight states, all the possible ones in a three flip-flop network. When filling in the diagram, we must take care to avoid adding other states and reuse those we already have, according to the input IN. The complete diagram can be seen in the next page.

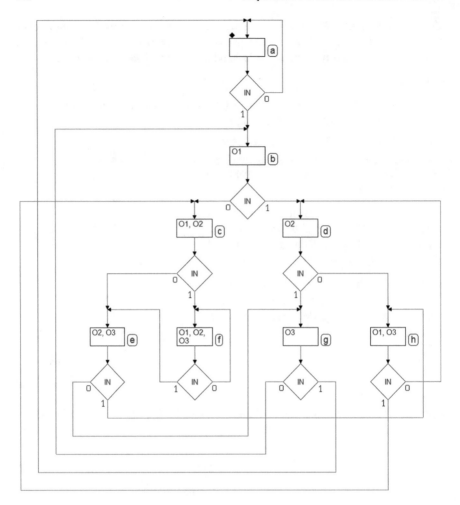

7.4   Synthesis of Synchronous FSM

In this section, we will discuss the *synthesis of synchronous FSMs*, a process that we can use to create a synchronous sequential network from its algorithmic description. A CAD system will normally allow us to synthesize automatically. Still, it is useful to tackle this subject systematically by examining the rules and concepts as if we needed to proceed manually.

Taking the *Mealy* model as an example, the issue of synthesis is reduced to the definition of two networks: the *Next State Combinational Network* and the *Output Combinational Network*.

We can see in the figure that both networks receive vector I (the inputs of the network) and vector X (the state of the network) in the input. The *next state combinational network* produces vector iX (the next state to be stored in the state register), the *output combinational network* produces vector U (the network outputs).

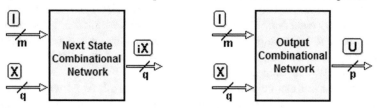

Inputs I and outputs U are defined by the network specifications. The number of variables that make up vectors iX and X will depend on the number of states foreseen by the algorithm at the project level (and by the type of flip-flop that is chosen for the state register). In the previous sections, we simply assigned a name to each state. To synthesize the network, it is essential to assign a *unique binary code* to each state. This process is called *state assignment*.

7.4.1 State Assignment

In a synchronous FSM, there are several ways to assign codes to the states. Some tend to reduce the number of circuit components, others to improve speed or reliability. Generally, designers try to find a compromise between circuit size and speed. The first technique that we will look at minimizes the number of state variables; with q variables, one can codify up to 2^q states.

The four-state counter, examined previously, is a *synchronous network with no inputs* (excluding the *Clock* and $\overline{Reset}$). The minimal number of variables needed to represent *four states* is two ($X1$ and $X0$). Taking *Moore's* model of the synchronous FSM as an example, this counter can be represented by the reduced model shown in the figure below.

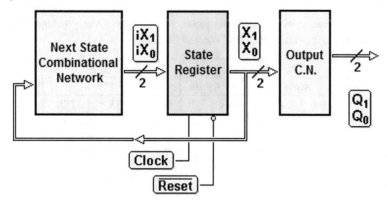

Due to the system's synchronous structure and the absence of inputs, the choice of state codes is completely *arbitrary*. With two state variables, we have four possible combinations and thus, 24 possible distinct assignments of the 4-state codes (24 = 4!, i.e., the number of possible *permutations* of the four combinations).

Given the freedom of choice, in our example we associate each state with the code that is the *binary combination of outputs generated by the state itself*. In other words, it is convenient to make outputs $Q1$ and $Q0$ correspond directly to the state variables $X1$ and $X0$ so that the output combinational network breaks down into two simple connections ($Q1 = X1$ and $Q0 = X0$).

State assignment can be represented in various, equally valid ways. The figure below left shows the state codes directly noted above every state block. The central figure shows the same assignment described by a *table*. On the right, a *map* is used.

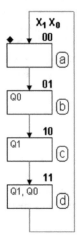

State	Code (X_1, X_0)
a	00
b	01
c	10
d	11

7.4.2   Describing an FSM with a State Table

The *state table* is a description that is formally *equivalent* to the ASM diagram and provides the very same information. For each state, it identifies the outputs and the next state in function of any relevant input. In the case of the previous machine, which has no inputs, the table is laid out as follows:

Current State					Next State		
State			Outputs		State		
Name	X1	X0	Q1	Q0	Name	iX1	iX0
a	0	0	0	0	b	0	1
b	0	1	0	1	c	1	0
c	1	0	1	0	d	1	1
d	1	1	1	1	a	0	0

The table is divided into two fields: the left describes the *current state* (name and code) and its *outputs*. The right-hand side shows the name and code of the *next state* for every possible current state.

Another example of the state table is derived from one of the previously described edge detectors (the one with four states, an input and an output). Let's assign the state codes directly on the diagram, as shown in the figure below. If we assume that the *input is synchronous*, the assignment will be *arbitrary*.

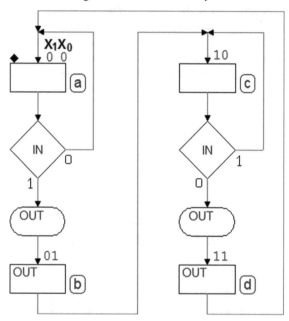

Let's derive the state table that corresponds to this ASM diagram. See the results below. Note that the table contains *one line for each path* connecting the state blocks so that it completely describes the evolution of the machine's states, as does the ASM diagram.

Current State					Next State	
State			Input	Output	State	
Name	X1	X0	IN	OUT	iX1	iX0
a	0	0	0	0	0	0
a	0	0	1	1	0	1
b	0	1	-	1	1	0
c	1	0	0	1	1	1
c	1	0	1	0	1	0
d	1	1	-	1	0	0

The table does not follow a timing sequence order, so it could be written in any order. Its information must be read row by row independently.

State by state, it lists the current outputs (conditional and otherwise) and the next state in function of the input value. For example, there are two rows for each of states (a) and (c) to describe how they depend on input *IN*.

Describing an FSM with a table is a good method for representing it on a computer. Yet, it is clear that the ASM diagram gives a better and more comprehensible vision of the MSF's algorithm, and so, it is this representation that we use here to better understand sequential network design.

In any case, this method has practical limitations when it comes to describing machines with many states since it becomes impossible to have an overview of it. This is when using a language to describe the hardware (eg., VDHL or VERILOG) is preferable.

7.4.3 State Table Synthesis

From the state table, we can extract the truth tables of the next state network and the output combinational network, shown here below:

Next State Network				
Current State		Input	Next State	
X1	X0	IN	iX1	iX0
0	0	0	0	0
0	0	1	0	1
0	1	-	1	0
1	0	0	1	1
1	0	1	1	0
1	1	-	0	0

Output Combinational Network			
Current State		Input	Output
X1	X0	IN	OUT
0	0	0	0
0	0	1	1
0	1	-	1
1	0	0	1
1	0	1	0
1	1	-	1

From these tables, we can synthesize these networks and draw the schematic of the circuits.

If we use the maps, these tables will give us two for the next state ($iX1$, $iX0$) and one for the output (OUT), as shown below:

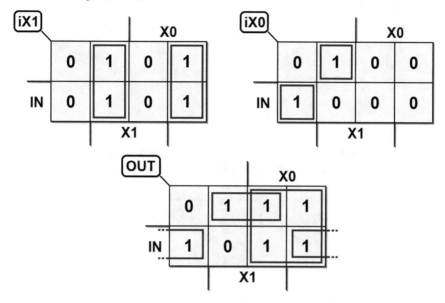

Alternatively, we can draw maps with entered variables. State variables are used as coordinates on the map while inputs (here, there is only IN) will be entered inside.

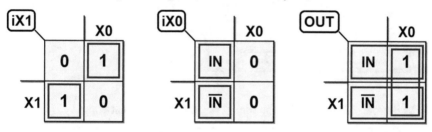

The synthesis of these maps gives us the equations that describe the functions we are looking for.

$$iX1 = (\overline{X0} \cdot X1) + (X0 \cdot \overline{X1})$$
$$iX0 = (IN \cdot \overline{X1} \cdot \overline{X0}) + (\overline{IN} \cdot X1 \cdot \overline{X0})$$
$$OUT = X0 + (\overline{IN} \cdot X1) + (IN \cdot \overline{X1})$$

We use the XOR gates and take a few steps to make these expressions more concise.

$$iX1 = X0 \oplus X1$$
$$iX0 = (IN \oplus X1) \cdot \overline{X0}$$
$$OUT = (IN \oplus X1) + X0$$

Here are the three functions represented in circuit form:

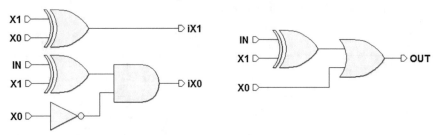

The figure below shows the whole network, complete with asynchronous initialization circuits (input $\overline{Reset}$ that acts on the $\overline{CL}s$ of the flip-flops):

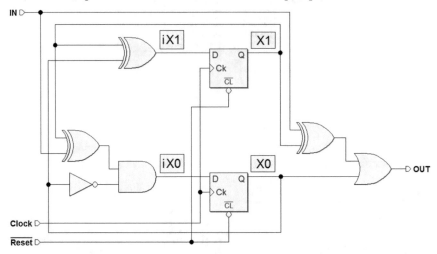

To be thorough, let's examine the timing simulation of this network:

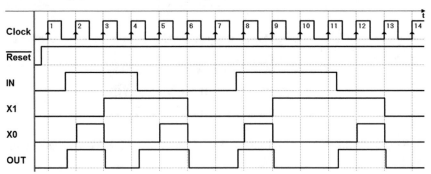

The timing diagram shows that, as expected, output OUT activates following the changes of input IN, without waiting for the rising edge of the $Clock$. Then, output OUT is kept active in the next state for one whole cycle.

7.4.4 Examples of Synchronous FSM Synthesis

Example 1 (2-Bit Up/down Counter)

In this example, we synthesize a synchronous FSM with a synchronous input and the terminations described in the figure below:

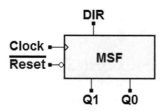

The ASM diagram of a *binary up/down counter* is shown in the figure at the right and was analyzed and defined on p. 271. The lack of conditional outputs makes this an example of a *Moore's* machine.

To do the synthesis, a code has been assigned to the states in the diagram (see at upper right-hand corners). The criteria were as follows: with four states, we can only use two state variables (X and Y).

Given that outputs $Q0$ and $Q1$ appear in all the possible combinations of the four states, we chose to align the state variables with the outputs:

$$Q0 = X \qquad Q1 = Y$$

The figure below shows a block schematic of the network that we want. We see the state variables X and Y (the outputs of the flip-flops) and variables iX and iY, which the next state combinational network must produce in function of state X, Y and input *DIR*.

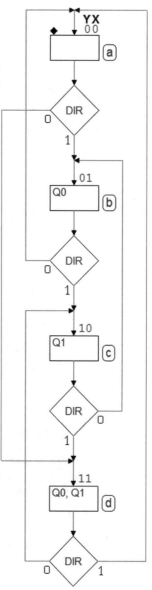

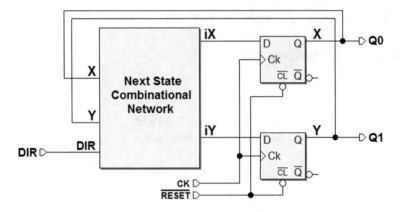

From the ASM diagram, we get the table of the next state (the output table is super-fluous here because of how it was defined), and then the maps:

State	Y	X	DIR	Next State	iY	iX
a	0	0	0	d	1	1
a	0	0	1	b	0	1
b	0	1	0	a	0	0
b	0	1	1	c	1	0
c	1	0	0	b	0	1
c	1	0	1	d	1	1
d	1	1	0	c	1	0
d	1	1	1	a	0	0

iX

		X	
1	1	0	0
1	1	0	0

DIR (left), Y (bottom)

iY

		X	
1	0	1	0
0	1	0	1

DIR (left), Y (bottom)

From this, we get the following. Notice the transformation of iY in the XOR tree is possible since the map is a *checkerboard*):

$$iX = \overline{X}$$

$$iY = \overline{DIR} \cdot \overline{X} \cdot \overline{Y} + DIR \cdot \overline{X} \cdot Y + \overline{DIR} \cdot X \cdot Y + DIR \cdot X \cdot \overline{Y}$$

$$= (\overline{DIR} \cdot \overline{X} + DIR \cdot X) \cdot \overline{Y} + (DIR \cdot \overline{X} + \overline{DIR} \cdot X) \cdot Y$$

$$= (\overline{DIR \oplus X}) \cdot \overline{Y} + (DIR \oplus X) \cdot Y = \overline{(DIR \oplus X \oplus Y)}$$

$$= DIR \oplus X \oplus \overline{Y}$$

Finally, we draw the logical schematic of the resulting network:

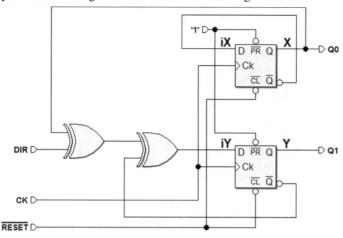

Example 2 (Up/down Counter with Maximum and Minimum Outputs)

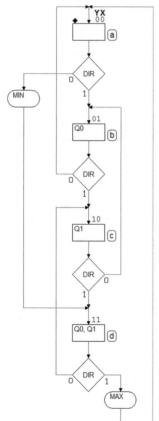

Here, we take the binary up/down counter from Example 1 and add two outputs to show when the maximum number is reached when counting *up*, or the minimum when counting *down*.

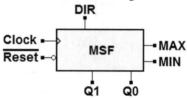

The ASM diagram at the left is identical in terms of the evolution of states. There are, however, two *conditional outputs*, so this is a *Mealy* model FSM. As we can see, *MIN* activates *while* we are in (a) assuming that $DIR = 0$. Thus, *MIN* shows the *terminal count* condition in the *down* direction. Likewise, *MAX* activates in (d) if $DIR = 1$ and shows the *terminal count* in the other direction.

The states are assigned in the same way as in Example 1 (variables X and Y) with aligned *state outputs*.

$$Q0 = X \qquad Q1 = Y$$

In the next page, we see the network block schematic that we want to synthesize. Note the *Mealy* model structure with outputs *MAX* and *MIN* that depend on input *DIR*.

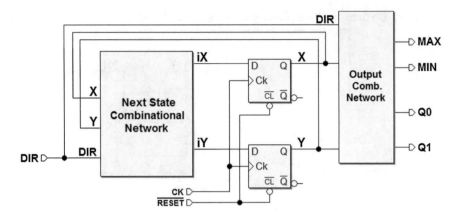

From the ASM diagram, we get the state table. Then, we get the maps:

State	Y	X	DIR	Nxt. St.	iY	iX	Q1	Q0	MIN	MAX
a	0	0	0	d	1	1	0	0	1	0
a	0	0	1	b	0	1	0	0	0	0
b	0	1	0	a	0	0	0	1	0	0
b	0	1	1	c	1	0	0	1	0	0
c	1	0	0	b	0	1	1	0	0	0
c	1	0	1	d	1	1	1	0	0	0
d	1	1	0	c	1	0	1	1	0	0
d	1	1	1	a	0	0	1	1	0	1

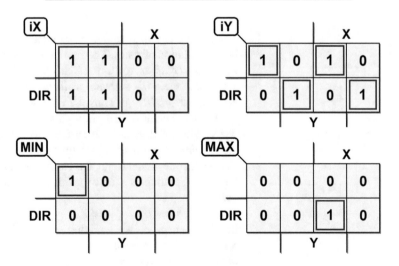

The maps of iX and iY are the same as in Example 1, while the maps of $Q0$ and $Q1$ are superfluous as we have defined them. We get the expressions:

$$
\begin{aligned}
iX &= \overline{X} & iY &= DIR \oplus X \oplus \overline{Y} \\
Q0 &= X & Q1 &= Y \\
MIN &= \overline{DIR} \cdot \overline{X} \cdot \overline{Y} & MAX &= DIR \cdot X \cdot Y
\end{aligned}
$$

After the synthesis, the resulting network is the same as in Example 1 plus the logic that generates outputs MIN and MAX:

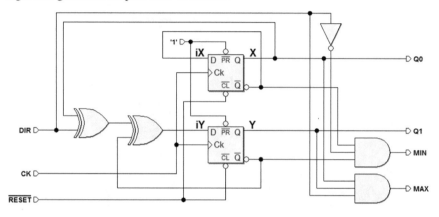

For practice, let's now try to redo the synthesis, this time using JK-PET flip-flops rather than D-PET. The network is described in the block schematic below:

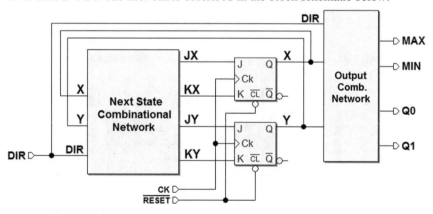

The state assignment and the output combinational network are the same as in the D-PET version. What changes is the next state network, which must generate four functions rather than two: JX, KX, JY, and KY. From the ASM diagram, we get the table of the next state and, from there, the maps.

State	Y	X	DIR	Nxt. St.	JY	KY	JX	KX
a	0	0	0	d	1	-	1	-
a	0	0	1	b	0	-	1	-
b	0	1	0	a	0	-	-	1
b	0	1	1	c	1	-	-	1
c	1	0	0	b	-	1	1	-
c	1	0	1	d	-	0	1	-
d	1	1	0	c	-	0	-	1
d	1	1	1	a	-	1	-	1

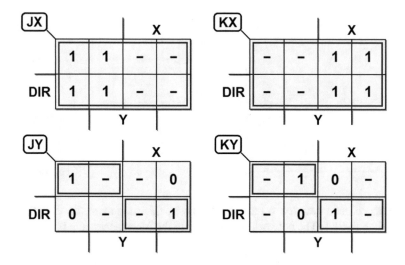

Alternatively, we could set the maps with the entered variable technique using X and Y as coordinates.

From the maps, we get the expressions that describe the four functions we are looking for:

$$JX = 1 \quad JY = \overline{DIR} \cdot \overline{X} + DIR \cdot X = DIR \oplus \overline{X}$$
$$KX = 1 \quad KY = \overline{DIR} \cdot \overline{X} + DIR \cdot X = DIR \oplus \overline{X}$$

After the synthesis with the JK-PET flip-flops, we get the network below:

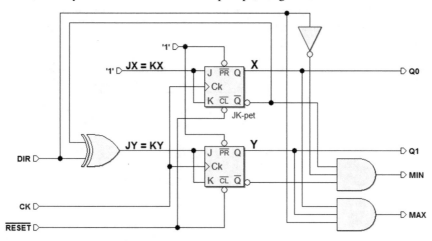

Let's examine the timing simulation of this network:

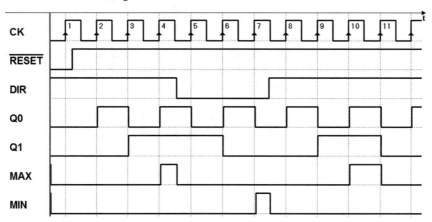

See, for example, what happens between edges 4 and 5 of the clock; the counter is asked to change count direction. Output *MAX*, which is active in the up count, is immediately deactivated when *DIR* changes since *MAX* is a combinational function of *DIR* (as well as of the state).

Example 3 (Sequence Detector)

Here, let's synthesize the *Sequence Detector* analyzed on p. 296. Its ASM diagram is shown again on the left, completed with the state assignment.

Given that input *LN* and the FSM are synchronous, code assignment can be arbitrary. We have chosen to use one of the state variables to directly generate output *OK*. We define the three state variables W, Y, and X with the assumption that $OK = W$. There are no conditional outputs in the diagram, so this synthesis gives us a *Moore* machine.

The block schematic of the network is shown here:

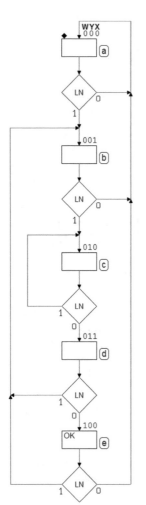
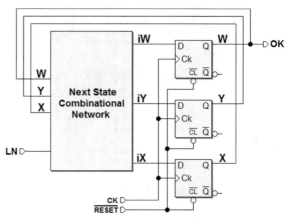

The network highlights the state variables W, Y, X, and the next state network outputs iW, iY and iX. Let's derive the table from the diagram.

State	W	Y	X	LN	Nxt. St.	iW	iY	iX
a	0	0	0	0	a	0	0	0
a	0	0	0	1	b	0	0	1
b	0	0	1	0	a	0	0	0
b	0	0	1	1	c	0	1	0
c	0	1	0	0	d	0	1	1
c	0	1	0	1	c	0	1	0
d	0	1	1	0	e	1	0	0
d	0	1	1	1	b	0	0	1
e	1	0	0	0	a	0	0	0
e	1	0	0	1	b	0	0	1
-	1	0	1	-	-	-	-	-
-	1	1	-	-	-	-	-	-

We get the maps from the state table. For comparison, we can use *ordinary* maps (on the left), or the more compact *entered variable* maps (on the right):

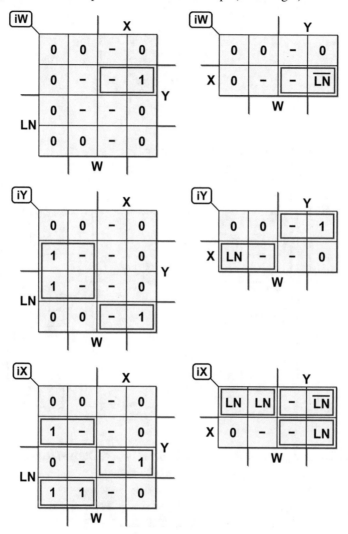

As we continue to read the maps, we get the expressions:

$$iW = \overline{LN} \cdot X \cdot Y$$
$$iY = \overline{X} \cdot Y + LN \cdot X \cdot \overline{Y}$$
$$iX = \overline{LN} \cdot \overline{X} \cdot Y + LN \cdot \overline{X} \cdot \overline{Y} + LN \cdot X \cdot Y$$

Finally, to be thorough (even though it is not necessary since $OK = W$ has already been defined), we have put the map of output OK below:

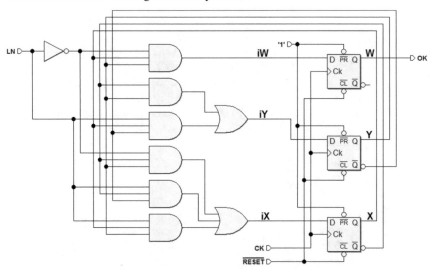

This is the network resulting from the synthesis:

Below, is the timing simulation that checks the network. The two sequences 1-1-0-0, which are recognized by the algorithm, are highlighted.

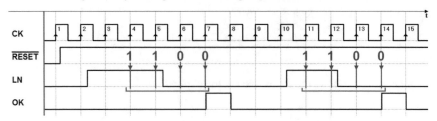

Example 4 (Push-Button Handling)

The figure below shows the ASM diagram from the *push-button handling* example (p. 306). The diagram has nine states, so we must use four-state variables (Z, W, Y, and X). The states are assigned arbitrarily since the FSM is synchronous and we assume inputs to be synchronous with the clock.

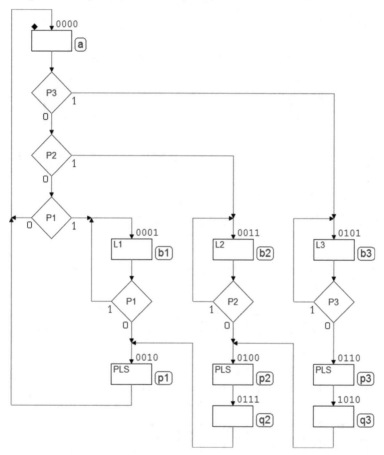

The reader should be aware that the assignment that was made leads to a very complex synthesis and network, as we will see in the next few pages. Perfectly equivalent networks with very different levels of complexity (and thus cost) can depend upon the specific assignation.

Thanks to the availability of CAD design systems, this type of choice and synthesis is done by automatic algorithms, where the designer typically only sets the rules and constraints at the beginning and checks system function at the end of the synthesis. Here, the example is shown as done *manually* to promote understanding of this subject.

We expect to get a network that, represented as blocks, has the look of the network in the figure below:

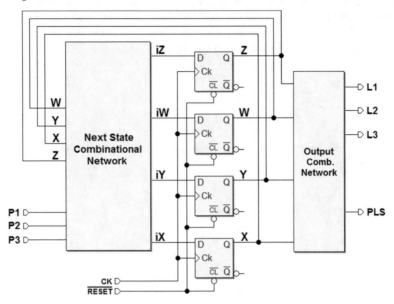

From the ASM diagram, we get the table of the next state, which defines iZ, iW, iY and iX of Z, W, Y, X, $P1$, $P2$, and $P3$:

State	Z	W	Y	X	P3	P2	P1	Next State	iZ	iW	iY	iX
a	0	0	0	0	0	0	0	a	0	0	0	0
a	0	0	0	0	1	-	-	b3	0	1	0	1
a	0	0	0	0	0	1	-	b2	0	0	1	1
a	0	0	0	0	0	0	1	b1	0	0	0	1
b1	0	0	0	1	-	-	0	p1	0	0	1	0
b1	0	0	0	1	-	-	1	b1	0	0	0	1
p1	0	0	1	0	-	-	-	a	0	0	0	0
b2	0	0	1	1	-	0	-	p2	0	1	0	0
b2	0	0	1	1	-	1	-	b2	0	0	1	1
p2	0	1	0	0	-	-	-	q2	0	1	1	1
b3	0	1	0	1	0	-	-	p3	0	1	1	0
b3	0	1	0	1	1	-	-	b3	0	1	0	1
p3	0	1	1	0	-	-	-	q3	1	0	1	0
q2	0	1	1	1	-	-	-	p1	0	0	1	0
-	1	0	0	-	-	-	-	-	-	-	-	-
q3	1	0	1	0	-	-	-	p2	0	1	0	0
-	1	0	1	1	-	-	-	-	-	-	-	-
-	1	1	-	-	-	-	-	-	-	-	-	-

From the table, we get the maps of iZ, iW, iY and iX. We must draw them with *entered variables* because we have too many variables ($P1$, $P2$, $P3$ and the four-state variables).

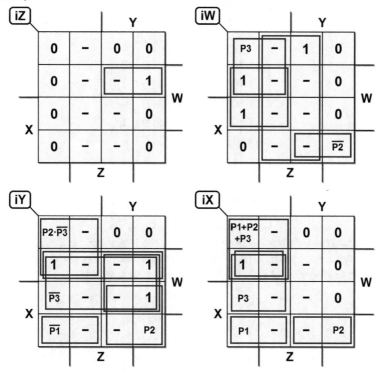

From the maps, we get the expressions:

$$iZ = \overline{X} \cdot Y \cdot W$$

$$iW = W \cdot \overline{Y} + Z +$$
$$P3 \cdot \overline{X} \cdot \overline{Y} + \overline{P2} \cdot X \cdot Y \cdot \overline{W}$$

$$iY = \overline{X} \cdot W + Y \cdot W +$$
$$P2 \cdot \overline{P3} \cdot \overline{X} \cdot \overline{Y} + \overline{P3} \cdot W +$$
$$\overline{P1} \cdot X \cdot \overline{Y} \cdot \overline{W} + P2 \cdot X \cdot Y$$

$$iX = \overline{X} \cdot \overline{Y} \cdot W + (P1 + P2 + P3) \cdot \overline{X} \cdot \overline{Y} +$$
$$P3 \cdot \overline{Y} \cdot W + P1 \cdot X \cdot \overline{Y} \cdot \overline{W} +$$
$$P2 \cdot X \cdot Y \cdot \overline{W}$$

Let's go back to the ASM diagrams and get the output table:

State	Z	W	Y	X	L3	L2	L1	PLS
a	0	0	0	0	0	0	0	0
b1	0	0	0	1	0	0	1	0
p1	0	0	1	0	0	0	0	1
b2	0	0	1	1	0	1	0	0
p2	0	1	0	0	0	0	0	1
b3	0	1	0	1	1	0	0	0
p3	0	1	1	0	0	0	0	1
q2	0	1	1	1	0	0	0	0
-	1	0	0	-	-	-	-	-
q3	1	0	1	0	0	0	0	0
-	1	0	1	1	-	-	-	-
-	1	1	-	-	-	-	-	-

Let's translate the information from the table to the four maps L3, L2, L1, and *PLS*. Here it is not useful to draw the maps with *entered variables* since the network's outputs are the function of only the four individual state variables.

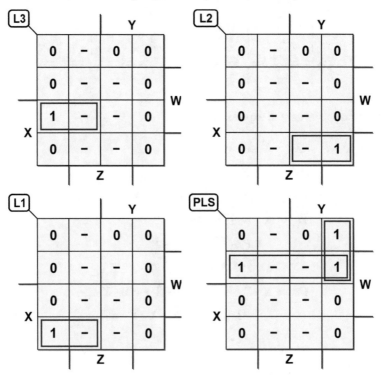

From the maps of the outputs, we get the expressions:

$$
\begin{aligned}
L1 &= X \cdot \overline{Y} \cdot \overline{W} \\
L2 &= X \cdot Y \cdot \overline{W} \\
L3 &= X \cdot \overline{Y} \cdot W \\
PLS &= \overline{X} \cdot W + \overline{X} \cdot Y \cdot \overline{Z}
\end{aligned}
$$

Finally, the synthesis gives us the following network:

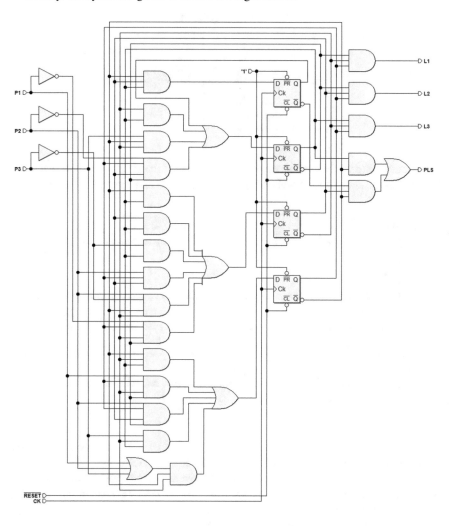

7.5   Time Behavior of Synchronous FSM

7.5.1   FSM with no Inputs

For simplicity's sake, let's start considering the synchronous FSM with no inputs (see the figure below) and for the moment, ignoring the outputs, focusing only on the logic of the next state.

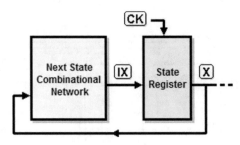

The timing diagram below represents the evolution of the states over time. Vector X represents the state of the FSM and its changes, which result from the action of the clock.

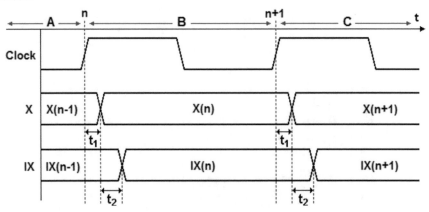

At the end of clock cycle A, the machine is in state $X(n-1)$, and the state register inputs $IX(n-1)$ for the next state $X(n)$ are already available.

At the rising edge n of the clock, the FSM moves to clock cycle B where the state vector $X(n-1)$ is *substituted* by that generated by the next state network $IX(n-1)$, giving us $X(n)$.

The timing diagram highlights delay t_1 between the edge of the clock and the instant vector X changes (this delay is caused by the propagation times of the memory elements that constitute the state register).

After delay t_1, state $X(n)$ is thus active and, thanks to the feedback, $X(n)$ is carried back to the input of the combinational network of the next state, which will make the new value of vector $IX(n)$ available after delay t_2.

Vector $IX(n)$ represents the new next state, which will be loaded in the state register on the next clock active edge (this further delay is due to the combinational network's propagation time). The entire sequence examined above will repeat with the next rising edge of the clock, when value $IX(n)$ is loaded onto the state register. On clock cycle C, this produces the next state $X(n+1)$... and so forth.

Notice that in order for events to occur as described, clock period T must be greater than the sum of the delays considered and of the setup time t_s of the state register $(T > t_1 + t_2 + t_s)$.

7.5.2 FSM with Synchronous and Asynchronous Inputs

There needs to be a distinction between *"synchronous"* and *"asynchronous"* FSM inputs. Remember that *synchronous inputs* change levels in *well-defined time intervals* that are *constant with respect to the active clock edge* so that the network can read them unambiguously.

This entails that they must be *stable* when the next state is loaded onto the state register. A classic example of a *synchronous input* is what comes from another sequential network that shares the same clock and therefore generates the signal with a constant delay relative to the clock (see figure).

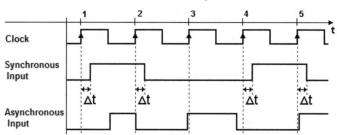

Asynchronous inputs, which can change at any moment, might be read by the network *while they are unstable* (i.e., when they are *changing*), thus producing state transition errors. An example of an asynchronous input is the signal from a manually operated push-button, which is by nature disengaged with the clock.

The figure above shows some *transitions* of an *asynchronous input* that are critical since they occur in correspondence with (or nearly so) an edge of the clock (at 2, 3, and 5).

The problem comes from the propagation times of the next state and state register networks. An input's change in value is generally propagated by the next state network with *slightly different delays* variable by variable.

If the input transition takes place too *close to the active edge of the clock*, due to the delay differences, some of the variations could be taken by the state register, while others might not. This would cause errors in state transitions, as explained in the following.

7.5.3 Synchronous Inputs

Let's now examine in detail the timing of a synchronous FSM with *synchronous* inputs. We will still leave the outputs aside (see figure below).

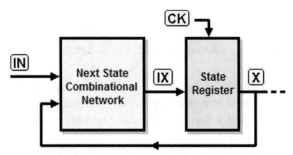

As we have seen before, the next state in this case is obtained by combining the current state X with inputs IN. The timing diagram below shows an example of the evolution of the states in presence of synchronous inputs.

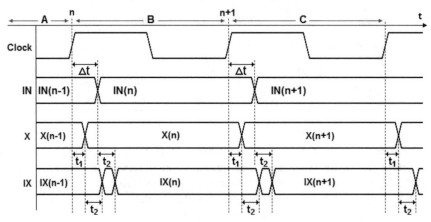

At the end of clock cycle A, the machine is in state $X(n-1)$, state register inputs $IX(n-1)$ and inputs $IN(n-1)$ are available. At the rising edge n of the clock, we move to clock cycle B where state vector $X(n-1)$ is substituted by that generated by the next state network $IX(n-1)$, giving us $X(n)$.

As with the network without inputs, there is a delay t_1 between the edge of the clock and the change of vector X (due to the state register). The new state $X(n)$ is brought by the feedback connection to the input of the combinational network of the next state. When propagation time t_2 is over, it will produce a new value for vector $IX(n)$, which is not yet valid, because *the inputs have changed* in the meantime.

The new value of inputs $IN(n)$ (occurring synchronously with the clock) give extra work to the next state combinational network. With propagation delay t_2, it will produce the definitive value of $IX(n)$, a function of the new value of the $IN(n)$ and the current state $X(n)$.

This sequence repeats likewise at every active edge of the clock. Notice that the synchronous nature of the inputs makes it possible to produce a vector IX that is *always stable* when read by the *next active clock edge*.

7.5.4 Asynchronous Inputs

In a synchronous FSM with *asynchronous* inputs, the state transition can occur incorrectly. The timing diagram below shows an example of two possible cases in relation to the *asynchronous* input change.

In the first example, the *asynchronous inputs* change in clock cycle B at a time that is still compatible with the networks' delay times (similarly to the case with the synchronous inputs).

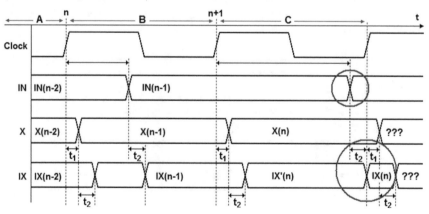

In the second example, the *asynchronous inputs* change too late in clock cycle C. As shown in the figure, the next state network produces vector $IX(n)$ almost at the same time of the active edge of the clock.

This edge forces the state register to load the new state, but some of the variations may arrive a bit *before* the clock edge and others *after*. Thus, the state register could load an incorrect vector and bring the machine *unpredictably* to *an unexpected state*.

A possibile solution is to use a *specific state assignment* so that *only one* state variable depends on the *asynchronous* input.

It is a common approach in the design practice to face the problem at its root, by synchronizing the asynchronous inputs (see pp. 186 and 187).

7.5.5 FSM Outputs (Moore's Model)

There are no specific observations to make about the outputs in a *Moore* model synchronous FSM since they are generated by the output combinational network as functions of the state (see figure below).

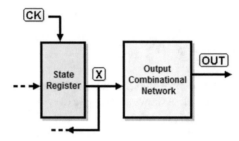

As shown in the timing diagram below, the outputs change when the state changes with a delay time $t_1 + t_3$ compared to the edge of the clock, where t_1 is the state register's propagation time and t_3 is that of the output network.

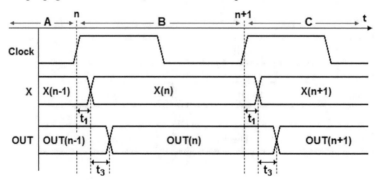

7.5.6 FSM Outputs (Mealy's Model)

In a *Mealy* model FSM, the evolution of the conditional outputs will generally depend on the behavior over time of the inputs, so myriad examples could be presented.

The figure below highlights the generation of conditional outputs as combinational functions of the state and the current inputs.

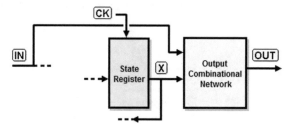

The timing diagram below shows us only a general representation. The conditional outputs are functions of the state and the inputs.

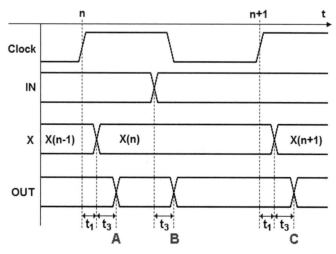

This means that the outputs can change either following the state transition (A and C) or a change in inputs (B) within the interval between two active edges of the clock, as shown in the figure below.

The figure shows that the output delay is measured against what caused its change. We can easily see that the signal path is $t_1 + t_3$ if it is referred to the clock, or just t_3 if referred to an input.

7.6   Exercises

7.6.1   Sequential Network Analysis in Terms of FSMs

Analyze the following synchronous sequential networks by designing an ASM diagram that functionally describes their behavior.

On the *Deeds* Web site, you will find an ASM diagram to be completed from each FSM as well as the schematic with the assigned network. The schematics are set up for you to insert your FSM next to the network given so that you can easily compare their behavior through the simulation.

Network 1

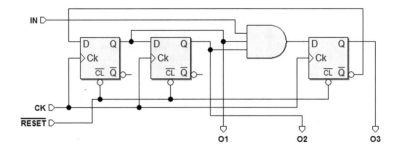

Network 2

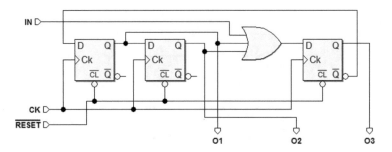

Network 3

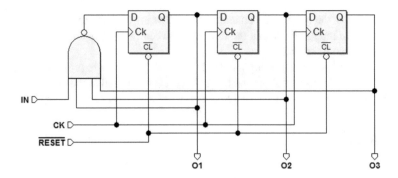

Network 4

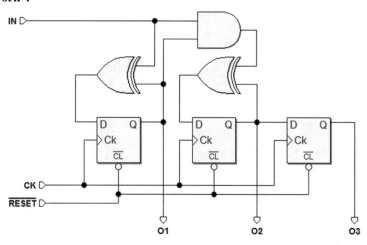

Network 5

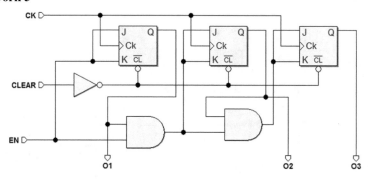

Network 6

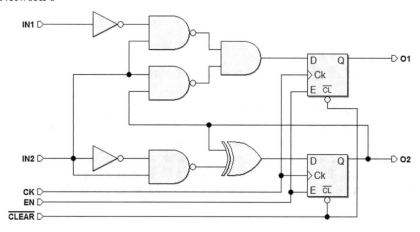

Network 7

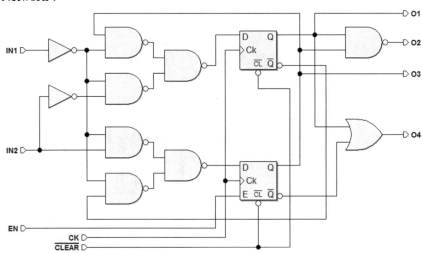

Network 8

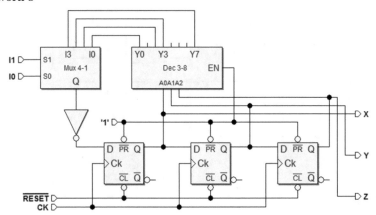

7.6.2 FSM Design Based on Textual Specifications

For each of the FSM design exercises below, you must draw the ASM diagram based on given specifications and complete the timing diagram indicating the state the FSM is in at every clock cycle. No synthesis is requested.

For each exercise, the *Deeds* Web site offers: an ASM diagram to be completed, a PDF file with the timing diagram template to fill in with pen and paper without using the simulator, and a schematic in which to insert your FSM into test its timing behavior.

Exercise 1

Using the ASM method, design a synchronous FSM with a synchronous input *DIR* and two outputs *Q1* and *Q0*. In binary, outputs *Q1 (MSB)* and *Q0 (LSB)* represent the decimal numbers 0–3. The device behaves like a non-cyclical binary up/down counter. Input DIR determines the count direction ($DIR = 1$, up count; $DIR = 0$, down count). Once it gets to the maximum or minimum, the device stops there. It can restart in the opposite direction by inverting the value of *DIR*.

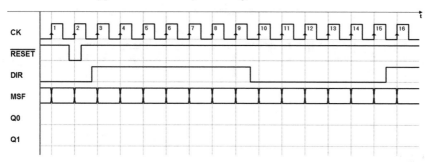

Exercise 2

Using the ASM method, design a synchronous FSM with an input *X*, synchronous with clock signal *CK*, and two outputs *D0* and *D1* that represent in binary code the number of 1*s* read at the input in the last three clock cycles.

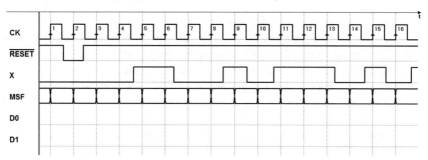

Exercise 3

Using the ASM method, design a synchronous FSM that can measure the delay between two signals *S* and *W* at the inputs according to the following specifications:

1. It has a master clock *CK* and receives in the input two *symmetrical square wave* signals *S* and *W* whose period is eight times that of the clock.
2. The level changes of inputs *S* and *W* are synchronous but the two signals can be delayed by 0, 1, 2 or 3 clock cycles from each other.
3. Four outputs *R0*, *R1*, *R2*, and *R3* show the delay amount of *W* versus *S*.
4. A synchronous *GO* signal with a duration of just one clock cycle commands the measurement sequence.

Keep in mind that the machine remains inactive until the *GO* signal arrives and that *S* and *W* can be kept valid only from the *GO* cycle (included); i.e., we can assume nothing about their previous timing position both with respect to the *GO* signal, and between themselves.

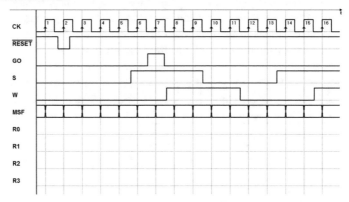

Exercise 4

Using the ASM method, design an FSM that receives a serial sequence of binary data *LIN* synchronous with the clock *CK* of the FSM itself. The machine has the outputs: *DAT* (data), *STR* (strobe), *ERR* (error), and *EOT* (end of transmission). It also has a synchronous input *RST* (restart). The *LIN* sequence is interpreted according to the following rules:

- A constant low level means absence of transmission.
- Two consecutive high levels indicate the presence of a data bit in the third position, which is copied onto output *DAT* contemporaneously with the activation of the *STR* signal (they need not be kept valid for more than one *CK* cycle).
- The level is always low after the data bit. If there is no low level, the FSM activates *ERR* and waits for the user to give an *RST*. The active-high *RST* will be considered only in the absence of transmission.
- An isolated high level marks the end of transmission, that the FSM signals by activating *EOT* and keeping it active until it receives another sequence.

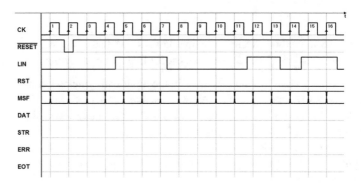

Exercise 5

Using the ASM method, design a synchronous FSM with an input X, synchronous with clock signal CK, and three outputs $S0$, $S1$, and WT (wait).

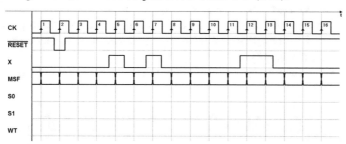

Input X receives sequences made of three bits, with the first one always at 1. The two sequences are always separated by at least three bits at 0. The two outputs $S0$, $S1$ report which sequence was received last, with a code chosen by the designer. That code must stay until the first bit of a new sequence is received. Output WT activates when the two outputs $S0$ and $S1$ are invalid.

Exercise 6

Using the ASM method, design a synchronous FSM that meets the following specifications:

1. It has three synchronous inputs coming from three push-buttons RDY (ready), GO, and STP (stop).
2. It generates two outputs RID (ride) and ERR (error).
3. Output RID is activated when the correct sequence of input signals is received, i.e., when the RDY push-button is pressed and then released and, after, the GO push-button is pressed. Output RID activates as soon as GO is activated, without waiting for the active edge of the clock. It is deactivated later when the STP push-button is pressed.
4. Output ERR is activated if the GO push-button is pressed before or at the same time as the RDY push-button. The ERR output will also be deactivated when the $STOP$ push-button is pressed.

Each output remains active for at least two clock cycles, regardless of whether the STP push-button is activated.

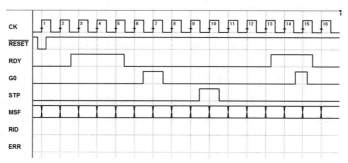

Exercise 7

Using the ASM method, design a synchronous FSM with a synchronous input *IN* and an output *OUT*. The input, which is normally at zero can be at 1 for intervals of 1, 2, or 3 consecutive clock *CK* cycles.

The output takes on the value 1 and keeps it for three clock periods if *IN* stays at 1 for one period, for two periods if *IN* stays at 1 for two periods, for one period if *IN* stays at 1 for three periods.

Assume that after the generation of the output is completed, input *IN* stays inactive for a few clock cycles.

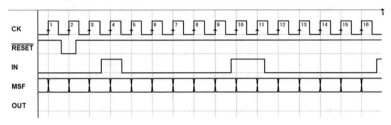

Exercise 8

Using the ASM method, design an FSM with:

1. A synchronous input *COM*;
2. Two groups of two inputs *R1*, *R0* and *D1*, *D0*;
3. An output *DPU*.

The device generates an aperiodic pulse *DPU*, whose duration and delay respect to the command signal are separately controllable by steps of the clock period. At every rising edge of COM, the system waits for a time of *r* clock cycles and then generates a pulse with duration of *d* clock cycles.

Inputs *R1* and *R0* define the extent of delay *r* (= 0, 1 or 2), while *D1* and *D0* define the pulse duration *d* (= 1, 2 or 3). Thus, the delay can be null and the pulse must always last at least one clock cycle.

The system only generates a new output sequence after the previous one is terminated.

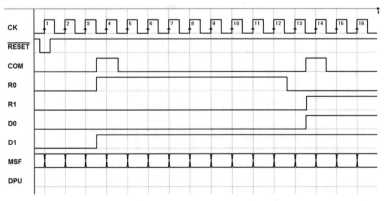

7.7 Solutions

7.7.1 Sequential Network Analysis in Terms of FSMs

The files of the FSMs represented here can be downloaded from the *Deeds* Web site on the *digital contents* pages of the book.

Network 1:   **Network 2:**

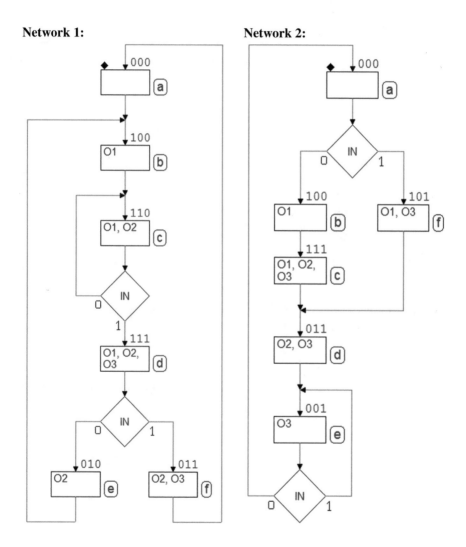

Network 3:   **Network 4:**

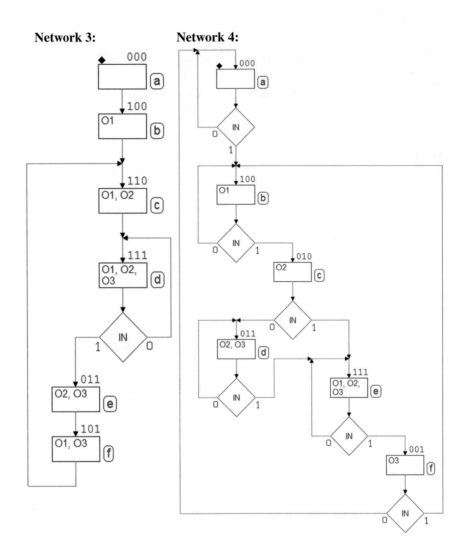

Network 5:

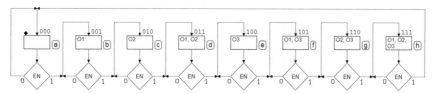

Network 6:

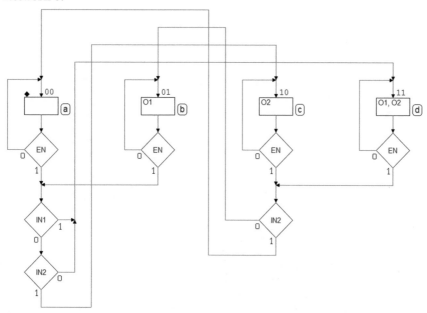

Network 7:

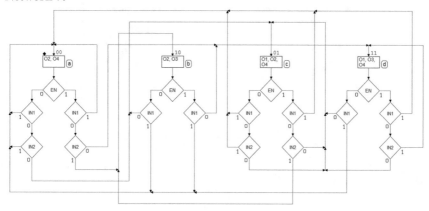

Network 8:

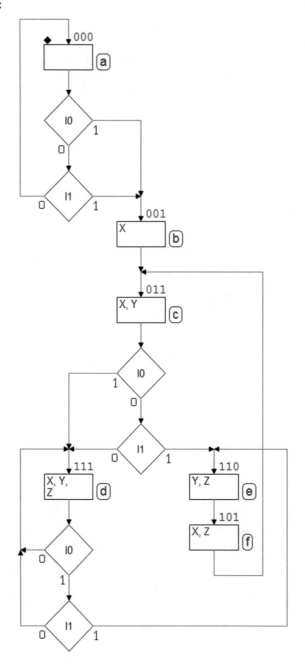

7.7.2 FSM Design Based on Textual Specifications

The solutions for each of the FSMs are also listed on the *Deeds* Web site as well as a downloadable circuit file, which can be used to check network behavior through a timing simulation.

Solution to Exercise 1:

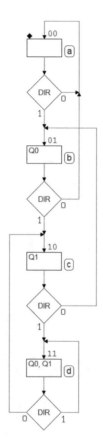

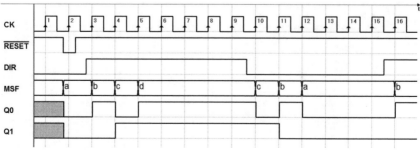

Solution to Exercise 2:

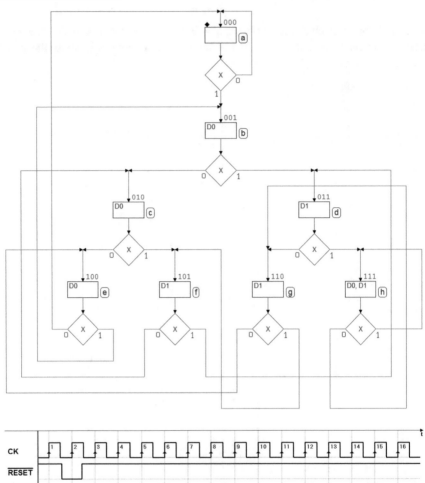

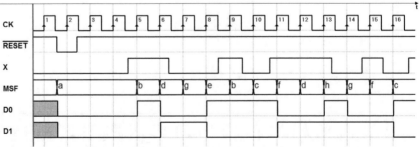

Solution to Exercise 3:

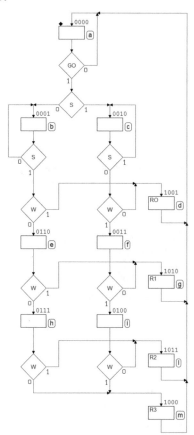

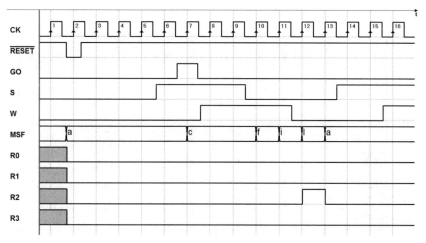

Solution to Exercise 4:

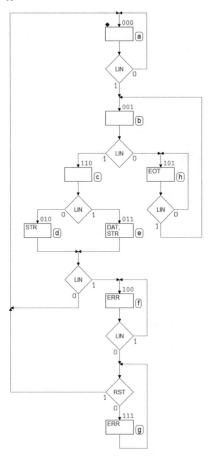

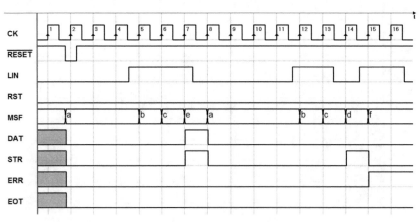

Solution to Exercise 5:

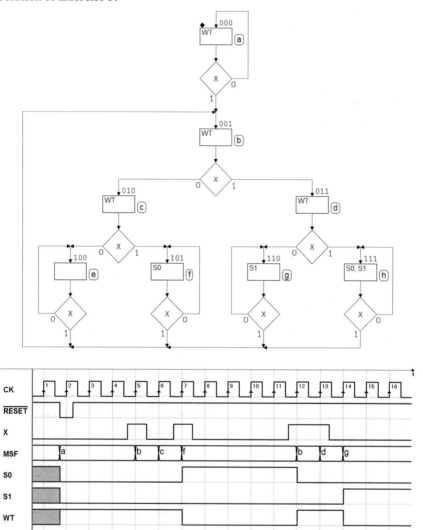

Solution to Exercise 6:

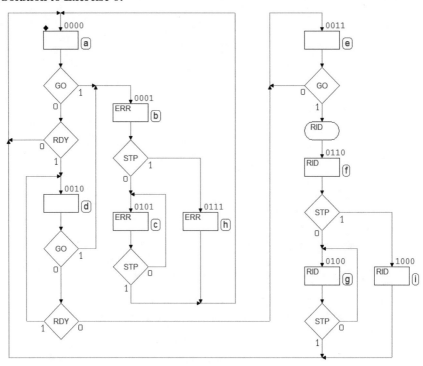

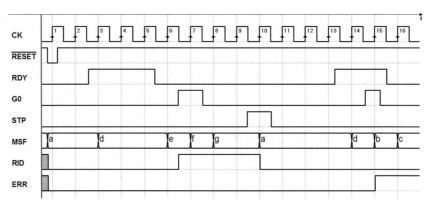

Solution to Exercise 7:

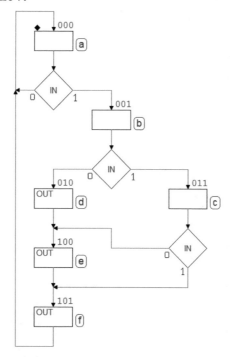

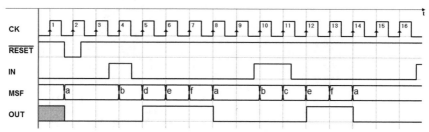

Solution to Exercise 8:

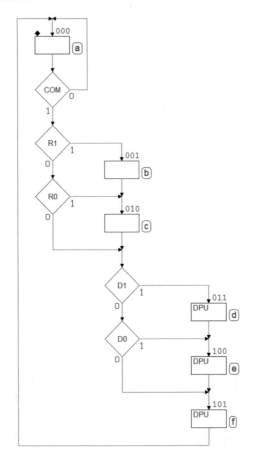

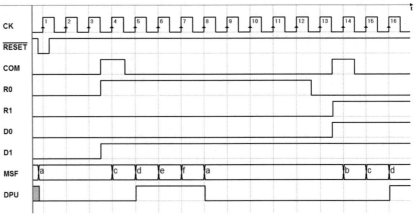

Chapter 8
The Finite State Machine as System Controller

Abstract The Finite State Machine can implement any algorithm, but it becomes over complicated when dealing with data. It is therefore convenient to include in digital systems other components that are more efficient to process and memorize data under the machines control. Such systems are called controller and datapath and described in this chapter. They optimize the sharing of duties between Finite State Machine and an external architecture. They are the first choice for the design of a wide variety of systems.

This chapter presents a collection of design examples of *synchronous digital systems* using an FSM as *system controller*.

8.1 Digital Systems

In previous chapters, we have seen that the FSM can model and implement various algorithms. A FSM describes a system in terms of the evolution of states and outputs as functions of inputs. From a theoretical perspective, the FSM can represent any discrete system that evolves over time moving from one state to another and has a finite number of inputs, outputs, and states.

Still, describing and designing a system in terms of *states*, while very general and versatile, can become overly complex and impracticable when the number of states becomes very large.

As we have seen with the shift register, the number of states in data management systems depends on all the possible configurations that the data can take on. For example, an 8-bit shift register will be described by an algorithm with at least 256 states. To describe a microprocessor in terms of states, one needs to only observe that a single 32-bit register (a microprocessor could contain many of them) would require 2^{32} states.

Regular digital structures for data management are available as blocks. They can be combinational (e.g., multiplexer, demultiplexer, encoders, decoders, and arithmetic circuits) or sequential (registers, counters, memories, and many others). Although the FSM can do arithmetic and logical operations, they can be done more effectively by dedicated devices.

© Springer International Publishing AG, part of Springer Nature 2019
G. Donzellini et al., *Introduction to Digital Systems Design*,
https://doi.org/10.1007/978-3-319-92804-3_8

A digital system can be optimized by dividing its tasks between a *"controller"* and *"dedicated components."*

The *controller* that we will design as FSM will manage the functioning of the system, while the *dedicated components* define its *"architecture,"* which interacts directly with the data.

The system's *controller* is sometimes called *"sequencer"* or *"timing generator."* The architecture is often called *"datapath."* Hereinafter, we will use the terms *controller* and *datapath*.

8.2   Open Control Systems

In the simplest version of this system, the *controller* gives commands to the *datapath* without receiving *any feedback*, as we can see in the figure below. The system's outputs can be generated by the *controller* or the *datapath*. Inputs can be dealt with by both subsystems.

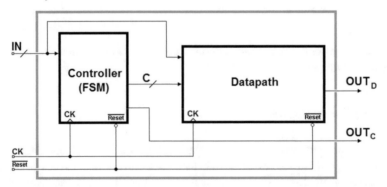

In the following, we will look at an example of this structure: a *2-bit serial receiver*, a version of the one based on the FSM alone that we have seen previously on p. 303.

The introduction of the *datapath* will simplify the FSM and make the system structure more easily *scalable*. In fact, in order to manage a serial signal containing *8 data bits*, a system based only on FSM would need at least *256 states*. Adding a *datapath* made up of as many flip-flops as serial bits would make the FSM only moderately more complex, when increasing the number of bits. In sum, the second approach raises the *complexity of the design linearly* along with the number of bits to manage, while the first approach raises it *exponentially*.

In further implementations of the serial receiver, we will show how an adequate *datapath* will make it possible to design a controller whose complexity is *independent* of the number of bits.

8.2.1 2-Bit Serial Receiver

Let's briefly summarize the specifications. We must design a *2-bit synchronous serial receiver*. The device must receive serial sequences on line *LN* and generate outputs *Q0, Q1, OK*, and *ERR*, as shown in the figure below. There, we reference the previous version with the FSM only. Note that in this figure the network's inputs and outputs appear as *active components*, as represented by *Deeds* during the simulation *by animation*.

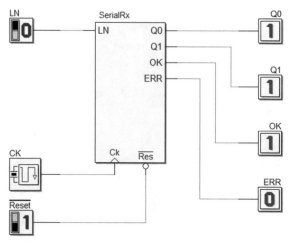

The format of the sequence received is summarized here verbally and also in the timing diagram below:

- The *bit time* is the clock period *CK*.
- The sequence begins with a *start bit* at 1.
- The sequence continues with the *two* data bits *D0* and *D1* (in order).
- The sequence ends with a *stop bit* at 0.

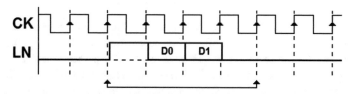

Notice that signal *LN* is synchronous with the clock *CK*, which is why we represent it with an approximate propagation delay after the clock rising edge. The system performs the following operations:

- Upon receiving a sequence, it makes the two data bits *D0* and *D1* available on outputs *Q0* and *Q1*.
- It checks that the *stop bit* is correctly at 0, and if so, it activates output *OK* and maintains it active until the next sequence arrives. If the stop bit is at 1, the receiver

only activates output *ERR* (Error) for the duration of one clock cycle and awaits a new sequence.

The figure shows the ASM of the previous version based on the FSM alone:

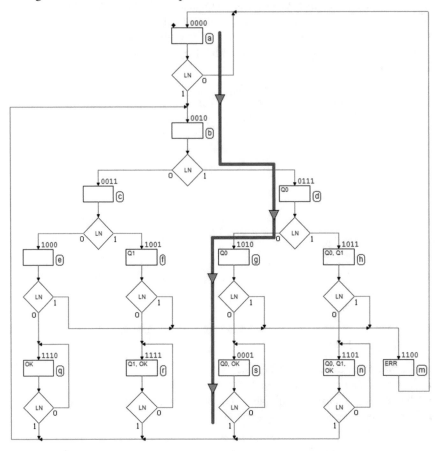

The path highlighted in red reminds us that in order for the FSM to *remember* which bits it received, it must *separate the paths*, since it cannot memorize the data in any other way. In the example, the FSM received a sequence where $D0 = 1$ and $D1 = 0$, followed by a regular stop bit. To keep the outputs active (here $Q0 = 1$ and $Q1 = 0$) until the next sequence, there must be a loop around the state that activates the outputs.

This requires a number of states that depend on all the possible data configurations. It is clear that extending the number of bits raises the complexity of the FSM (13 states in this example, which go up to 25 if the serial signal contains three bits and to 49 with four bits; 97 with five bits, etc.).

Thus, using this design structure is impractical for real-life cases (a typical serial signal contains eight bits).

The situation changes radically when we introduce a *datapath* that can memorize data, consisting in two E-PET flip-flops.

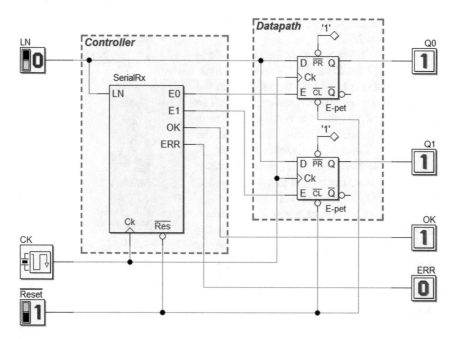

Notice the initialization input $\overline{Reset}$, which acts both on the *controller*, forcing the FSM into the *reset state*, and on the sequential components of the *datapath*. We will find this type of connection in all the upcoming designs excluding specific cases, which will be specially marked. Let's begin to draw the ASM diagram (see the figure below). The FSM waits for the *start bit* in state (a), as in the previous approach.

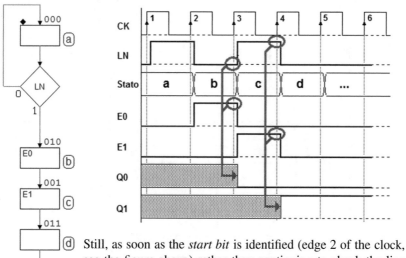

Still, as soon as the *start bit* is identified (edge 2 of the clock, see the figure above) rather than continuing to check the line cycle by cycle, this version *delegates* the job of memorizing the bits of the sequence to the flip-flops.

Input D of the flip-flops is actually connected directly to line LN.

The FSM only needs to *enable* the two flip-flops by activating lines $E0$ and $E1$ *at the right time*, that is in states (b) and (c), as suggested in the timing diagram seen in the previous figure. The highlighting shows that the value of the line is copied onto the corresponding flip-flop on the active edge of the clock in presence of enable $E0$ or $E1$. It is assumed that their previously memorized value is unknown and that the sequence received is $D0 = 0$ and $D1 = 1$.

In the clock cycle between edges 4 and 5, while the FSM is in state (d), the serial sequence presents the *stop bit*. Its value determines whether to generate OK (until the next sequence) or ERR for one cycle and then returning to await another sequence (see the complete ASM diagram on the right).

In the timing simulation below, the first sequence is correct, while the second presents an erroneous stop bit (due to line disturbances) causing ERR to be activated.

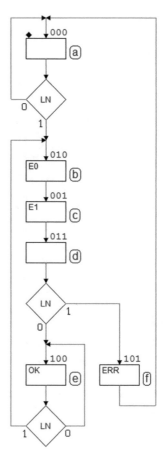

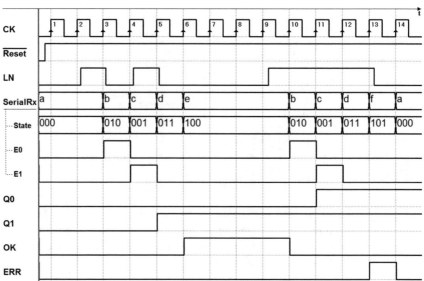

To extend the receiver to more bits, we only need to introduce more states like (b) and (c) into the ASM diagram and, obviously, change the *datapath* by adding the necessary flip-flops.

8.3 Feedback Control Systems

FSM generally needs to have *feedback* information from the *datapath*.

The *feedback* connection that carries this information as FSM inputs greatly increases the system's possibilities.

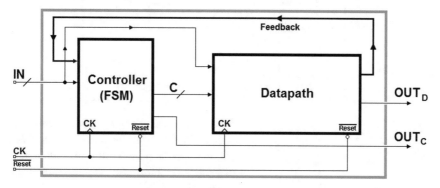

8.3.1 2-Bit Serial Receiver and Transmitter

This design is an example of a system in which the FSM acquires signals generated by the *datapath* as inputs and uses them to re-transmit the serial data in another format.

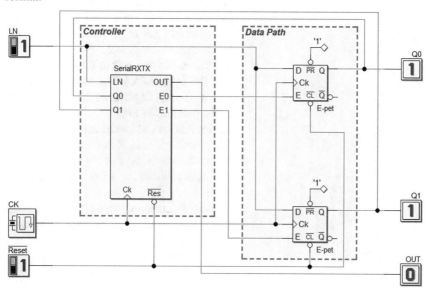

At first glance, the schematic resembles the previous example in that it uses two E-PET flip-flops to store serial data received through the *LN* line. This system, however, has an output *OUT* that re-transmits the serial signal in a different format.

The connection between flip-flops' outputs *Q0* and *Q1*, and the FSM inputs with the same name allows the FSM to know the values of the bits received, which is necessary for re-transmission.

In *idle state*, lines *LN* and *OUT* are at 0. The device waits for a bit packet on line *LN* in the known format (*start bit, D0, D1, stop bit*).

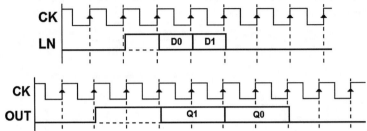

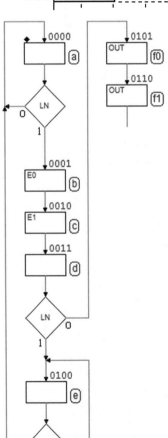

When it is done receiving, the device transmits a similar sequence on *OUT* (*start bit, Q1, Q0, stop bit*) but with a bit time that is two times (two clock cycles) the one of the sequence received.

As we see in the figure above, this operation halves the *bit rate* (the number of bits transmitted in a second).

The transmission on *OUT* should only occur if the sequence received on *LN* is correct (the *stop bit* must be at 0). If it is not, the device generates no output on *OUT*, but rather *checks LN* to go back to zero before awaiting a new sequence.

Let's draw the first part of the ASM diagram (see the figure aside). The receiver saves the data on the flip-flops, as we have already seen in the previous example.

Notice that the FSM waits in state (e) until the line goes back to 0, if the *stop bit* is wrong. If, however, it is right as expected, the FSM begins to generate the serial signal on output *OUT*.

The *OUT* activation in the consecutive states (f0) and (f1) means that the *start bit* is transmitted on *OUT*, with the duration of two clock cycles.

To generate the serial sequence as defined, the FSM needs to know the values of the serial data received, now contained in the flip-flops. Thus, the FSM re-reads outputs $Q1$ and $Q0$ in order and, based on their value, generates pairs of states with or without the activation of the output OUT.

Notice the inverted order of transmission as required (before $Q1$, then $Q0$).

Finally, the FSM generates the *stop bit* in state (st). Since state (a), which has no output, occurs after (st), there is no need to duplicate (st).

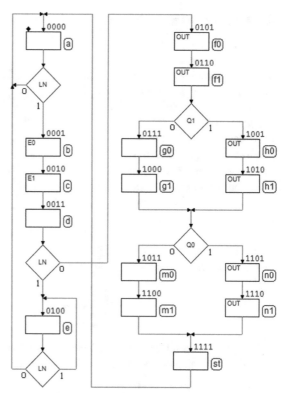

In the following timing diagram, the sequence received $1 - 0 - 1 - 0$ produces sequence $11 - 11 - 00 - 00$ in the output.

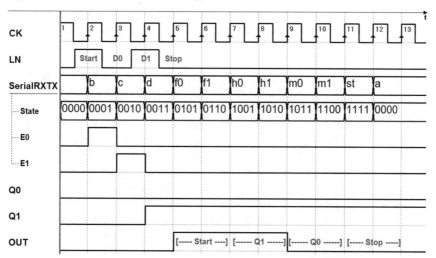

8.3.2   Pulse Generator

This design is another example of a system made in two different ways: using only the FSM and with the *controller - datapath* structure.

On detection of $0 \to 1$ transition on the input *TRG*, the *pulse generator* produces a *pulse* on output *OUT* (a high signal with a duration that is multiple of the clock cycle).

In this first version, we use only the FSM (see the figure aside).

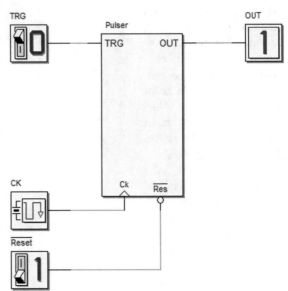

The pulse duration is fixed and lasts only six clock cycles, as seen in the timing track below, which shows an example of *TRG* command activation, too.

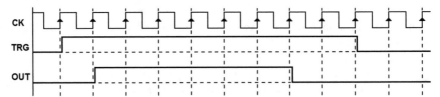

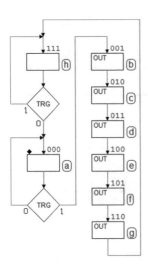

The ASM diagram does not pose any particular difficulty. In the *reset state* (a), the FSM waits for *TRG* to go to 1.

When this occurs, the FSM goes to state (b) and the following ones, generating $OUT = 1$. The pulse duration is equal to the number of consecutive states where *OUT* is active.

When pulse generation is finished, in (h) we wait for *TRG* to go back to 0.

The *wait state* (h) is necessary. If it were skipped, and if *TRG* were still at 1 when it went back to state (a), the FSM would immediately generate a new pulse.

The specifications, however, require this to happen on the *rising edge* of *TRG*.

Finally, this is the timing diagram produced by the simulator when input *TRG* duration is short.

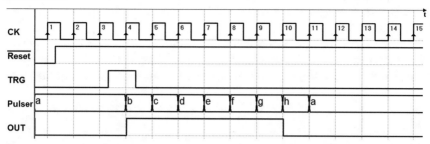

Below, with *TRG* very long, the FSM will wait until *TRG* it goes back to 0.

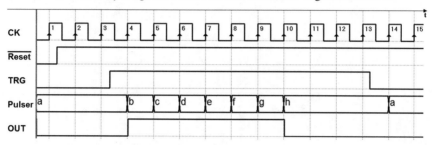

Let us now look at the second version based on the *controller–datapath* structure. We add to the system a "Cnt4" counter (see below).

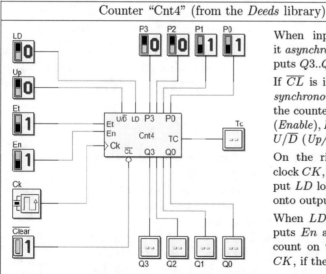

Counter "Cnt4" (from the *Deeds* library)

When input $\overline{CL}$ is at 0, it *asynchronously* forces outputs $Q3..Q0$ to zero.

If $\overline{CL}$ is inactive, the other *synchronous* inputs control the counter: LD (*Load*), En (*Enable*), Et (*EnableTc*) and $U/\overline{D}$ (*Up/Down*).

On the rising edge of the clock CK, the active-high input LD loads inputs $P3..P0$ onto outputs $Q3..Q0$.

When LD is not active, inputs En and Et enable the count on the rising edge of CK, if they are both at 1.

The count is *up* if $U/\overline{D} = 1$, otherwise it is *down*.
Output TC (*Terminal Count*) activates to 1 when $Et = 1$ and the counter's outputs reach the value 1111 (if the count is *up*) or the value 0000 (if the count is *down*). TC always equals 0 if $Et = 0$.

Before continuing, it can be useful to get familiar with the behavior of the counter "Cnt4" by simulating a test circuit as the one of previous figure, available also in the *digital contents* of the book. We suggest to check the operation modes of up/down count, enable, load, and clear.

As we can see in the figure below, the *datapath* is made up of a *counter* that keeps trace of the number of clock cycles where *OUT* must remain active. This extension allows us to implement more flexible specifications, like the ability to program pulse duration.

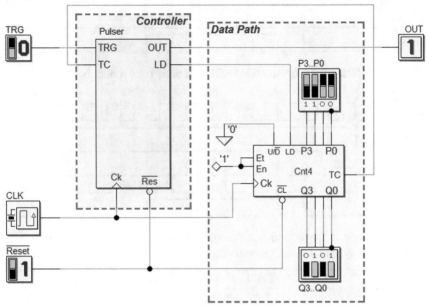

Counter is set to count *down* ($U/\overline{D}=0$) and always *enabled* ($EN=ET=1$). The FSM defines the start number of the count controlling the line *LD* and checks its end by reading *TC*. In the ASM diagram (see figure at right), we wait for command *TRG* in state (a).

In the waiting state (a) we activate *LD* to force the counter to load the number set on the switches (lines *P3..P0*).

Note that activating *LD* in (a) we freeze the count, thus reducing also the network's energy consumption (a real circuit uses a certain amount of energy each time lines change levels).

Exiting state (a), the counter will start to count. However, given that command *LD* is synchronous, the counter does not start yet in the transition between (a) and (b), since *LD* is still active on that rising edge of the clock.

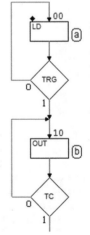

In the figure below, the red arrows indicate clock edges 2, 3, and 4. These are the instants when the number $P3..P0$ is loaded onto the counter, including edge 4, when the FSM moves to state (b). This means that the count will effectively begin only on edge 5.

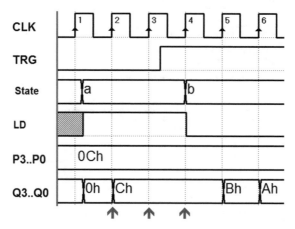

State (b) repeats until the counter signals the end of the count. At each edge of the clock, the counter decreases its value until it *gets to zero* and activates TC as in the timing diagram below:

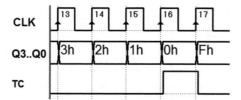

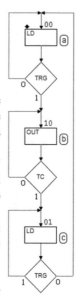

As a result, the FSM leaves state (b). Notice a detail that in this case poses no problem: edge 17 of the clock allows the FSM to leave state (b) but since the counter is enabled, the down count continues (from 0000 to 1111).

We finish the ASM diagram by adding state (c) where output OUT is no longer active and waits for TRG to go back to zero if, it is still at 1, as in the previous version of the generator. In state (c), we also activate LD to freeze the count.

Finally, we carry out the complete system simulation in *Deeds*. An analysis of its behavior (see the timing diagram on the next page) will allow us to answer detailed questions such as the following:

How many times state (b) has been repeated?

How many clock cycles have occurred since the counter was loaded?

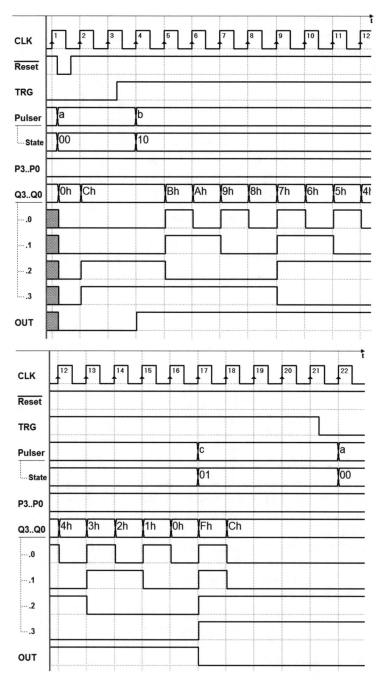

As the diagram shows, OUT lasts 13 clock cycles, in response to a 12 set on counter inputs $P3..P0$ (the 0 is counted as well).

Finally, notice that the system can be easily modified to generate a longer pulse by simply using a counter with the right number of bits, without modifing the FSM. The system shown below uses a 16-bit counter (the "Cnt16" from the *Deeds* library):

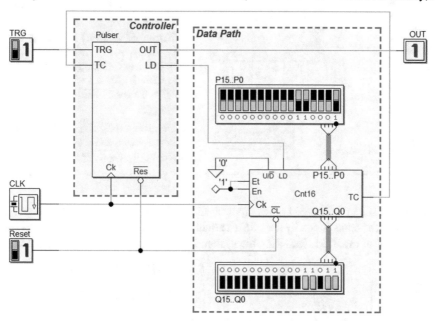

This FSM is exactly the same, but the new circuit can generate pulses with durations of up to $2^{16} = 65536$ clock cycles.

8.3.3 8-Bit Serial Receiver

The two projects here introduce two variations of the serial receiver we have seen before: to memorize bits, they use a *shift register* rather than flip-flops. The serial sequence has the same protocol as what was used for the 2-data-bit receiver, except that it contains eight, from D0 to D7. The specifications remain the same regarding the activation of *OK* or *ERR* when the sequence has been received.

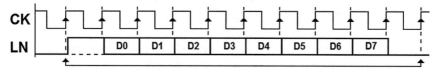

Let's now look at the first of these two new versions of the receiver. The *datapath* is made up of only a shift register, the "SiPo8" component of the *Deeds* library (see the figure in the next page).

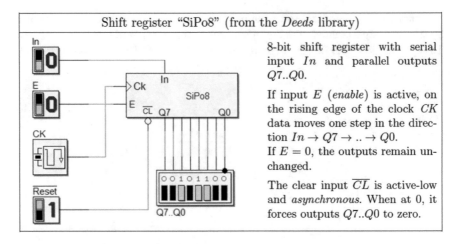

As we can see in the figure below, input *IN* of the *shift register* is directly driven by line *LN* with no mediation by the FSM. The shift enable *E* of the register is controlled by line *EN* of the FSM. Resetting the system initializes the FSM and the register.

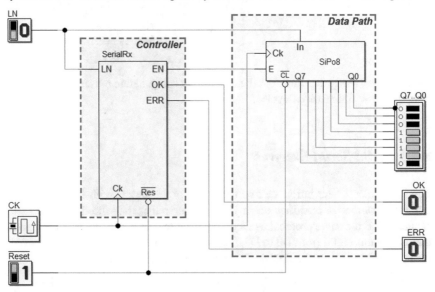

To correctly design the FSM, one must have a clear understanding of the timing of the signals in play. Before drawing the ASM diagram, it is often useful to make an outline of signal evolution over time. The figure below gives an example that assumes a serial sequence that transports bits $D0..D7 = 01001101$, in this order.

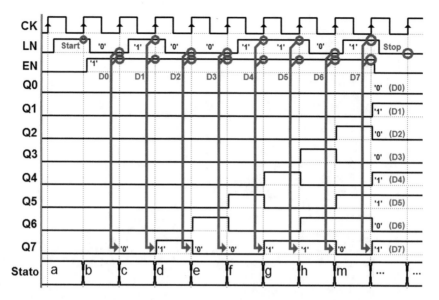

We must load the bits onto the shift register one by one.

As we see in the timing diagram above, we must activate *EN* as soon as the start bit is identified, and keep it active for eight clock cycles.

The bits enter into *Q*7 and they shift toward *Q*0, at every clock edge.

When we deactivate *EN* after the data bits on the line are terminated, all the received bits will be stored in the outputs *Q*0..*Q*7 of the register, available *in parallel*.

Thus, after waiting for the start bit in state (a), we will activate *EN* from the following state (b) (see the ASM diagram).

To keep it active for eight cycles, we must insert a total of eight states where *EN* is active (b)..(m).

Finally, we check the *stop bit* and generate *OK* or *ERR* according to specifications as in previous exercises, thus completing the ASM diagram (see the figure in the next page).

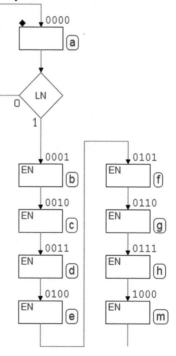

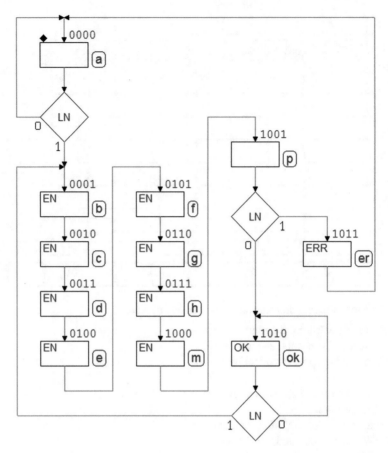

Now, with a complete timing simulation, we can verify that the system functions correctly.

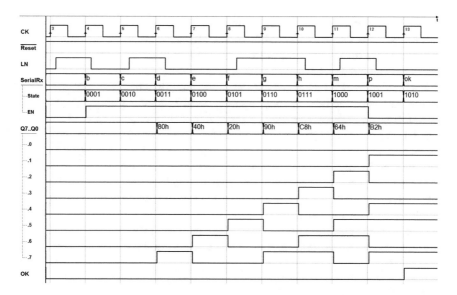

The version of receiver just examined uses an external device to memorize the serial data but still uses the FSM to count the *number of bits*. Obviously, it is possible to delegate this function to the *datapath* by using a counter set for the down count from 7 to 0.

The counter is a "Cnt4," used in a previous example (see the *Pulse Generator* on p. 372). The new system schematic now contains two elements in the *datapath*, a register and a counter, and optimizes task distribution between the *datapath* and the *controller*.

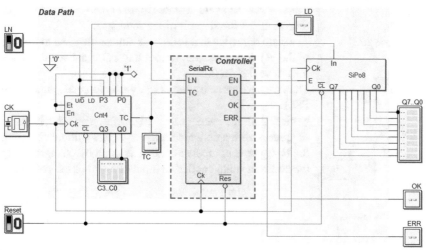

The beginning of the ASM diagram is the same as the previous version (see the figure at the right). In the idle state (a) the machine waits for the start bit in line *LN* and initializes and freeze the count through line *LD*.

When the start bit is detected, the FSM moves to state (b) where deactivates *LD* and activates register input *EN*, as in the previous case. Here, the difference is that it uses a single state rather than introducing as many states as there are bits to insert in the register.

The counter, in any case, is enabled ($En = Et = 1$). As soon as *LD* is deactivated, the count will decrease at each following rising edge of the clock, starting from 0111 (the value set on the inputs $P3..P0$, see the schematic).

In state (b) data bits are memorized one by one onto the register and counted by the counter, while the machine waits for *TC* to be activated.

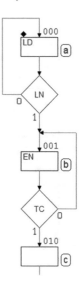

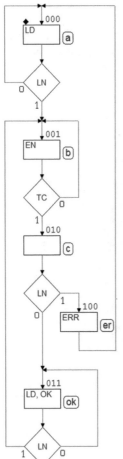

When the count gets to 0000, the counter activates *TC*, so the FSM leaves the cycle at state (b) and goes to (c) at the right time to check the stop bit of the sequence.

The state sequence that follows is almost identical to the one of the previous example, except that *LD* is activated in state (ok) as well as in (a).

Thus we have created a remarkably simple, easily analyzable FSM.

Notice that the *very same algorithm* can check serial sequences with a definable number of bits without changing the *datapath's* components. This is done by changing only the constant value applied to inputs $P3..P0$ of the counter.

The timing simulation of the complete design now allows us to check every aspect of the relation between the evolution of states and the behavior of the register and the counter (see the figure below). Specifically, we can observe the following details.

- The counter decrements from 7 to 0, starting from clock edge 5 (that is *at the end* of the cycle where the FSM is in state (b) for the first time).
- The machine stays in state (b) until *TC* is read at 1 (on edge 12).
- The counter, after getting to zero (0000), continues to count cyclically (1111, 1110...) until the FSM activates *LD* in state (ok), so it is re-loaded to 0111 on edge 14.

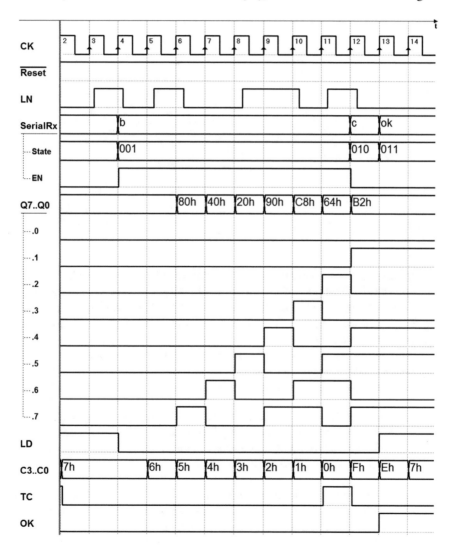

8.3.4   Light Dimmer

The system in the figure is a *light dimmer* that controls the intensity of a light bulb
by means of four switches (*ON*, *UP*, *DN* e *OFF*).

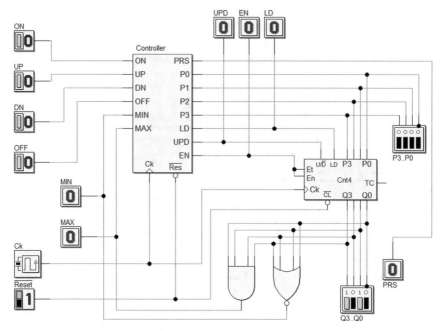

The circuit generates the output lines $Q3..Q0$ that encode 16 possible values of light
intensity, starting from 0000 (fully off) to 1111 (fully on). Note that the light bulb
will be driven by a device that converts the code value $Q3..Q0$ in an analog quantity,
which is fed to the light bulb. This device is not part of our design, and it is not
represented in the schematic.

The *switches* generate a *low* level in idle state and *high* while they are pressed. They
should be considered *ideal*, that is having no electromechanical contact *bouncing*.
Pressing and then *releasing* each switch determine the system's behavior, as follows.

> *ON*   → *Turns on* the lamp to the maximum($Q3..Q0 = 1111$).
> *UP*   → *Increases* light intensity by one unit.
> *DN*   → *Decreases* light intensity by one unit.
> *OFF* → *Turns off* the lamp($Q3..Q0 = 0000$).

Output *PRS* serves as a signaling device and is active while a switch is pressed.
The output number $Q3..Q0$ must not be incremented if it has reached the *maximum*
value or decremented if it has reached the *minimum* value.

This system uses the counter "Cnt4" more completely than in previous examples
(see *Pulse Generator* on p. 372 or *8-bit Serial Receiver* on p. 377), in that here it
must be *loaded*, *enabled*, and set for the *up or down* count.

The two logical gates have the counter outputs $Q3..Q0$ as inputs: the AND activates *MAX* when the count gets to the maximum, while the NOR activates *MIN* when it gets to the minimum.

As for reading the switches, they should be given the same considerations as in the *Push-button Handling* example on p. 306. The structure of the first part of the ASM diagram is the same, the simultaneous control of the pressure on the push-buttons and the wait states that follow.

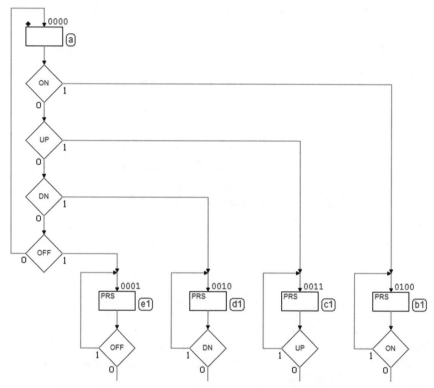

The diagram branches off into four separate paths corresponding to push-buttons *OFF*, *DN*, *UP*, and *ON*. Based on specifications, output *PRS* is active in states (e1), (d1), (c1), and (b1), where the FSM is when a push-button has been pressed and it is waiting for it to be released.

Let's fill in the four paths of the diagram keeping in mind how it must work when the push-buttons are released. For now, let's focus on the paths related to the release of *OFF* and *ON*. The figure below shows that in states (e2) and (b2), we send the counter the command to load the data present on counter inputs $P3..P0$ by activating *LD*.

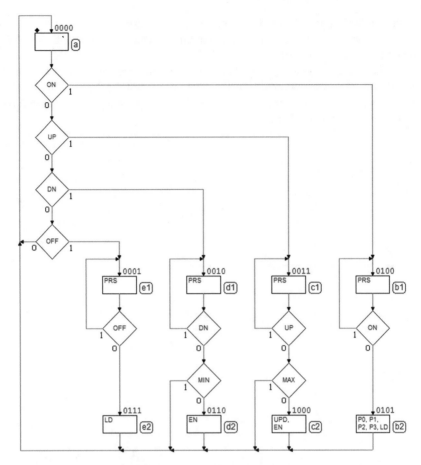

To achieve this, the FSM sets $P3..P0 = 0000$ in (e2) while it assigns $P3..P0 = 1111$ in (b2). So, in (e2), the counter will be loaded to the minimum value (light off) and in (b2) to the maximum (light fully on).

Notice that lines $P3..P0$ and LD are set at the same time in the same state. Given that LD is synchronous, the actual loading takes place at the edge of the clock that makes the FSM leave the state and return to (a).

Let's look at the diagram in relation to the decrement and increment of the number (push-buttons DN and UP). The decrement of the counter takes place in state (d2) and the increment in state (c2), after the MIN and MAX check, because the number must not be changed if it is respectively at the *minimum* or *maximum* value.

In these states, EN must be activated to enable the count and UPD to determine its direction. In state (d2), UPD is set at 0 (*to count down*), while in (c2) it is set at 1 (*to count up*).

Note that the count occurs on the positive edge of the clock; therefore, the increment or decrement will take place *on exit* from states (d2) and (c2) where it is enabled, at the same time when the machine goes back to (a).

Below a timing simulation of the system. Note the loading, increments, and decrements discussed above.

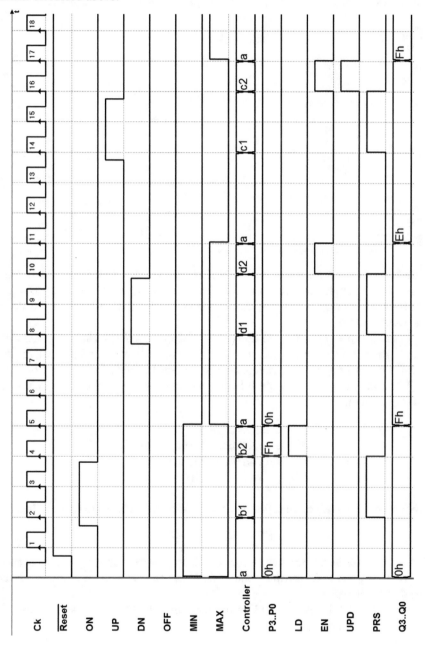

8.3.5   *Combination Lock*

This system commands the opening of a door with an electric lock through a *combination* the user inserts. The lock is commanded by output *OPN* and opens when the input push-buttons *PA*, *PB*, *PC*, and *PD* are pressed in a specific sequence.

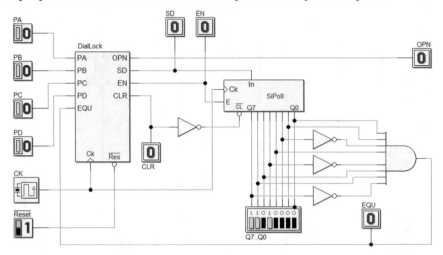

The *datapath* is made up of an "SiPo8" shift register (seen in *8-bit Serial Receiver* on p. 377) and a simple combinational logic that reads its outputs $Q7..Q0$ and activates the FSM's input *EQU* when they take on the value 01101011_2 (the internal code of the *combination*).

Register inputs *In* and *E* are driven by FSM outputs *SD* (*serial data*) and *EN*, respectively. Notice that, unlike the previous cases, $\overline{Reset}$ is applied only on the FSM, while the register's $\overline{CL}$ command is generated by the FSM's output *CLR*.

When the user *presses* the push-button and then *releases* it, the system inserts a 2-bit code in the shift register.

$$PA \ \rightarrow \ \text{Loads code } 00.$$
$$PB \ \rightarrow \ \text{Loads code } 01 \, (\text{first1, then0}).$$
$$PC \ \rightarrow \ \text{Loads code } 10 \, (\text{first0, then1}).$$
$$PD \ \rightarrow \ \text{Loads code } 11.$$

The sequence of push-buttons (the combination) we have chosen to open the lock is:

$$PD - PC - PC - PB \, .$$

When the data in the register is equal to that produced by this input sequence (01101011_2), the FSM activates output *OPN* for one clock cycle and then clears the data in the register.

Once the lock is open, it will be mechanically locked again when the door is closed. We assume that:

- Pressing a push-button activates the corresponding line to 1 at the FSM input, while releasing it puts it back to 0.
- Two push-buttons are never pressed at the same time.
- Enough time passes between releasing a push-button and pressing the next one so that the system concludes its operations.

As we can see in the figure below, the section of the ASM diagram that handles the push-buttons was previously introduced (as in *Push-buttons Handling* on P. 306 and in the *Light Dimmer* on p. 384).

Notice that here, when the system is reset, the FSM goes to state (a), which *precedes* the push-button control cycle. In (a), *CLR* is active and clears the register (for reasons that will become clear later).

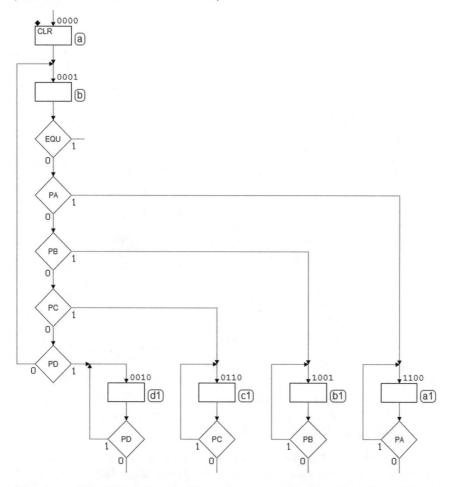

After the waiting for the release of the push-buttons, the states that follow will *load onto the register* the two bits corresponding to the pressed push-button.

The figure below shows an almost complete ASM diagram. States (d2) and (d3) have been inserted along the path related to the release of push-button *PD*. They load the pair of bits 11 onto the register by activating *EN* and *SD* for two clock cycles.

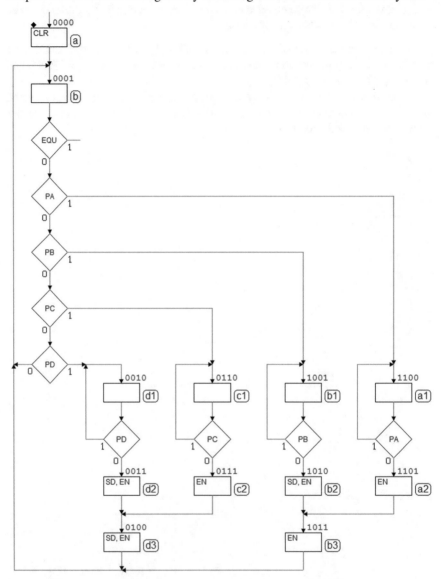

Likewise, we load a 0 onto the register in state (c2). Then, we take advantage of the existing state (d3) to load the next 1. We do the same in states (b2), (b3), and (a2). Here, we see another example of how we can use the same state many times along different paths.

Now, let's finish the diagram according to the specifications regarding opening the lock (see below).

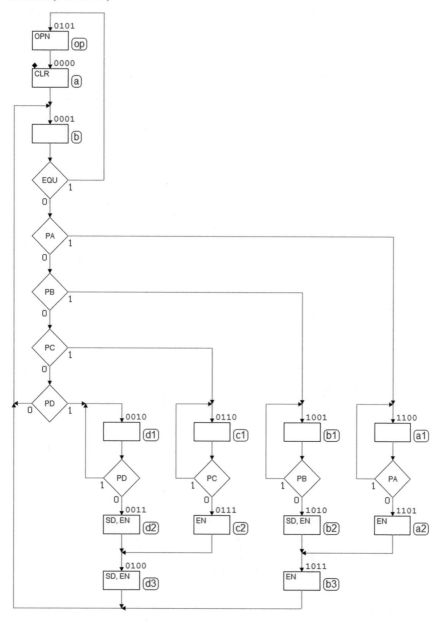

Notice that the check on *EQU* is inserted in the same cycle that checks the push-buttons; this is for simplicity's sake. When *EQU* is 1, the FSM moves to state (op) activating *OPN* for one cycle, as per specifications, and then goes to state (a), which clears the register. Without this step, once the right combination was set, *OPN* would stay active forever.

Finally, notice that (a) is also set as a *reset state*. This is because the register is not cleared directly by the system reset, as previously explained. Rather, it needs to be cleared before entering the loop waiting for the push-button.

The timing simulation below highlights the FSM outputs *SD* and *EN* in relation to the values loaded on the shift register.

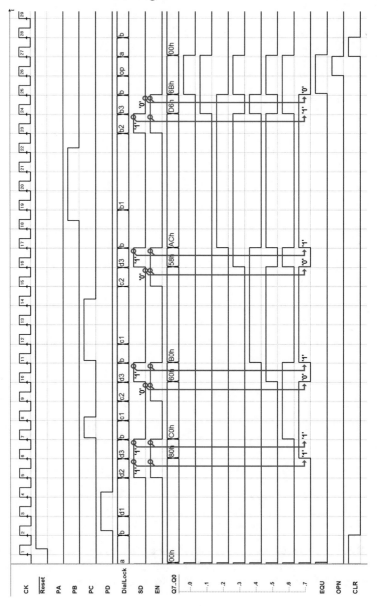

8.3.6 Automatic Drink Dispenser

We are designing a digital system to manage a (simplified) automatic drink dispenser. It is made up of a synchronous FSM plus a *datapath* made of a *4-bit arithmetic circuit* that includes an "Add4" *adder*, a "PiPo4" *parallel register*, and a "Cp4" *magnitude comparator*.

Both the FSM and the register are initialized by the system $\overline{Reset}$.

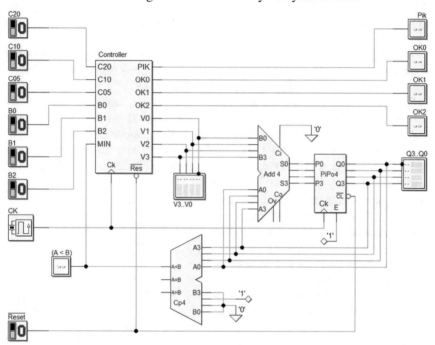

The dispenser offers *three types* of drinks that all cost *20 cents*. The dispenser waits for a user to insert the money to pay for the drink. Inputs $C05$, $C10$, and $C20$ show when a 5, 10, or 20 cent coin is inserted, respectively. When this happens, they produce a high *pulse* on the corresponding line for *one clock cycle* (when idle, the lines are low).

The system calculates the total value of the coins inserted and memorizes it in the "PiPo4" register. By convention, a unit of $Q3..Q0$ corresponds to *5 cents*: for example, $Q3..Q0 = 0100_2 = 4_{10}$ corresponds to $(4 \cdot 5) = 20$ cents.

The register's input is connected to the output of the "Add4" adder. One of the adder's inputs is connected to the register's output, while the other is directly provided by the FSM through lines $V3..V0$. As a result, instant by instant, the adder generates the sum of the number on the register and the number provided by the FSM.

The FSM adds the *value* of the *inserted coin* to the register, presenting it to the adder for one clock cycle. Given that the register's enable input E is always active, it will be necessary for the FSM to keep lines $V3..V0$ at zero for the rest of the time to keep the calculated total unchanged.

The comparator compares the total $Q3..Q0$ accumulated on the register with the price of the drink ($= 0100_2 = 4 = 20$ cents, established on inputs $B3..B0$ of the comparator). Of its three outputs $A < B, A = B, A > B$, only one $A < B$ is read by the FSM at input MIN ("*minor than*").

When *at least* 20 cents are inserted, the dispenser activates PIK to ask the user to choose the drink by pressing one of the three push-buttons $B0$, $B1$, or $B2$ (PIK lights up the push-buttons). Push-buttons $B0$, $B1$, and $B2$ are normally at the logical value 0 and go to 1 when they are pressed.

When one of the buttons is pressed, the FSM deactivates PIK and activates one of the outputs $OK0$, $OK1$, or $OK2$ (corresponding to the drink the user picks) for one clock cycle, thus commanding the dispensing of the drink.

Finally, the dispenser *collects* the cost of the drink but *credits* the next user for any amount inserted *over* the 20 cent cost of the drink. To do this, the FSM must subtract 20 cents from the value stored in the register, and does so by generating *two's complement* of the code of the value of the drink on lines $V3..V0$ for one clock cycle. After this operation, the register will contain the credit to be used by the next customer.

Further specification: assume all the input signals to be synchronous with the clock, that it is impossible to insert more than one coin at a time and that enough time passes between inserting one coin and the next that the system can function properly.

Let's start drawing the ASM diagram (see aside). We wait in (a) for a coin to be inserted.

Depending on the coin's value, it activates $V2$, $V1$ or $V0$ for one clock cycle. Thus, it provides the binary numbers 0100_2, 0010_2 or 0001_2 (corresponding to 20, 10 o 5 cents) to the adder.

In one of states (b), (c) or (d), the adder generates the sum of the number $V3..V0$ and the number on the register.

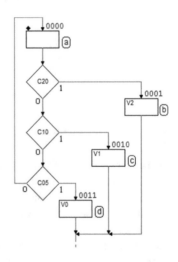

The register memorizes the sum on the *next positive edge* of the clock, so it is necessary to add to the ASM state (e) to wait for one clock cycle before checking MIN (see the complete ASM diagram below).

We go back to (a) if the money inserted is lower than the cost of the drink. When the sum reaches or exceeds the 20 cent cost, *MIN* goes to zero and the FSM to state (f) where it activates *PIK* to ask the user to choose a drink and waits for him/her to press one of the push-buttons *B0*, *B1* or *B2*.

Depending on which button is pressed, *OK2*, *OK1*, or *OK0* is activated for one clock cycle to command the corresponding drink to be dispensed.

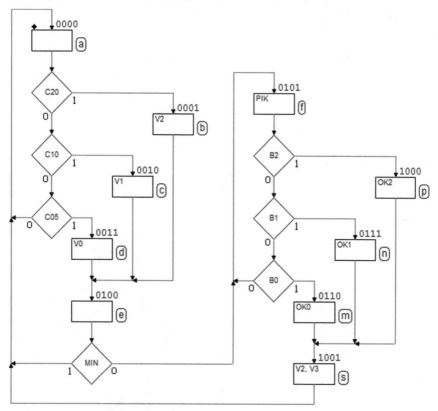

Finally, to close the diagram, we make it so that any residual credit is available for the next user. To subtract 20 cents from the register, the FSM sets $V3..V0 = 1100_2$ in state (s) (two's complement of the code for 20 cents).

The system's timing simulation in the next figure highlights the adding operation of the value set by the FSM on $V3.V0$ and the value on the register at that moment.

Notice the last addition where the cost of the drink is subtracted from the total (35 cents $= 7_{10} = 7_h = 0111_2$) by adding two's complement to it (-20 cents $= -4_{10} = C_h = 1100_2$).

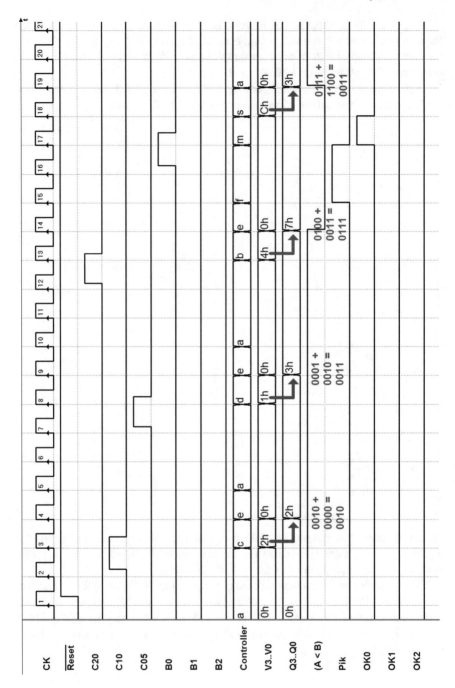

8.3.7 *Programmable Square Wave Generator*

We anticipate that here the *network structure* will not be provided as in previous examples but must be designed according to specifications.

We are designing a synchronous (*controller–datapath*) digital system that generates a two-level periodic signal (hereinafter called *"square wave"*). The system must have a group of eight input lines, which we will call *TH*, a second group of input lines *TL* and a single output line *SW*. Upon activation, the system generates a square wave on *SW* without waiting for any command. The high part of the square wave lasts for as many clock cycles *CK* as those set by *(TH + 1)*; the low part has the same duration as the number set by *(TL + 1)*. The generation of *SW* begins with the high part of the waveform.

The steps to take to carry out a design of this type are:

(a) Define the system as a functional block, highlighting inputs and outputs.
(b) Choose the components to use in the datapath.
(c) Design the datapath with the connections between the controller and the datapath components and the inputs and outputs of the whole system.
(d) Draw the ASM diagram of the FSM that describes the controller.
(e) Perform a timing simulation of the whole system by choosing input sequences that help demonstrate the system's functionality set.

Step (a)
It is fundamental in this phase to correctly identify the system's inputs and outputs based on the given specifications. Here is a summary of what was defined in the text:

1. *Inputs*:

   (a) *TL* (8 bit) sets the duration of the low part of the periodic signal in pure binary.
   (b) *TH* (8 bit) sets the duration of the high part.

2. *Outputs*:

   (a) *SW* (1 bit) periodically commutes between the logical value 0 and 1.

The system, represented as a single functional block, is as follows:

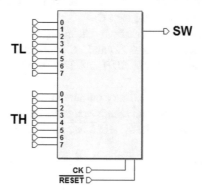

Step (b)

To choose the datapath components, we must know which functions are required to create the system. Remember that the controller must be used as such, not as a register, a counter, a comparator or to memorize and/or process numerical data sets, for example.

To create the final, complete system, the controller must be the *"director"* of combinational, arithmetic, or memorization components. In this example, the system must acquire *TL* and *TH* and use them to measure the number of clock cycles where *SW* must remain low or high.

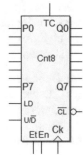

The best component for this job is an *8-bit universal counter* like the "Cnt8" from the *Deeds* library. We have already encountered a smaller version of it (for example in the *Pulse Generator* on p. 372).

We can pre-load it with a number and then, with the *count down* enabled, we can wait for *TC* (*Terminal Count*).

So, according to the specifications, the components needed are two universal counters of this type, one for *TL* and one for *TH*.

Step (c)

This step is deeply connected to step (b). As explained above, two counters enabled for counting down are needed. They will be loaded at the right time by the controller (with the values of *TL* and *TH*). The *TCs* will be evaluated by the controller to check if the set time has passed.

Now, it is important to know how to connect the two counters' inputs and outputs to the controller and the system. Specifically:

- The inputs *P7..P0* of the two counters must be connected to the *TL* and *TH* input groups, as explained before.
- The counters must count down so let's set input $U/D = 0$.
- The input *LD* of the counters must be commanded by the controller so that they are loaded at the right time. It is possible to keep Et and En active all the time so that the counters work always and *LD* is the only necessary control. Then, the controller will be simply waiting for *TC*.
- The output *SW* can be driven directly by the controller.
- The schematic includes signals as *LD* and *TC* of the two counters, *LDL* and *TCL* for the counter connected to *TL*, *LDH* and *TCH* for the counter of *TH* and the output *SW*.
- It is convenient to add a few auxiliary outputs into the system schematic, just to visualize the *relevant signals* and *counter outputs*, during the simulation.
- Notice that the clock *CK* and the system $\overline{Reset}$ must be shared by the counters and the controller.

After these evaluations, the final schematic is shown here:

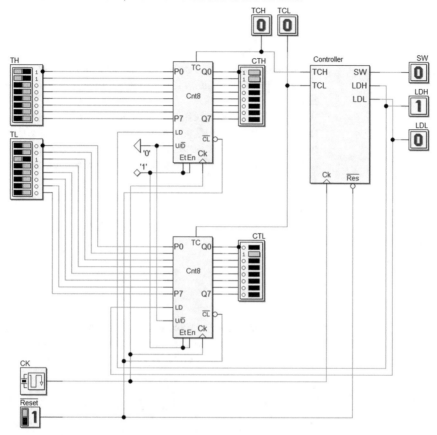

Step (d)
Once points (a), (b), and (c) are defined, point (d) should not pose particular difficulties in that the controller should implement the logic specified in describing the datapath design.

Notice that when the system's $\overline{Reset}$ is activated, the two counters go to zero. When $\overline{Reset}$ is released, if *LDH* and *LDL* are not active, the counters will start to count. It can be useful to activate them in the *reset state* (rs), as shown at the right.

In this way, we start pre-loading the counters with the *TL* and *TH* values. To satisfy the specifications, the system needs to start generating $SW = 1$, so we activate also *SW* in the reset state.

We no longer activate *LDH* in the next state (w1) in order to leave the counter free to count, and we wait for its terminal count *TCH*.

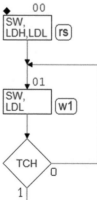

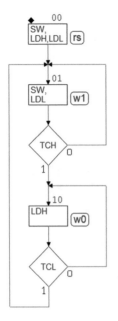

This way, we keep output SW at 1 for $(TH + 1)$ clock cycles (+1 because we count also the zero).

To be ready to start the count for the low part of SW, we maintain pre-loaded in (w1) the related counter, activating LDL.

When TCH signals that count has ended, it is time to change the output SW value to 0, so we introduce another state (w0). In this new state, the counter of the low part is free to count (LDL is 0) and the FSM stands by for the terminal count TCL.

Then, we finalize the ASM diagram closing the loop to state (w1), on the activation of TCL.

The output sequence on SW will be repeated indefinitely.

Step (e)

To test the system, we define a timing trace that includes the activation of reset and the setting of TL and TH (in the example, they are set to 3_{10} and 2_{10}, respectively). The output square wave on SW is low for four clock cycles and high for three.

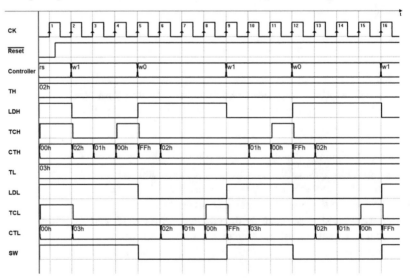

As we can see in the simulation, the first part of the signal on output SW is wrong (it is kept high for a longer time than needed). Starting from the subsequent cycles, however, the system produces the square wave that was set by the inputs TH and TL.

8.3.8   Christmas Light Systems

Design a system for Christmas lights that controls three strips of red, green, and blue lights. Each strip should light up in sequence for a time controllable by a 4-bit number. The *network structure* must be designed according to the specifications above.

Here too, we should read the project specifications carefully and follow the same five steps (a), (b), (c), (d), and (e) suggested on p. 397.

Step (a)

In step (a), we must define clearly inputs and outputs of the system. Reading the specifications above, we understand that the machine must have 12 inputs and three outputs that we identify as follows.

1. *Inputs*:

   (a) *DR*, four bits, lighting time of the red light (in binary);
   (b) *DB*, four bits, lighting time of the blue light;
   (c) *DG*, four bits, lighting time of the green light.

2. *Outputs*:

   (a) *R* (one bit), red light control (on for *DR* clock cycles);
   (b) *B* (one bit), blue light control (on for *DB* clock cycles);
   (c) *G* (one bit), green light control (on for *DG* clock cycles).

The system, as a single functional block, will appear as follows:

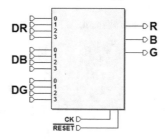

Step (b)

The system acquires *DR*, *DB*, and *DG* to use them to measure the number of clock cycles the output signals *R*, *B*, *G* should be kept active for. We could use three counters, as in the previous example.

Another possible architecture is to use a single 4-bit counter that will be pre-loaded to the correct value by multiplexers connected to system inputs *DR*, *DB* and *DG*.

We need a counter "Cnt4" (see on p. 372) and four multiplexer "Mux4-1" (p. 51), shown on the right.

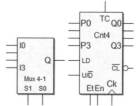

Step (c)

Following the previous analysis, we should decide how to connect the elements of the system:

- The counter's inputs $P3..P0$ must be connected to the three groups of inputs DR, DB, and DG by the multiplexers.
- We use four multiplexers, one for each bit of DR, DB, and DG, as described in the following:

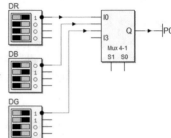

The bits in position 0 of DR, DB and DG will be connected to the inputs $I0$, $I1$ and $I2$ of the first multiplexer, respectively, as shown in the figure. The same will be done for the other bits, through the other multiplexers. This way, the outputs Q of the four multiplexers will copy DR, DB or DG, based on how their selection inputs $S1$, $S0$ are set (at 00, 01 and 10, respectively).

- The counter must count down, so U/D is set at 0. We set also $Et = En = 1$, so that we control the counter using the input LD only.
- Let's take DR as an example: the controller sets the selection lines $S1$, $S0$ at 00 and then activates LD. The counter generates TC after $DR + 1$ clock cycles (since the count also includes zero).
- The controller generates directly the outputs R, B, and G.
- Clock CK and $\overline{Reset}$ are common to the counter and the controller, as discussed in the previous example.

Below is the final schematic of the system.

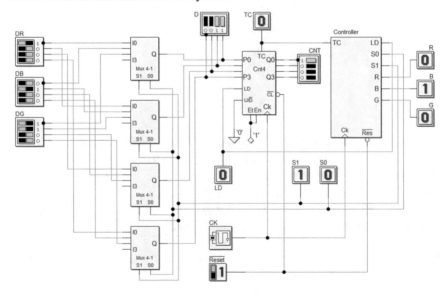

Step (d)

As we can see in the figure, in the *reset state* (rs), we activate only the counter's input *LD*. This means in the same state (rs) we are setting $S1, S0 = 00$ on the multiplexers, so that *DR* is routed to the counter pre-load inputs *P3..P0*.

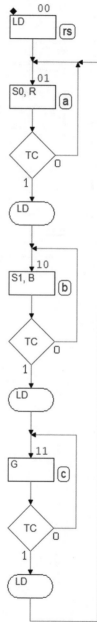

Note that the *reset state* (rs) is not included in the main algorithm loop (it is used only to initialize the counter when the FSM starts working).

On entering the following state (a), the counter will be loaded with the number on *DR* (the lighting time of the red light). In state (a) we activate output *R*, and wait for *TC* from the counter.

The counter's input *LD* is activated in state (a) only when the counter reaches zero. When this happens, on entering the following state (b) the counter is loaded again, but this time with the number *DB*. In fact, in state (a) we have prepared $S1S0 = 01$ (declaring only *S0* in the block), to route the multiplexers accordingly.

State (b) is similar in its operation to state (a), but now *B* is active and we set the multiplexers to present *DG* to the counter's pre-load inputs.

In the same way, in state (c) we activate *G* and set the multiplexers to re-load again the counter with the number on *DR*. The FSM loops back to state (a) when the lighting time of the green light has elapsed.

Step (e)

To simulate the complete system (see the timing diagram on the next page), we set inputs $DR = 1, DB = 2$ and $DG = 3$ in addition to the usual initial reset sequence. Values have been chosen to obtain a reasonably short graphic.

Notice how the outputs related to lights *R*, *B* and *G* activate in a cyclical sequence for 2, 3 and 4 clock cycles, respectively.

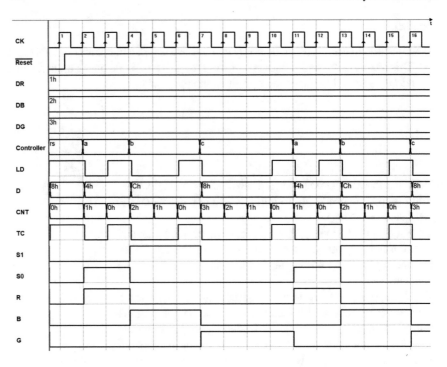

8.4 Design Exercises

8.4.1 Design of the Controller of a Given Datapath

For each of the following exercises of digital systems design, the schematic of a complete *controller–datapath* architecture is supplied.

You should define the controller's ASM diagram according to the specifications and complete the timing diagram (on the opposite page). Remember to indicate the state of the FSM at each clock cycle. The synthesis of the FSM is not requested.

For each exercise, the Web site offers:

(a) A trace of the FSM to design where state variables, inputs, and outputs are pre-defined.
(b) A PDF file with a suggested timing diagram template, to fill in on paper without using the simulator.
(c) The network schematic where you can insert your FSM to check its behavior through the timing simulation.

Exercise 1 The system shown in the figure below contains the controller, a "Cnt8" counter and two logical gates.

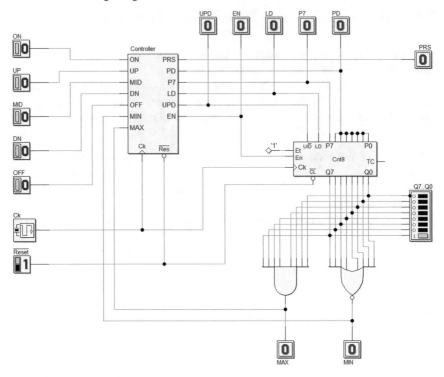

The system implements a *binary number generator*, controlled by five push-buttons *ON*, *UP*, *MID*, *DN*, and *OFF*. Output $Q7..Q0$ is taken directly from the counter.

Pressing and *then releasing* each push-button determine the system's behavior. Push-buttons' functions are:

ON → Sets number $Q7..Q0$ at the highest value.
UP → Increments the number by one unit.
MID → Sets number $Q7..Q0$ to the intermediate value $(= 10000000_2)$;
DN → Decrements the number by one unit.
OFF → Sets number $Q7..Q0$ to the lowest value.

In the idle state, the push-buttons are at 0, while pressed they are at 1.

The controller activates output *PRS* when one of the push-buttons is pressed. Number $Q7..Q0$ does not have to be increased or decreased if it has reached the highest or lowest value, respectively.

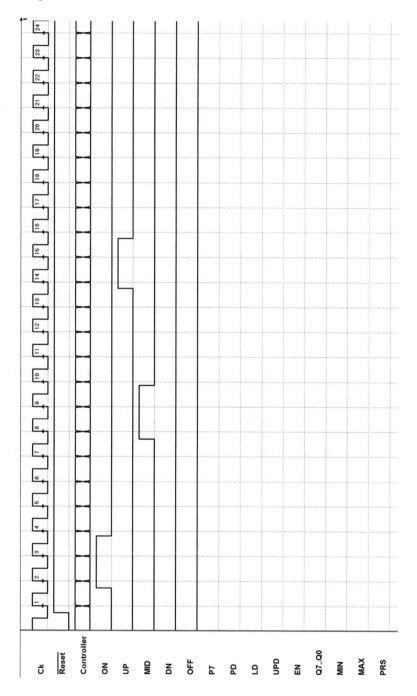

Exercise 2 The system shown in the figure below is made up of a controller, a "Cnt4" counter and an E-PET flip-flop.

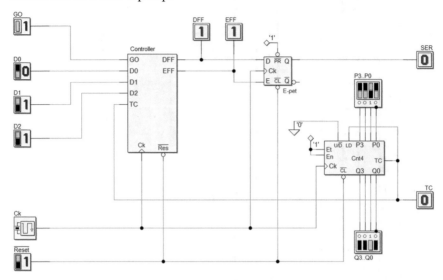

The system, a *serial transmitter*, must generate on output *SER* a 5-bit packet with one start bit at 1, three data bits (order: *D0*, *D1*, and *D2*), and a stop bit at 0.

Duration N of each bit of the packet (the bit time) is P time the clock period *CK*. P is the number set at the counter's inputs $P3..P0$.

To carry out the system's timing analysis, assign $P = 2$.

The data to transmit are available on the controller's inputs *D0*, *D1*, and *D2*. A falling edge on input *GO* starts the packet's generation.

Determine the relation between N and the number P that depends on your own specific solution.

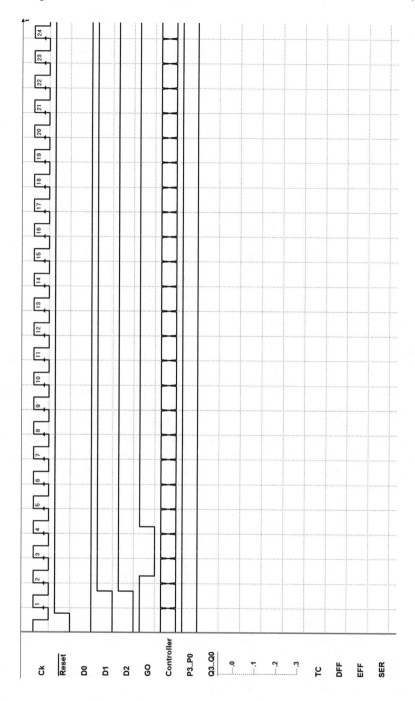

Exercise 3

The system shown in the figure below includes a controller that receives data from a serial line and manages a network made of two E-PET flip-flops, a register, and a few arithmetic circuits.

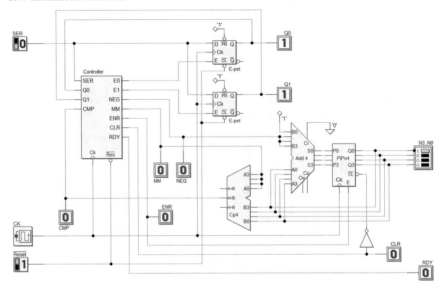

On *SER*, the controller receives a 4-bit pack with one start bit at 1, two data bits (order: $D0$ and $D1$), and a stop bit at 0. The duration of the bit time is one clock period; the two data bits codify an operation that the system carries out on outputs $N3..N0$.

If an incorrect stop bit ($= 1$) is received, no operation is carried out and the system waits for *SER* to go back to zero before waiting for the next packet. When a correct one (stop bit $= 0$) is received, the system carries out the following operations according to the value of $D1$ and $D0$.

D1	D0	Operation
0	0	No operation (NOP)
0	1	Clear output $N3..N0$
1	0	Increment output $N3..N0$ by one
1	1	Decrement output $N3..N0$ by one

Output $N3..N0$ does not increase when it has reached the highest value nor does it decrease when it has reached the lowest. Output *RDY* is activated for one clock cycle when any command (except NOP) is received.

Notice that the FSM memorizes bits $D1$ and $D0$ on flip-flops $Q1$ and $Q0$, so that their values can be reused by the FSM itself.

The FSM output *NEG* selects two different values on the input $B3..B0$ of the adder. The FSM line *MM* allows to choose two different numbers to be compared with the output number $N3..N0$.

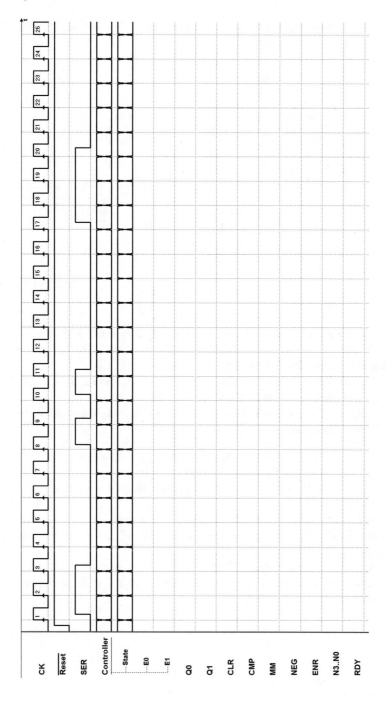

Exercise 4 The system shown in the figure below is composed by of a controller, a register, and a few arithmetic circuits.

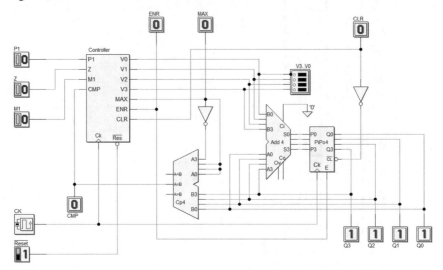

The system is a binary number generator controlled by three push-buttons $P1$, Z, and $M1$. Output $Q3..Q0$ is taken from the register "PiPo8." The binary number in the output is signed (two's complement code).

In the idle state, the push-buttons are at 0; while pressed, they are at 1. The push-buttons are assumed to be ideal with no mechanical bounce.

Pressing and then releasing each push-button determine the system's behavior. Push-buttons' functions are defined as follows:

$$P1 \;\rightarrow\; \text{Increments the number } Q3..Q0 \text{ by one.}$$
$$Z \;\;\;\rightarrow\; \text{Clears the number } Q3..Q0.$$
$$M1 \;\rightarrow\; \text{Decrements the number } Q3..Q0 \text{ by one.}$$

The number $Q3..Q0$ must not be incremented if it is at the maximum positive value nor decremented if it is at the minimum negative value. When the system is initialized, the number $Q3..Q0$ must be cleared.

Notice that the FSM outputs $V3..V0$ set the values to add on the adder inputs, while output MAX allows to compare the number $Q3..Q0$ with the maximum and minimum values.

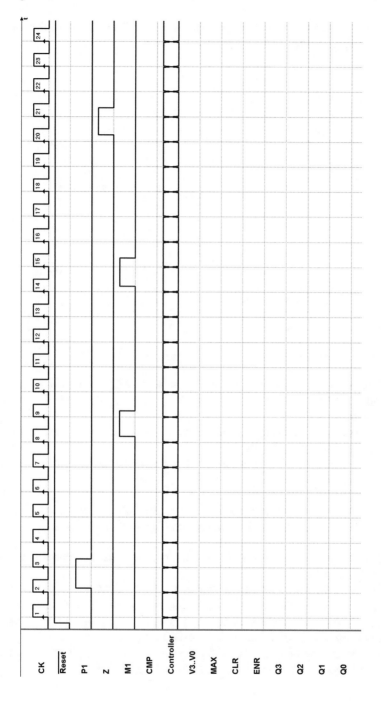

Exercise 5 Consider the digital system shown in the figure below. It is made up of a controller, a register, and a few arithmetic circuits.

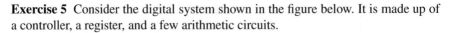

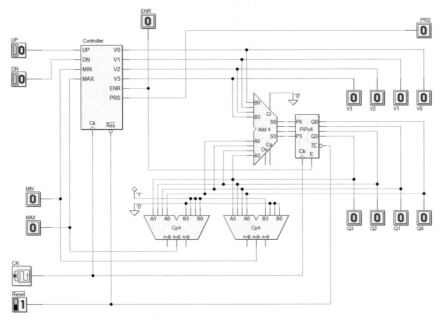

The system implements a *binary number generator* controlled by two push-buttons *UP* and *DN*. Output $Q3..Q0$ is taken from the parallel register "PiPo4." The output binary number is signed (two's complement code).

In idle state, the push-buttons are at 0 and at 1 while they are pressed. The push-buttons are assumed ideal with no mechanical bounce.

When the push-buttons are pressed and then released, the system carries out the following functions:

UP button → increments the number $Q3..Q0$ by one.
DN button → decrements the number $Q3..Q0$ by one.

The system ignores the pressing of the push-buttons if its time duration is shorter than 3 clock cycles.

The two comparators receive the value of output $Q3..Q0$ on inputs $A3..A0$, and they compare it with the maximum and minimum values set on $B3..B0$. The number $Q3..Q0$ must not be incremented if it is at the maximum (positive) value or decremented, if it is at the minimum (negative) value.

The system activates output *PRS* for the time a push-button is pressed. Assume that the two push-buttons can never be pressed at the same time.

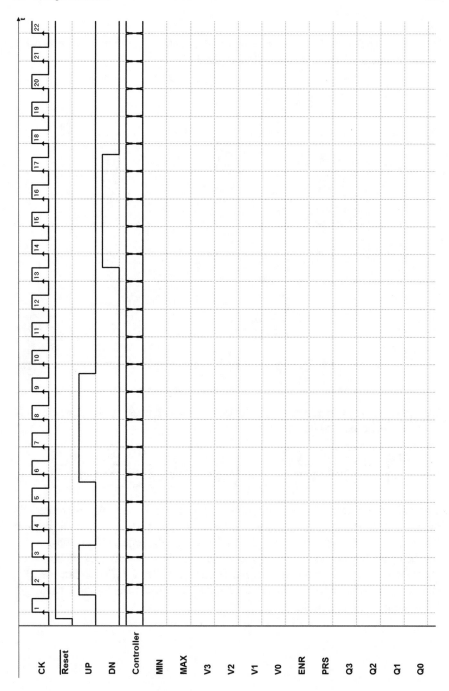

Exercise 6

The system shown in the figure is made up of a controller, a "SiPo4" shift register, an 8-bit counter "Cnt8," and a few logical gates.

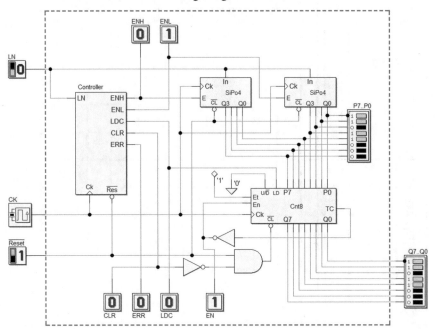

On input *LN*, the system receives a standard serial signal made up of a *start bit*, eight *data bits*, and a *stop bit* (see below).

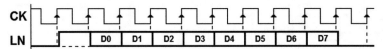

The controller commands the shifting of the serial data into the two registers with the order that we can infer from the schematic (the low part $D3..D0$ on the register at the right of the schematic, the high part $D7..D4$ on the register at the left).

The controller checks the *stop bit* at the end of the serial sequence. If correct, it loads the counter with the byte received through the lines $P7..P0$. If, however, the stop bit is not valid, the counter is cleared and the controller signals *ERR*, keeping it active until *LN* returns to zero.

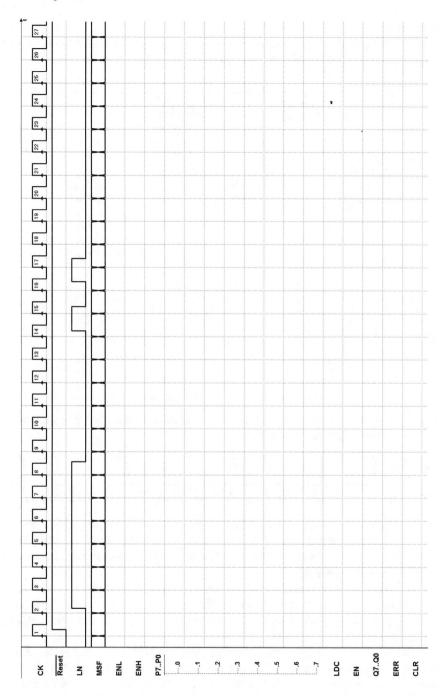

Exercise 7

Consider the digital system shown in the figure below. It is made up of a controller, an 8-bit counter "Cnt8," a "PiSo8" shift register, a D-PET flip-flop, and a few logical gates.

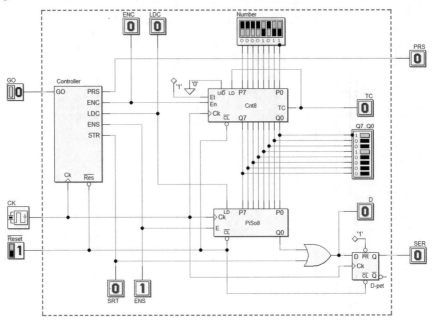

When push-button *GO* is released, the system transmits on *SER* a serial signal made up of a start bit, eight data bits, and a stop bit.

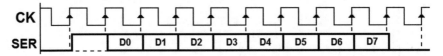

The eight data bits to transmit are taken from the counter's outputs. When push-button *GO* is not pressed ($GO = 0$), the counter is enabled through line *ENC*. For the rest of the time, the count is disabled. When push-button is pressed ($GO = 1$), the FSM activates line *PRS*.

When the push-button is released, the controller starts transmission by loading the counter's outputs onto the shift register through line *LDC*. Notice that line *STR* makes it possible to generate the start bit for one clock cycle and that serial output *SER* is generated by the flip-flop.

The controller commands serialization through line *ENS*, which enables the shift register. When the generation of *SER* is over, the system checks push-button *GO* again. For the purposes of the timing analysis, assume that *Number* (see schematic) equals 0Bh.

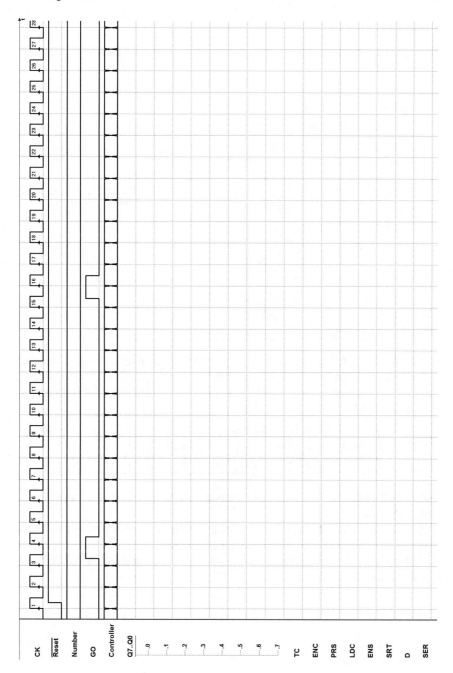

Exercise 8

The digital system shown in the figure below is made up of a controller, a counter, and a shift register.

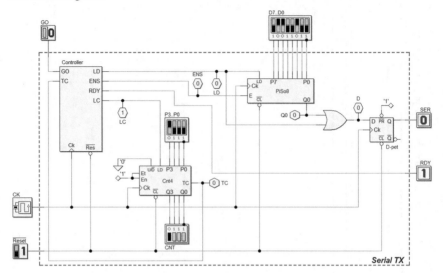

The system is a synchronous serial transmitter whose *bit time* is equal to the clock period. Transmission is launched on the rising edge of input *GO*. The packet transmitted on *SER* is made up of a *start bit* at 1, the eight *data bits* $D0..D7$ (in that order), and a *stop bit* at 0.

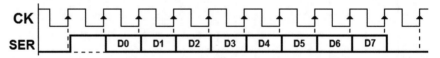

Output *RDY* (ready) is activated when the system does not transmit but waits for the launch command.

The counter is used by the controller to count the number of transmitted bits in order to end operations after the packet is transmitted.

Define the number $P3..P0$ required by your specific project (in the figure, 0111_2 is just an indication).

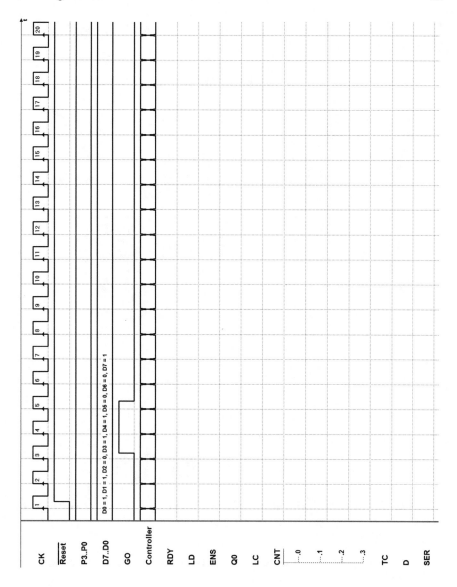

Exercise 9 The system shown in the figure below is made up of a controller, a counter, a comparator, and a parallel register.

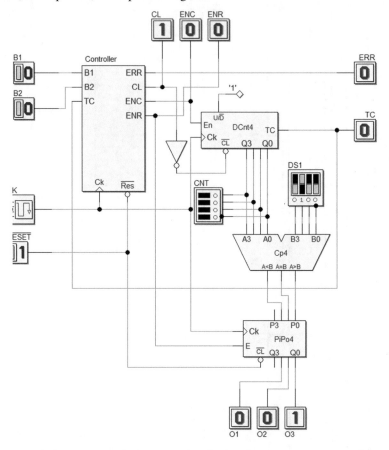

The system's inputs are push-buttons *B1* and *B2* (when pressed are at 1). The FSM checks how long they have been pressed at the same time.

Specifically:

- It activates output *O1* if *B1* and *B2* are pressed at the same time for less than four clock cycles.
- It activates output *O2* if *B1* and *B2* are pressed at the same time for exactly four clock cycles.
- It activates *O3* if *B1* and *B2* are pressed at the same time for more than four clock cycles.

If *B1* and *B2* are pressed for more than 16 clock cycles, the system activates *ERR* until the two push-buttons are both released.

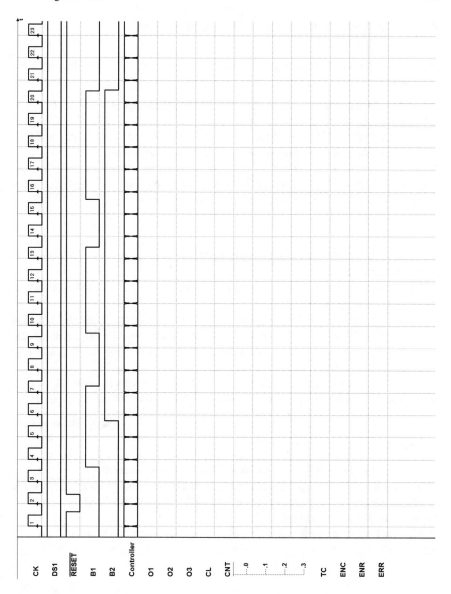

8.4.2   Design of a Controller–Datapath System

To carry out a design of this type, as suggested on p. 397, it is advisable to follow these steps:

(a) Define the system as a functional block, highlighting inputs and outputs.
(b) Choose the components to use in the datapath.
(c) Design the datapath with the connections between the controller and the datapath components and the inputs and outputs of the whole system.
(d) Draw the ASM diagram of the FSM that describes the controller.
(e) Perform a timing simulation of the whole system by choosing input sequences that help demonstrate the system's functionality set.

Exercise 1

Design a *serial transmitter* of the measure of a *pulse duration*, by using the *controller–datapath* structure.

The system waits for line *IM* to go to 1 and then begins to count how many clock cycles the signal *IM* stays at 1.

At the first clock edge after *IM* goes back to zero, the system sends a 6-bit packet on the *PKG* channel. It consists of a *start bit* at 1, *four data bits* that represent the pulse duration in terms of clock cycles (an error of $+/-1$ is tolerated), and a *stop bit* at 0.

If the pulse lasts longer than 15 clock cycles, the machine activates output *ERR* until *IM* goes back to zero. It sends no data pack but waits for the next pulse.

(Design tips on p. 427)

Exercise 2

Design a *kitchen timer* by using the *controller–datapath* structure.

When push-button *ST* (at 1 when pressed) is released, the device waits for the number of clock cycles externally set on inputs *P7..P0* (eight bits) and then activates an output *BEP* for *three cycles*.

The clock's frequency is $1Hz$, and the time must be controlled between 2 and 256 seconds. The actual time can differ by a maximum of one second from the defined time.

(Design tips on p. 428)

Exercise 3

Design a *serial–parallel converter* by using the *controller–datapath* architecture.

The system has an input *SER* and outputs *D0..D5* and *RDY*. It waits for the serial data in the format {1, *D0*, *D1* .. *D5*, 0} and transfers it into parallel format on the six outputs *D0..D5*. If the *stop bit* check is positive, it activates *RDY* and keeps it until it receives new serial data.

If the *stop bit* is incorrect, the system simply goes back to waiting for new serial data.

(Design tips on p. 429)

Exercise 4

Design a *serial–serial repeater* by using the *controller–datapath* structure.

On its input *IN*, the system receives a serial packet in the format {1, *B0*, *B1*, *P*, 0} where *P* represents the *parity* (XOR function) of *B0* and *B1*.

The system acquires *B0* and *B1* on two JK-PET flip-flops and uses an XOR gate to check that the parity of *B0* and *B1*, which are saved on the flip-flops, corresponds to bit *P* of the serial packet, which was transmitted. If it does correspond and the stop bit is correct, it transmits the pack {1, *B0*, *B1*, 0} to output *OUT*.

If the parity check is negative or the stop bit is incorrect, there is no transmission and the system goes back to waiting for another input packet.

(Design tips on p. 430)

Exercise 5

Design a *parallel–serial converter* by using the *controller–datapath* architecture.

The system has five inputs (*GO* and *D0..D3*) and three outputs (*SER*, *BSY* and *RDY*). When *GO* appears, the system generates a serial packet made up of: {1, *D0*, *D1*, *D2*, *D3*, and 0}.

Signals *D0..D3* are only guaranteed to have the correct value when *GO* is received. *GO* lasts for one clock period.

The system activates *BSY* during the transmission of the packet and generates a pulse with a duration of one clock cycle on *RDY* just after the packet is transmitted.

(Design tips on p. 431)

Exercise 6

Design a digital system that works as an *accumulator* that generates, on the output, the sum (4 bits) of the current number and the one at the input.

The system has five inputs:

D2..D0   The data the accumulator must add (3 bits with no sign).
NEG   If active, it commands to change the sign of the number
         on D2..D0 before adding it.
GO   If active, it commands the execution of the sum.
The system has six outputs:

Q3..Q0   The content of the accumulator.
OVF   To be activated in case of *Overflow*.
CO   To be activated in case of *Carry*.

(Design tips on p. 432)

Exercise 7

Design a *controller for a microwave oven* by using the *controller–datapath* structure.

The system has two 1-bit inputs and one 8-bit input.
Inputs:

GO   Starts cooking for time *TCC*
TCC   8-bit input to set cooking time
OPN   Active while the door is open
Outputs:

COK   When active, the oven heats
BEL   Signals the end of cooking time and lasts one clock cycle
The user sets the cooking time and presses the push-button. The signal *GO*, which is active for one clock cycle, starts the cooking for time *TCC*.

If the door is open, the oven does not work; when it is closed, it begins to heat. If the door is opened before the end of the set cooking time, it stops cooking and starts again when the door is closed.

(Design tips on p. 433)

8.5 Design Tips

8.5.1 Design of a Controller–Datapath System

Here, you will find some hints to help you complete the design exercises starting on p. 424.

Exercise 1

We recommend using the counter "DCnt4" and logical gates of your choice.

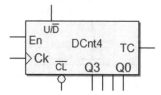

We recommend defining a timing trace like this one below:

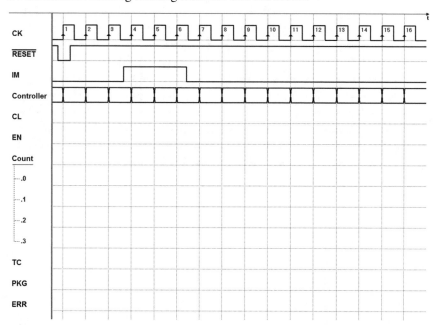

Exercise 2

You may use the counter "DCnt8" and any logical gate.

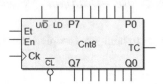

We recommend defining a timing trace like this one below:

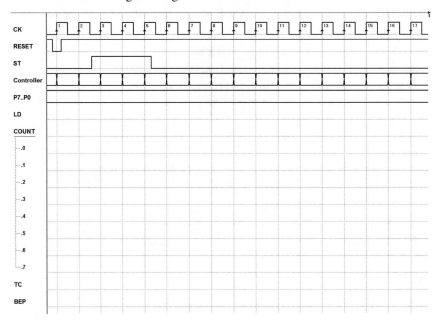

Exercise 3

We suggest using the components "Sipo8" and "Cnt4" as well as any logical gate.

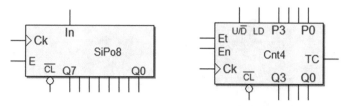

We recommend defining a timing trace like this one below:

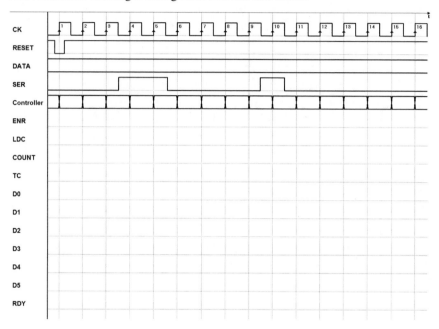

Exercise 4

We suggest using two JK flip-flops and an XOR gate for the datapath.

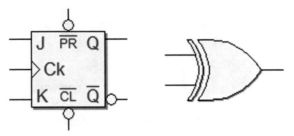

We recommend defining a timing trace like this one below:

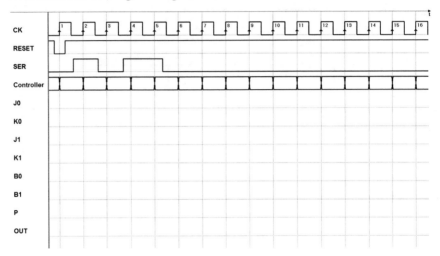

Exercise 5

For the datapath, we recommend using the components shown below (keep in mind that we can also design the system without a counter).

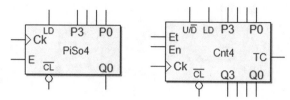

We recommend defining a timing trace like this one below:

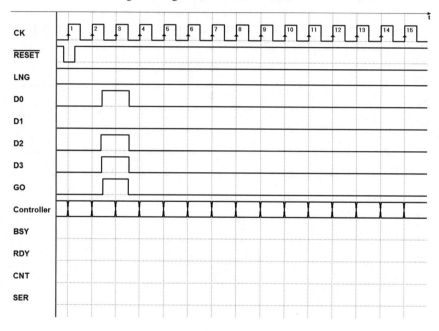

Exercise 6

We recommend using the components shown below:

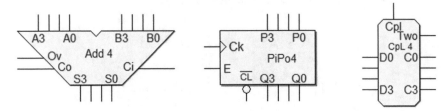

We recommend defining a timing trace like this one below:

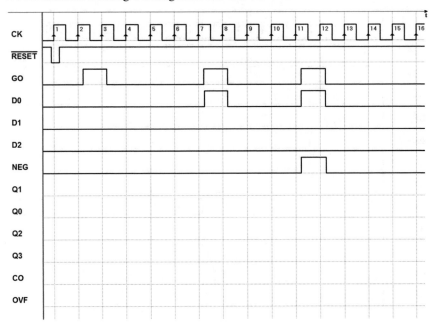

Exercise 7

For the datapath, we recommend using the component shown below:

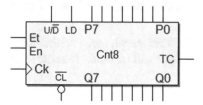

We recommend defining a timing trace like this one below:

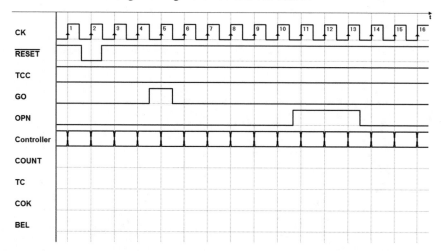

8.6   Solutions

8.6.1   Design of the Controller of a Given Datapath

We can download the MSF files shown here and their complete circuit schematics from the Web site of *Deeds*. It will be useful to analyze the solutions by using the timing simulation.

Solution of Exercise 1:

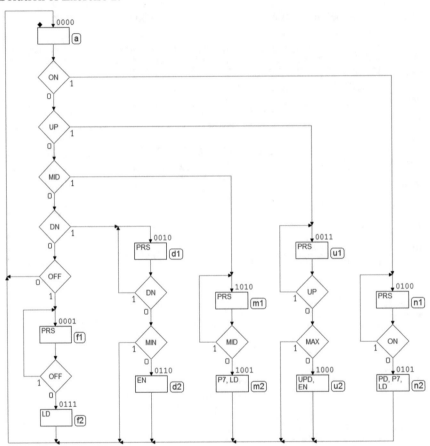

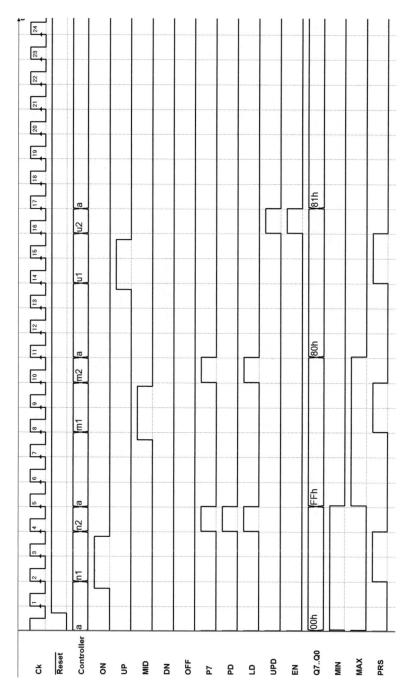

Solution of Exercise 2:

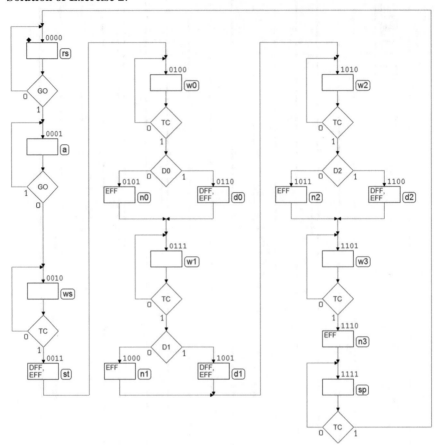

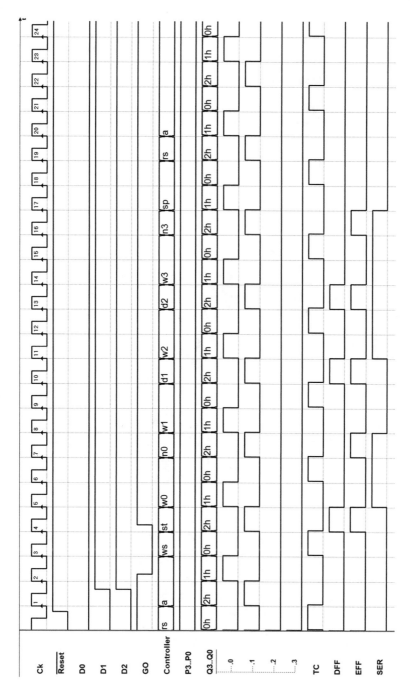

Solution of Exercise 3:

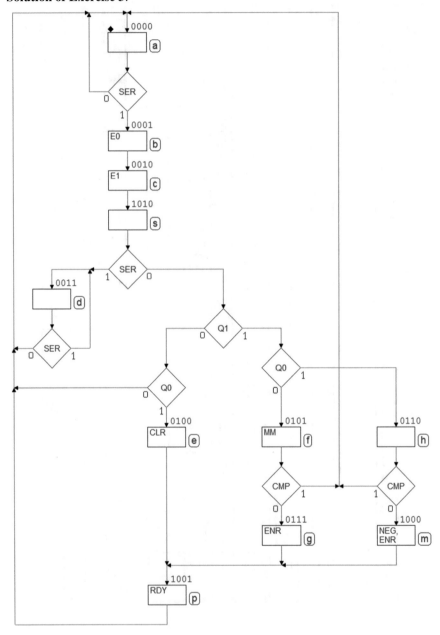

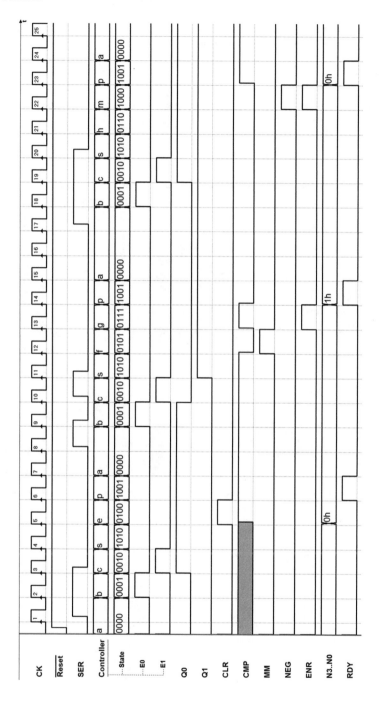

Solution of Exercise 4:

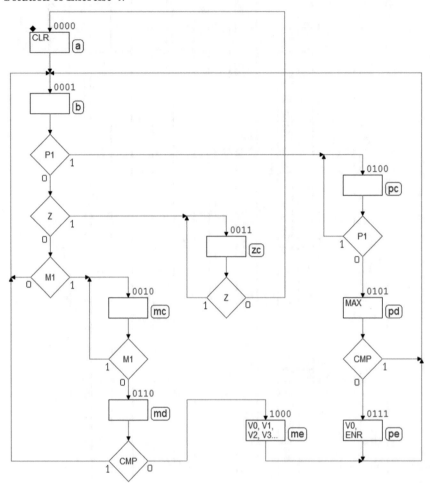

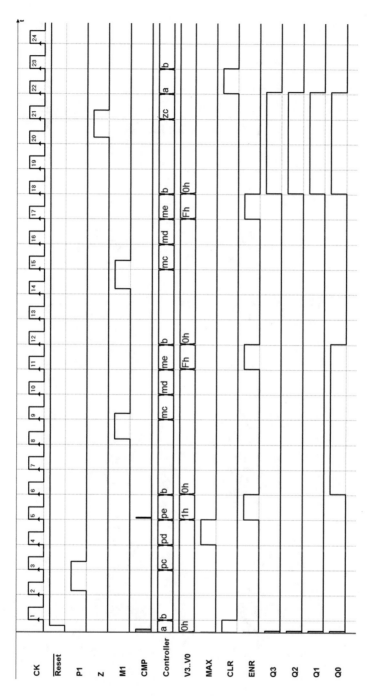

Solution of Exercise 5:

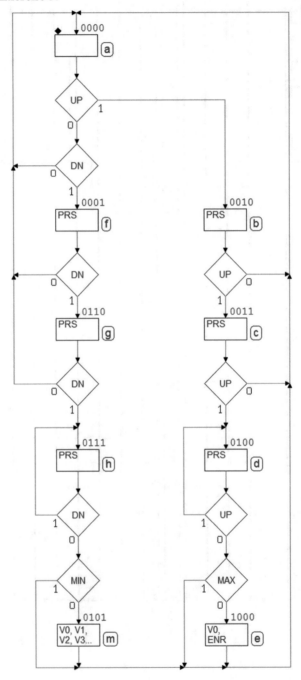

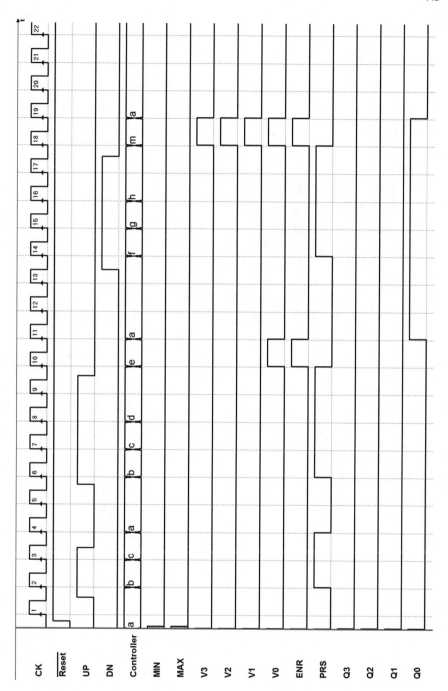

Solution of Exercise 6:

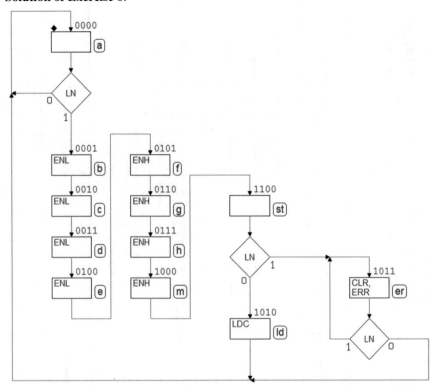

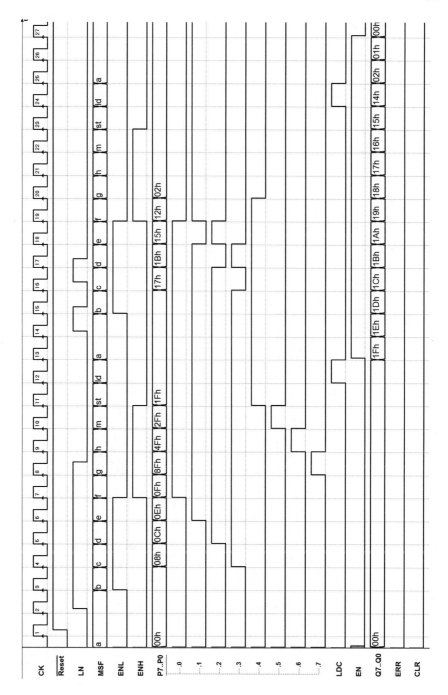

Solution of Exercise 7:

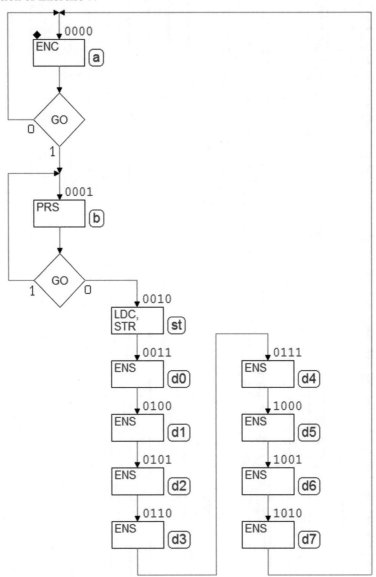

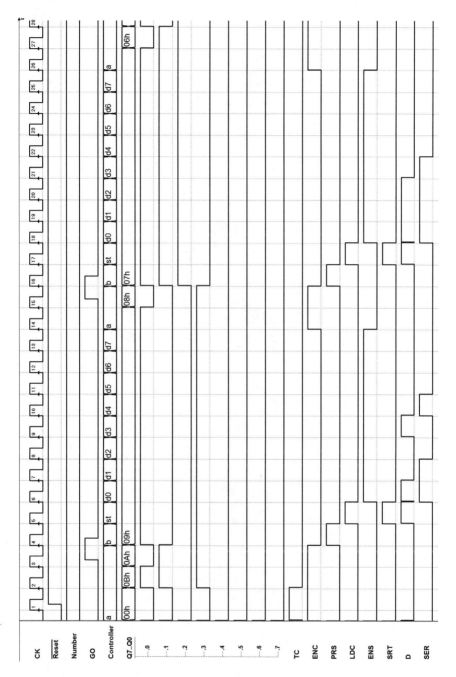

Solution of Exercise 8:

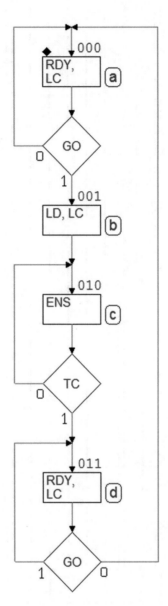

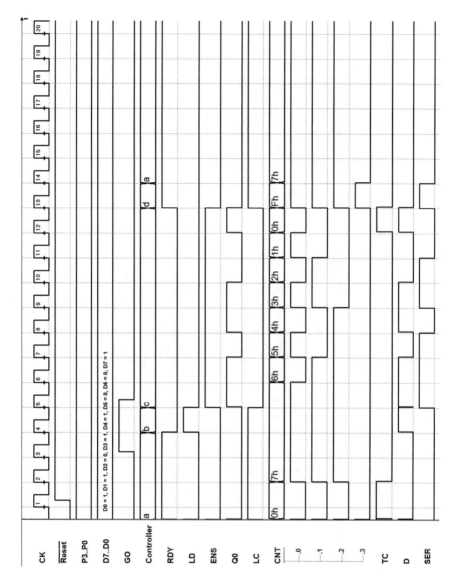

Solution of Exercise 9:

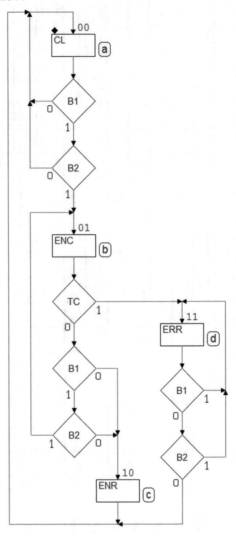

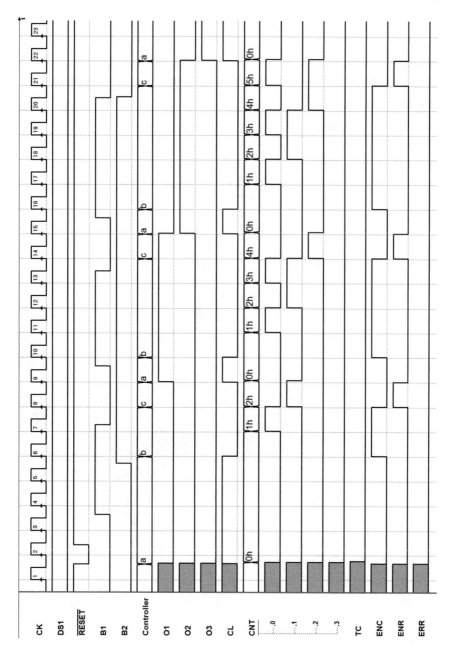

8.6.2   Design of a Controller–Datapath System

The Web site of *Deeds* has all the files corresponding to the figures shown here (circuit schematics and FSMs) so that the solutions can be checked by simulation.

Solution of Exercise 1

The system's inputs and outputs:

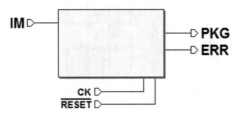

The network schematic (controller + datapath):

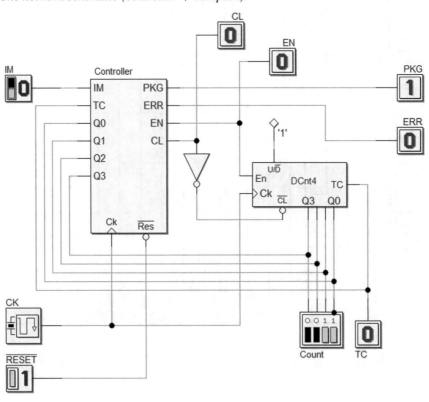

Diagram of the states of the controller:

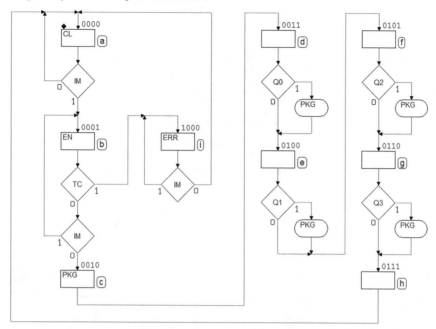

Timing simulation:

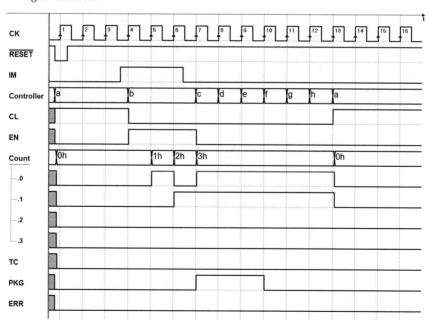

Solution of Exercise 2

The system's inputs and outputs:

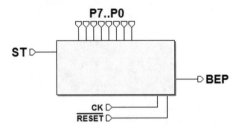

The network schematic (controller + datapath):

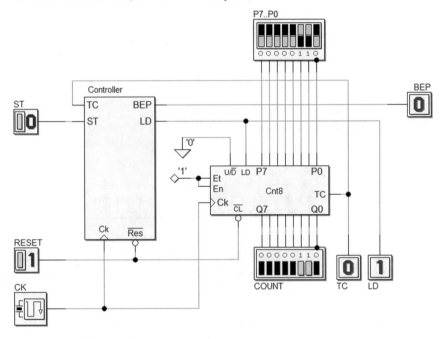

Diagram of the states of the controller:

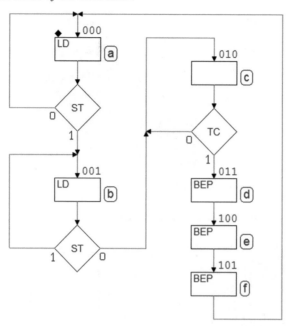

Timing simulation:

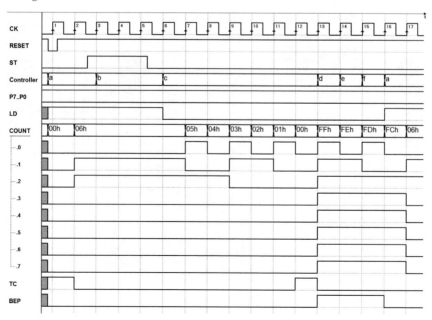

Solution of Exercise 3

The system's inputs and outputs:

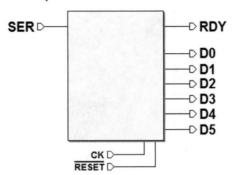

The network schematic (controller + datapath):

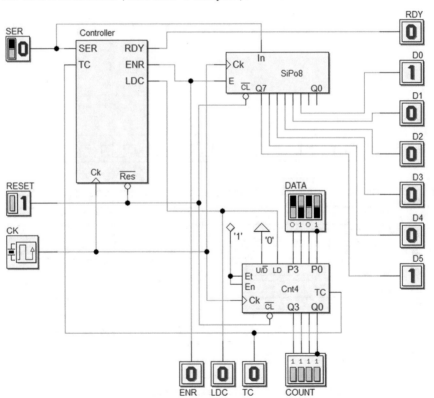

Diagram of the states of the controller:

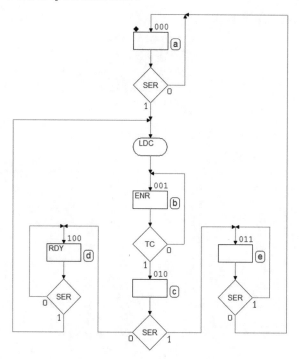

Timing simulation:

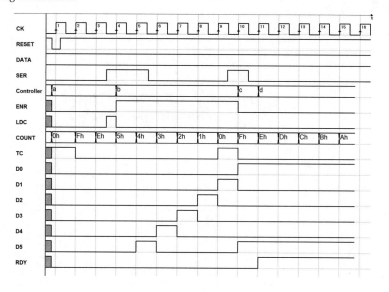

Solution of Exercise 4

The system's inputs and outputs:

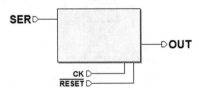

The network schematic (controller + datapath):

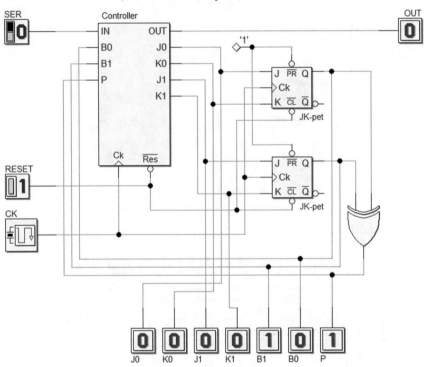

Diagram of the states of the controller:

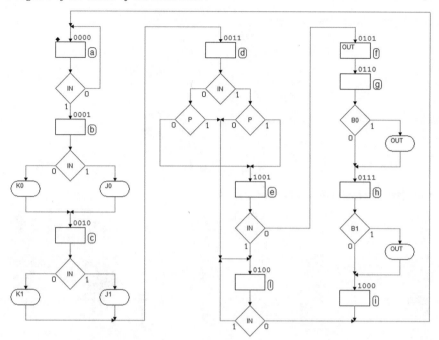

Timing simulation:

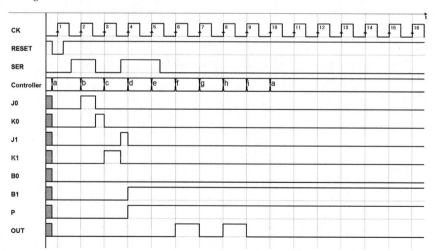

Solution of Exercise 5

The system's inputs and outputs:

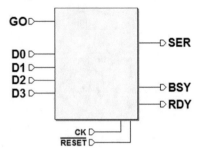

The network schematic (controller + datapath):

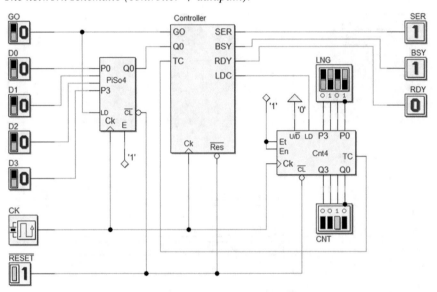

Diagram of the states of the controller:

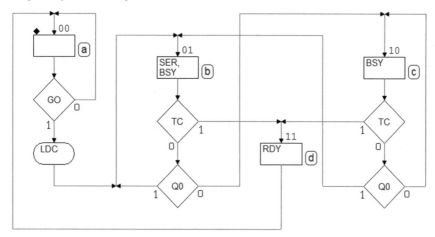

Timing simulation:

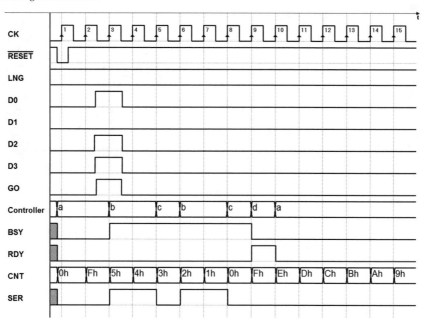

Solution of Exercise 6

The system's inputs and outputs:

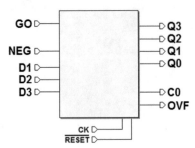

The network schematic (datapath only):

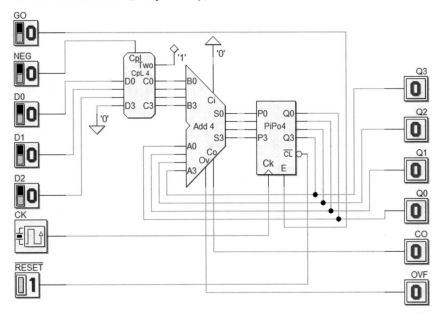

Solution of Exercise 7

The system's inputs and outputs:

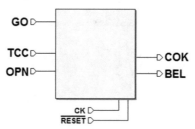

The network schematic (controller + datapath):

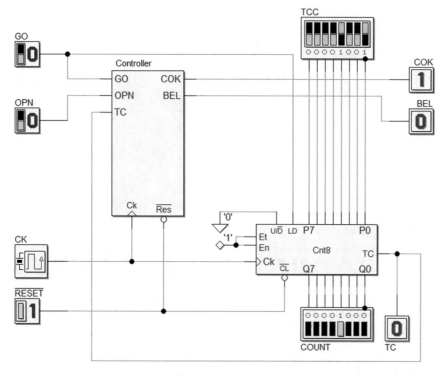

Diagram of the states of the controller:

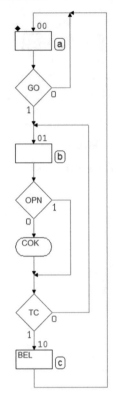

Timing simulation:

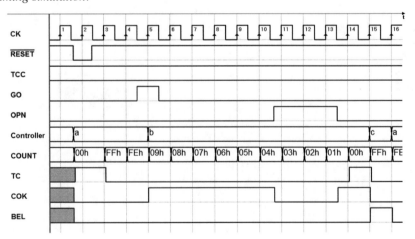

Chapter 9
Introduction to FPGA and HDL Design

Abstract The last chapter deals with the practical implementation in hardware of systems similar to the ones presented in previous chapters and tested by simulation only. The devices that host the projects are Field-Programmable Gate Arrays (FPGAs), inserted on commercially available boards and managed by Deeds and proprietary tools. A short description of the devices and the associated tools is presented. An original, hands-on introduction of the VHDL hardware description language is included. A few exercises of digital system design and prototyping complete the chapter.

The reader that has successfully followed the book is now familiar with the issues at the bases of digital systems and able to practice with their design and simulation. Projects and examples presented in the previous chapters were targeted more to understanding and less to practical implementation, since the former is an essential skill for the designer, upon which the latter is based. Furthermore, circuit implementation is strongly dependent on the state of the art of microelectronic technologies and subject to a rapid evolution and inevitable obsolescence.

The networks presented in this chapter are similar, as for their approach and complexity, to the ones already studied, with the difference that the work will go all the way to physical implementation and testing.

9.1 Field-Programmable Gate Arrays

Physical implementation will be based on components called Field-Programmable Gate Array (FPGA). The figure in the next page shows the visual appearance of two FPGA, by *Intel/Altera FPGA* (ex *Altera Corporation*), on the left, and by *XilinX*, on the right.

© Springer International Publishing AG, part of Springer Nature 2019   465
G. Donzellini et al., *Introduction to Digital Systems Design*,
https://doi.org/10.1007/978-3-319-92804-3_9

An FPGA is a chip that contains a large quantity of basic logical elements, such as gates and flip-flops (and quite a lot of more complex circuits) that can be wired together to form the system thanks to a matrix of switches. The connections are created by using development tools provided by the FPGA's manufacturers, downloaded in the chip and kept alive in a memory.

FPGAs are the youngest child within the large family of programmable logic devices (PLDs), a term that designates all the chips that can be programmed, i.e., specialized for a given application by establishing or changing their inside connections, during production or in the field. PLD must not be confused with devices, as microcomputers, whose hardware is fixed and programming means execution of external instructions. PLD has changed deeply, since the 1980s, design and implementation of complex systems.

FPGAs have a great value for the educational field, too, since they lend themselves very well for the fast and inexpensive realization of working prototypes of systems designed for learning purposes.

9.1.1   System Prototyping and FPGA

In a not too distant past, circuit prototyping implied the connection (by soldered wires) of many discrete components. That process was extremely time consuming and very sensitive to mistakes in the connection or bad contacts with the wires to a point that it was not easy to understand if the malfunctioning of the system was due to design mistakes of faulty connections.

It was common the use, in the laboratories, of solderless breadboards, with a fixed grid of holes partially connected together, in which students inserted components, usually in the form of integrated circuits (ICs) and established connections through wires. The ICs made available gates, flip-flops, and a wide variety of combinational and sequential blocks: the circuit implementation was therefore the same, or very close, to the hardware structure designed with *Deeds*.

For instance, below is a picture of the breadboard of an 8-bit parallel to serial converter, where the system is built with standard ICs implementing the functions of gates, flip-flops, registers and counters.

The problems with breadboarding digital systems are the same mentioned when presenting the traditional prototyping. There are a few advantages: it is easier to change connections and the risk that students burn their finger with the soldering iron disappears. The problem of faulty contacts is even worse than in the previous method. Solderless breadboards are still useful for rapid prototyping of system whose core is a FPGA or a microcontroller: the board can host simple interface or ancillary circuits around the core.

Nowadays are available FPGA-based prototyping board, especially suited for educational purposes. They include several input/output interfaces, allowing the implementation of system prototypes without the addition of components outside the board.

9.1.2 FPGA Board Examples

A wide variety of FPGA boards are commercially available, with performances continuously evolving. They are targeted to different applications, from simple and inexpensive boards ideal for educational purposes to complex, high-speed boards for professional designs. It is not our intention to go into the details of the boards. The experimenters or the designers can find on the market the most suitable for their application, paying attention to their performances, available software and, last but not least, budget.

As an example, we present here a general description of a few FPGA boards, suitable for the implementation of the systems developed in the book. It is worth to notice that each of them has the capacity to host much more complex systems and, therefore, to allow a natural transition toward professional design. The figure below shows the board DE0-CV, produced by *Terasic/Altera*. The DE0-CV is supported by *Deeds* environment.

On the board are available, as the picture shows, push-buttons, switches, LEDs, seven-segment displays, connectors, and other devices. The heart is the FPGA, the big black square in the center of the board, a device from *Intel/Altera FPGA*, member of the family "Cyclone® V FPGA." The chip contains a matrix of more than 40,000 logic units (the basic FPGA block that will be explained later in this chapter), 60,000 flip-flops, and a microprocessor *ARM CortexTM* dual-core, the same used in many mobile phones.

In the next page is another example: the ARTY S7-50 board, produced by *Digilent*, using an FPGA chip from the *Spartan®-7* family, produced by *XilinX*. The FPGA chip is placed at 45 degrees with respect to the board's boundaries.

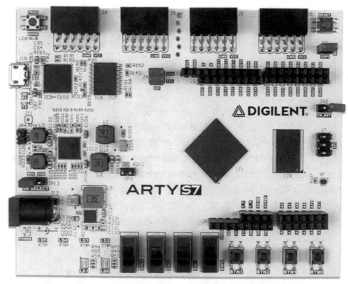

As in the previous board, push-buttons, switches, LEDs, and other interfaces are available. In particular, there are connectors designed to host the input/output and additional boards (shields) originally designed for *Arduino* microcontrollers. The FPGA contains more than 52,000 basic combinational blocks and more than 65,000 flip-flops. No microprocessor is included.

Note: it is possible to program the FPGA to implement a processor, by using the chip's resources (logic elements and flip-flops). This is called "soft processor" since it is assembled by software. It behaves exactly as an "hard" processor, i.e., one built as such on the silicon.

The last example is a very inexpensive FPGA board available online and supported by *Deeds*. It is based on an *Intel/Altera FPGA* chip "Cyclone® II." In addition to the four connectors placed around the chip, three LEDs and a push-button are available. The two 10-pin connectors on the left are used for programming the FPGA chip.

To implement our projects, which usually need more input/output devices, we must connect push-buttons, switches, displays, etc., to the four connectors. The chip is large enough to implement small microcomputers, like the ones available with *Deeds*. The resulting "soft processor" would use about one half of the 4,600 logic units and only 400 of the 4,600 flip-flop.

9.1.3   FPGA Architecture

FPGAs' manufactures make available a wide variety of devices, classified by *families* varying by complexity and targeted to different application fields. Typical examples are audio/video signal processing, radar, automotive systems, and, generally speaking, all the applications that require high performances but do not have the volume to justify the cost of a full custom chip. In spite of the variety presented by the families, all FPGA devices have in common a basic architecture. An FPGA is essentially a large matrix of logic blocks, arranged by rows and columns, as in the figure below.

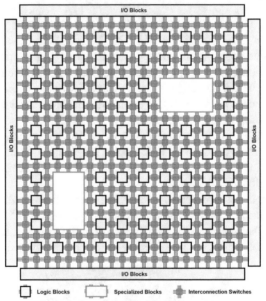

Each block contains one or more flip-flops and combinational networks. A matrix of programmable connections is spread through the chip, using the largest share of its area. Connections are, again, arranged by rows and columns: at every crossing electronic switches allow the individual junction of rows and columns. Such structure provides the interconnection among the blocks. Local submatrices may be available to improve the speed of communication among blocks physically close together.

Inside the matrix special blocks targeted to specific functions may be available, such as read/write memories (RAM), arithmetic circuits (very often multipliers), and others.

The matrix is surrounded along the four edges by other logic blocks (I/O Blocks), in charge of the interface of the chip with external devices.

The previous figure shows only the elements of the chip that are available for design, while hides the ones in charge of programming (configuring) logic blocks and connections, made of a very large number of flip-flops connected to form a shift register, as in the figure below.

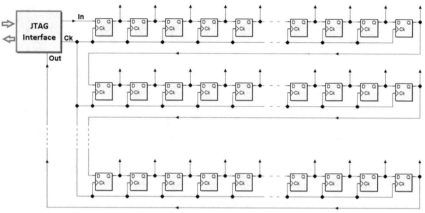

A particular synchronous serial port (JTAG interface, that we explain in the next section) is in charge to write the flip-flops during the FPGA programming phase. In normal operation, the flip-flops are not accessible.

The FPGA must be re-configured each time at power-up, since, as we know, a flip-flop cannot maintain information when power is off. Therefore, at power-up, a non-volatile memory in the board transfer programming information through dedicated pins. From the above, we understand that the FPGA may be re-programmed to perform a different function, using the JTAG interface.

Logic Block

In the next figure, the schematic of a simplified FPGA *logic block* is represented. Basically, we have an E-type flip-flop, edge-triggered, driven by a combinational logic network.

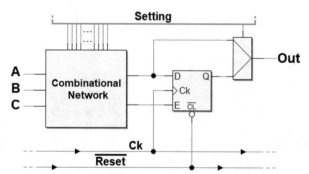

The combinational network's operation is controlled by the configuration flip-flops (top of the figure), as well as the multiplexer on the right that allows the option of storing in the flip-flop the output of the combinational network.

It is worth to have a closer look at the combinational network, which is implemented as a Lookup Table (LUT). In Sect. 2.6.6, we have seen how to use a multiplexer as a configurable combinational network, by feeding into its inputs the values of the desired function, in the present case provided by the configuration flip-flops.

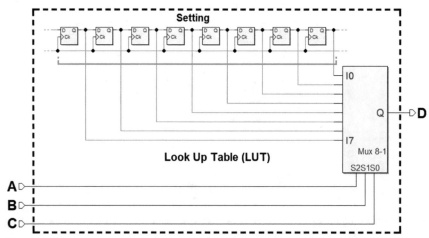

The multiplexer copies in its output D, according to the combination of A, B, and C, the values stored in the flip-flops, implementing the function simply by reading the LUT, which is defined at the time of FPGA configuration and does not change during its operations.

Note that a logic block could be more complex than the one just presented, containing also a full adder, XOR gates, other flip-flops, or multiple LUTs.

9.1.4   JTAG Programming

JTAG is the acronym of the consortium (Joint Test Action Group) that defined, at the end of the 1980s of last century a standard protocol for the functional test of integrated circuits, which later became the IEEE 1149.1 (*IEEE Standard Test Access Port and Boundary-Scan Architecture*). In the following, we will refer to it as JTAG (the term *Boundary-Scan* is sometimes used).

The version of the protocol released in 1994 added the possibility of programming memories, microcontrollers, and other devices. In addition, it allowed to perform the functional verification of the firmware and the possibility to activate automatic testing (Built-In Self-Test), defined by the component's manufacturer. A standard language to access components (*Boundary Scan Description Language*), by using JTAG, has been developed.

Nowadays, JTAG is the only procedure to access electronic systems, such as cell phones, tablets, wireless access points, and the like, for testing and troubleshooting.

In synthesis, the standard provides the possibility of blocking the normal operation of a system and disconnecting its clock, to switch to a modality in which the JTAG interface takes control of all the components' pins and the testing and programming circuits that may be inside the system itself.

The physical JTAG interface is composed of a limited number of standard connections. The simplest set allows to communicate with the circuit using a few lines, as shown in the following figure.

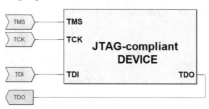

Pin	Name	Function
TCK	*Test Clock*	Data Clock Pin
TMS	*Test Mode Select*	Mode Control and Operation Selection
TDI	*Test Data In*	Serial Data Input Pin (toward the device)
TDO	*Test Data Out*	Serial Data Output Pin (from the device)

All JTAG's signals are serial and synchronized by clock TCK (usually in the range 10–100 MHz). The activation of TMS signals to the system to enter the JTAG-compliant mode. Then, through the same line it is possible to perform the operation requested, using a state algorithm, the "JTAG State Machine" (not described here). The standard defines an optional Test Reset control (TRST) also, but its functionality can be obtained via TMS control, and often it is not used, as in the example above.

Moreover, the standard allows the series connection of the pin TDI and TDO of more than one device ("Daisy Chain" connection) in order to access all the JTAG-compliant devices in a board (shown in the next figure). An example of the power of this method is the possibility of performing a "Chain Integrity Test". Each JTAG-compliant device has its own ID code. All the ID codes can be read and checked against the ID of the design project, to verify if the JTAG chain is working as designed.

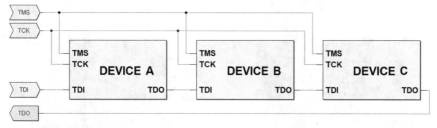

FPGA Programming

Many programmable devices, such as FPGA and PLD, are not designed to be JTAG-compliant for testing purposes only. They use JTAG for their programming.

It must be stressed that FPGA is programmable after their insertion on the system's board. This fact provides several advantages, such as to simplify the programming phase, avoid the use of external programmers, update, and modify the networks implemented by the FPGA. This feature makes FPGA systems ideal for the implementation of prototypes, experimental circuits, including educational ones.

There are several standards for the JTAG physical interface. The simplest uses a 10-pin connector, as shown in the figure below that refers to the FPGA board with the *Intel/Altera FPGA* chip "Cyclone® II" already described.

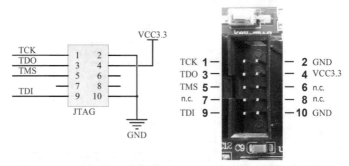

In the next figure, an example of JTAG programmer with its 10-pin cable (right) and a standard USB cable for connection to the PC. The software is always provided by the FPGA manufacturer.

Often, especially in the most advanced boards, the programmer is built in with the board, available through a dedicated USB interface. This is the case of the first two boards described before.

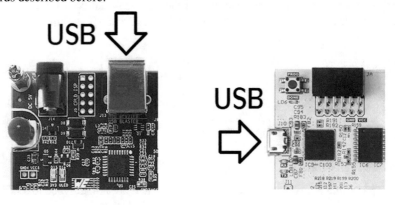

9.1.5 FPGA Development Tools

FPGA manufacturers offer (for a fee) proprietary development tools, targeted to professional designer and to the implementation of complex systems. The same manufacturers make available, usually free of charge, reduced versions of the same tools, usually targeted to the education field.

The obvious purpose is to publicize their products to future designers and, therefore, to influence their choices when they will be working. Plenty of documentation of such software tools is provided by the manufacturers.

In the following, we present a general view of what is available, free of charge, on the net. All the tools we mention offer similar features, such as schematic editor, source code editor, compiler, pin manager, optimization tools, and programmer and allow to design digital systems using logical schematics or Hardware Description Languages (HDL), such as Verilog, VHDL, or System C.

The large amount of features available, which make the professional's work more productive, produces in the beginner the impression of a difficult to manage complexity. In the following, we will see how *Deeds* allows to use FPGAs for a fast prototyping of our projects without going into the technicalities of the FPGA tools.

About boards based on *XilinX* devices, at the moment of writing two free tools are available: *Vivado® Design Suite HL WebPACK™* and *ISE® WebPACK™*. The screenshot below shows *Vivado®*, which is targeted to the most recent families of devices.

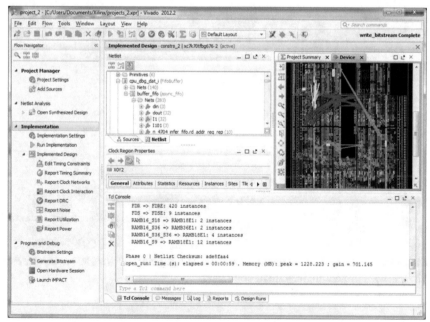

Below is *ISE®*. It supports project development on less recent FPGAs:

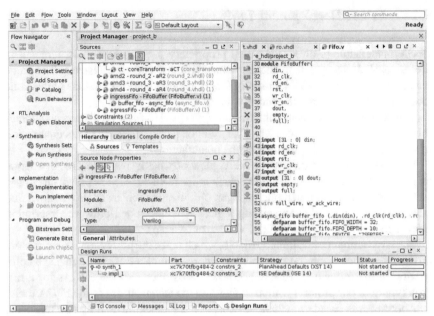

Intel/Altera FPGA provides two free tools: *Quartus® Prime Lite Edition*™ and *Quartus® II Web Edition*™. The next screenshot shows the mail window of the first one. It supports the most recent *Intel/Altera FPGA* families:

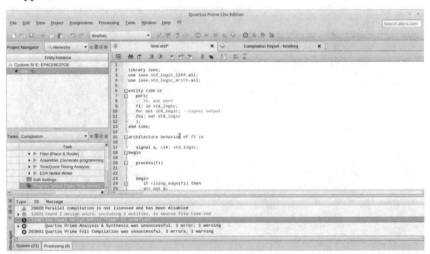

The older versions of the same software are called *Quartus® II*. Below is its main window with project management commands:

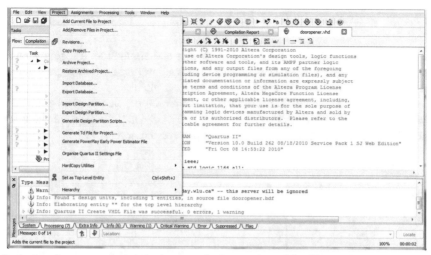

In the last few years, due to the increase in the number of families and the complexity of the chips, FPGA producers have put more efforts in developing and maintaining the tools that support the new families than in assuring their compatibility with older chips. In general, the less recent boards require the use of older version of development software. It is therefore necessary to pay attention in the choice of chips and tools, by studying the documentation available in the manufacturers' Web site.

Utilities called *Software Selector* associate the FPGA family to the corresponding software version. In the following figure, from the *Intel/Altera FPGA* Web site, we see that the *Cyclone® II* family is supported by a version *13.0 - ServicePack 1* (and older) of *Quartus® II Web Edition™*:

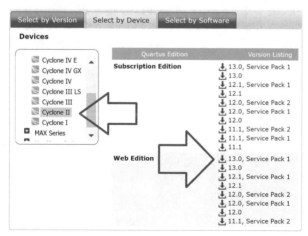

9.1.6   Deeds Support for FPGA

In the project development process, after design and simulation, a logical step is to test the network in a physical system. *Deeds* allows a fast implementation of prototypes on several FPGA boards commercially available.

To examine the path that goes from project to prototype, we use as an example a *Pulse Generator* similar to the ones seen on p. 372 in Chap. 8. We implement the project on the *Terasic/Altera* DE0-CV board presented before.

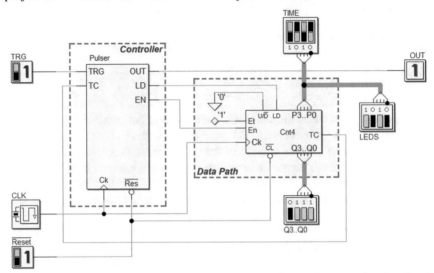

In this version, counter's outputs are visualized on a LED array, and pulse duration is set using 4 switches, with the aid of another LED array that shows the number while it is set. Controller's functionality has been extended by the addition of the line *EN* to enable counting. The FSM, in the reset state (rs), waits until the *TRG* input goes to zero.

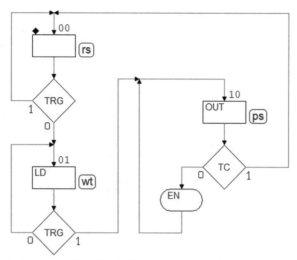

On next state (wt), the machine loads the counter and waits for a rising edge of *TRG*, after which goes to state (ps) that generates the pulse on *OUT*. In (ps) the counter in enabled until *TC* is activated and the FSM returns to the initial state.

"Test On FPGA" Window

In the following, we go through all the steps of the process (in several projects available on the *Deeds*'s Web site most of the settings are already included in the files).

Let's assume we are done with the behavioral verification of the project, obtained by simulation, and are ready for the prototype implementation. We open the *"Test On FPGA"* window (see the *Tools* menu item).

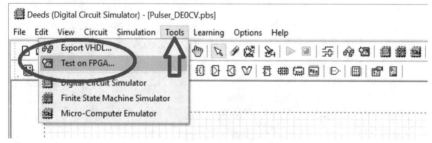

In the *"Test On FPGA"* window firstly we select a FPGA board from the list box (blue arrow in the next figure):

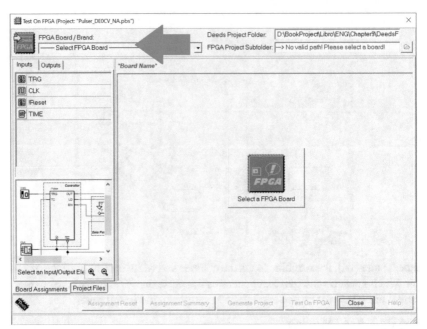

After the selection of the board, in our case the DE0-CV (green arrow in the figure below), we see the picture of the physical board in use (red arrow).

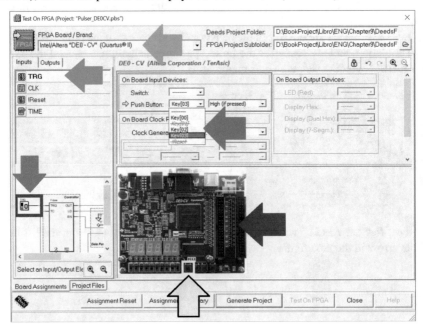

Resource Assignment

We need to associate the network's input and output components with the devices available on the FPGA board. We select, one by one, the input/output components of our network, either by clicking on the schematic (brown arrow) or in the list (blue arrow).

In the example of the previous figure, we point to the input *TRG* and the system shows only the board's resources that are compatible with our selection. In this case, we choose to associate to *TRG* the push-button "Key[03]".

While scrolling the list (gray arrow), each physical device is identified on the board inside a yellow frame (yellow arrow).

Assignment Summary

After the association of all inputs/outputs of our network to the corresponding physical resources of the board, we can check the "Assignment Summary" by clicking the button with the same name in the bottom of the window.

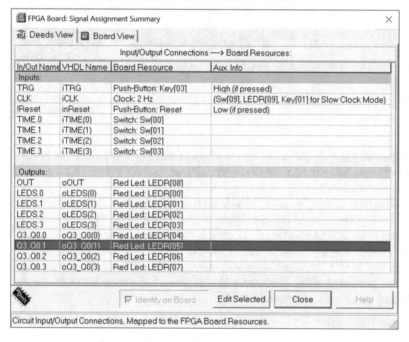

Clock Frequency Setting and Slow Clock Test Mode

The clock frequency on the board (Fc) can be defined at wish, independently of the one chosen in *Deeds*'s schematic. A frequency divider (not shown here) allows to scale downward the native clock frequency (50 MHz in the case of the DE0-CV board).

In our *Pulse Generator* project, the output pulse can last up to 16 clock cycles. Therefore, for visual testing, $Fc = 2$ Hz allows to generate pulses up to 8 seconds.

Deeds inserts a frequency divider by 25,000,000 between the board clock generator and the clock input of our network.

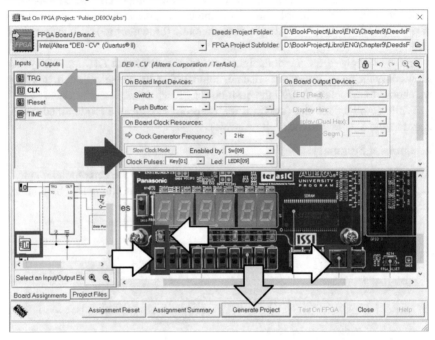

It is possible to slow further down the clock to observe *step by step* the behavior of the network, using the *"Slow Clock Mode"* (see the area pointed by the red arrow, in the figure above).

We set the switch "Sw[09]" to activate it and the LED "LEDR[09]" to visualize the clock pulses sent to the circuit under test. Push-button "Key[01]" will be the manual command to generate clock pulses, one by one.

If switch "Sw[09]" is at '0' during the test, the clock works with the regular 2 Hz frequency. If at '1' the clock stops and, at every push of "Key[01]" a complete clock cycle is generated. If the same button is kept pressed, the pulse is repeated at the rate of about two cycles per second.

Project Generation

The time has come to translate our network, with all its associations, into a *Project* that can be opened in the tool corresponding to the FPGA used.

In our case, the board is the DE0-CV, so the tool is *Quartus® II Web Edition*™. We use the button "Generate Project" (see the yellow arrow in the figure above) to start the process.

The window now looks as it is shown in the following figure.

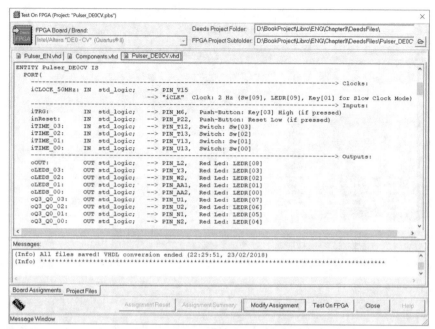

Deeds generates several files that represent the translation of our network in VHDL code. VHDL will be introduced later in this chapter: for the moment is enough to know that the files describe in a textual format connections and behavior of all the components in our network.

When the *Project* has been generated a dialog windows appears:

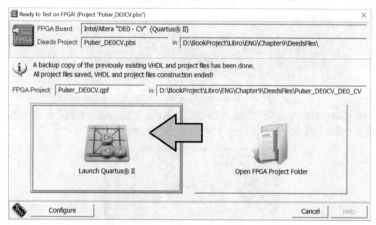

The window confirms that the generation of the Project has been successful and allows either to launch *Quartus® II* (green arrow) or simply to access the project's file folder.

Quartus® II will open our project, as seen in the figure in the next page.

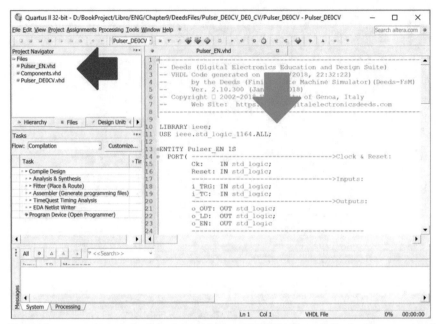

The red arrow points to the list of the generated VHDL files, while the blue to one of the files opened in the editor. It would be possible, at this point, to modify the VHDL code generated automatically by *Deeds*.

In fact, FPGA boards offer a wealth of possibilities from which students could take advantage, using the code generated by *Deeds* as a starting point. For our purposes, instead, it is enough to click "Start Compilation" to proceed with the FPGA configuration process.

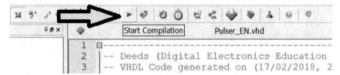

When the compilation is finished, a message reports a certain number of *warnings*. Warnings are not errors and, in our case, can be ignored.

Chip Programming

In the last phase, we must download the result of compilation on the board, using the "Programmer" module:

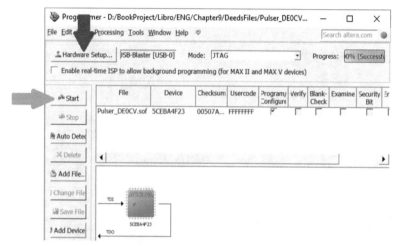

The green arrow points to the "Start" button that commences downloading. The red arrow points to the command that establishes communication between the PC hosting the tools and the FPGA board.

The VHDL code generated in this example will be explained in the next section, that also will introduce the basic elements of the language.

9.2 Introduction to VHDL

As stated in the preface of this book, we presented so far digital design basics using a traditional schematic entry approach. This choice favors an intuitive and visual understanding of concepts and circuits and allows a "conscious" transition to *Hardware Description Languages* (HDL).

In fact, HDLs are the current industry standard to describe and design digital systems. A very large and complex digital system can be efficiently designed with a top-down methodology using HDLs. Schematics are still used at the board level, where it is necessary to describe the wiring to the other parts of the system, or to the external connections.

A HDL is a programming language that allows us to describe digital circuits, either in a behavioral or structural way. HDLs are often used for simulation.

Very high-speed integrated-circuit Hardware Description Language (VHDL) and Verilog are the most widespread and are supported by the majority of CAD tools available. Currently a great and growing attention is reserved to System C, mainly a set of C++ class libraries that provide the necessary modeling code to describe systems. Other HDLs to be mentioned are JHDL (*Java HDL*) or Active-HDL (from *Cypress Semiconductor*).

In this section, we introduce VHDL, starting from the example seen just before. Our presentation of the language is far from exhaustive since VHDL is used for many purposes and here we see only one of them, the *"VHDL for synthesis"*.

9.2.1   VHDL Code from Deeds

We refer to the schematic seen in the previous section (see below).

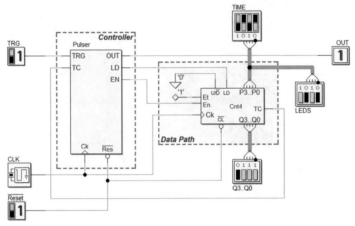

Apart from the connections (from inputs, to outputs and between blocks), the network contains two components, a controller (designed as FSM) and a 4-bit counter. Now, let's look at the files generated by *Deeds*, by going back in the chapter to the window that follows.

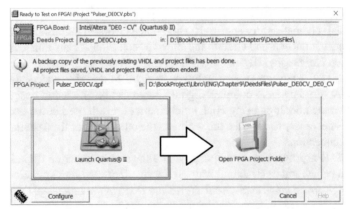

This time we choose to press the big button on the right ("Open FPGA Project Folder") to examine in more detail the files generated within the project.

Pulser_DE0CV.qsf	18/02/2018 12:43	4 KB
Pulser_DE0CV.qpf	18/02/2018 12:43	1 KB
Pulser_EN.vhd	18/02/2018 12:43	4 KB
Pulser_DE0CV.vhd	18/02/2018 12:43	16 KB
Components.vhd	18/02/2018 12:43	23 KB
ReportMessages.txt	18/02/2018 12:43	4 KB

The first two, in text format, define all the project's parameters, such as the name of the FPGA chip and the correspondence between the input/output connections of our network and the FPGA physical connections (pins). The last one is a log file that serves for debugging purposes. We concentrate our interest on the files with extension ".vhd" (the complete code is available on in Appendix C).

The "Components.vhd" contains the behavioral description of all the components included in the *Deeds* project (except FSMs and microcomputers). In our case, excluding the FSM (the controller), and the I/O objects, only one component should be in this VHDL file, the 4-bit counter.

However, the file contains also another component (named *"ClockScaler"*, not described here in detail), which is generated by *Deeds* to implement the *"Slow Clock Mode"*. As introduced before, this is the component that provides to the network a clock whose frequency is scaled down to the value defined in the settings.

9.2.2 Counter

A quick look into the VHDL file allows us to see the following code, which describes the terminals of the counter.

```
 1  ENTITY Counter4b IS
 2    PORT ( Ck  : IN  std_logic;
 3           nCL : IN  std_logic;
 4           LD  : IN  std_logic;
 5           ENP : IN  std_logic;
 6           ENT : IN  std_logic;
 7           UD  : IN  std_logic;
 8           P3  : IN  std_logic;
 9           P2  : IN  std_logic;
10           P1  : IN  std_logic;
11           P0  : IN  std_logic;
12           Q3  : OUT std_logic;
13           Q2  : OUT std_logic;
14           Q1  : OUT std_logic;
15           Q0  : OUT std_logic;
16           Tc  : OUT std_logic );
17  END Counter4b;
```

VHDL uses the *design entity* concept, and, in our example, "Counter4b" is an *entity*. This piece of code represents the *entity declaration unit* of the component and describes its external "PORT" interface, including all the terminals by name and type. For instance, *Tc* is defined as output ("OUT") and of type *std_logic*. The same type applies to all inputs and outputs of this example.

The type *std_logic* is used to define a digital signal that can assume the 0 and 1 values. While this may look obvious, really it is not since this type can assume nine different values, useful for simulation (for instance, X = Unknown).

The *entity declaration unit* must be followed by the *architecture body unit*, which represents the internal description of the *design entity*. It can describe its behavior, its structure, or a mix of both. For instance, in the following code example, we define an entity with two inputs (A, B) and one output (U).

```
 1  ENTITY MyNand IS
 2     PORT ( A,
 3           B:  IN  std_logic;
 4           U:  OUT  std_logic );
 5  end MyNand;
 6
 7  ARCHITECTURE behavioral OF MyNand IS
 8  BEGIN
 9    U <= not (A and B);
10  END behavioral;
```

The architecture part is described between the couple of keywords BEGIN .. END. The word "behavioral" is completely arbitrary (it should be the *private name* of the architecture, but it is not useful in this context). Instead, it is often used by designers to identify the style of the description (*behavioral* or *structural*).

With *behavioral description*, we mean that the entity is described by a function or an algorithm, without dealing directly with the components and connections (as in the example, where a Boolean function defines the output U). A *structural description*, instead, is the textual translation of connection among the blocks that form the logic network, in a way equivalent to traditional schematics. In any case, the two styles can be mixed, when convenient.

Let's return to our 4-bit counter, and consider the following code, extracted from its architecture description. We see an example of process:

```
 1  ARCHITECTURE behavioral OF Counter4b IS
 2  BEGIN
 3    Count4b: PROCESS ( Ck, nCL, ENP, ENT, UD )
 4    variable aCnt: unsigned ( 3 downto 0 );
 5    BEGIN
 6                    -- omissis...
 7
 8    END PROCESS;
 9  END behavioral;
```

A process could resemble, at a first glance, a function as the ones that we can write in C, JAVA, or other *procedural languages*. The resemblance is only apparent: a process describes the entity's behavior as a *parallel process*.

There is not a *main program* that calls the processes: they are *"launched"* when at least one of the signals defined between parentheses (*sensitivity list*) changes. In our case, the process is executed if the value of at least one of inputs *Ck, nCL, ENP, ENT*, or *UD* changes, as it happens in a physical network.

Let's examine the code of the process that describes the 4-bit counter.

```
 1  Count4b: PROCESS( Ck, nCL, ENP, ENT, UD )
 2  variable aCnt: unsigned( 3 downto 0 );
 3  BEGIN
 4    if    (nCL = '0') then   aCnt := (others =>'0');
 5    elsif (nCL = '1') then
 6      if (Ck'event) AND (Ck='1') then
 7        if    (LD = '1') then   aCnt := (P3 & P2 & P1 & P0);
 8        elsif (LD = '0') then
 9          if  (ENP = '1') and (ENT = '1')then
10            if    (UD = '1') then
11              if (aCnt < "1111") then  aCnt := aCnt + 1;
12              else  aCnt := (others =>'0');
13              end if;
14            elsif (UD = '0') then
15              if (aCnt > "0000") then aCnt := aCnt - 1;
16              else  aCnt := (others =>'1');
17              end if;
18            else  aCnt := (others =>'X');  -- (UD: Unknown)
19            END IF;
20          elsif not((ENP ='0') or (ENT ='0') ) then
21            aCnt := (others =>'X'); -- (ENP: Unknown)
22          END IF;
23        else  aCnt := (others =>'X');  -- (LD: Unknown)
24        END IF;
25      END IF;
26    else  aCnt := (others =>'X'); -- (nCL: Unknown)
27    END IF;
28    --
29    Tc <= ENT and ((aCnt(3) and aCnt(2) and
30                    aCnt(1) and aCnt(0) and UD) or
31                  (not(aCnt(3) or   aCnt(2) or
32                    aCnt(1) or   aCnt(0) or UD)));
33    --
34    Q3 <= aCnt(3);
35    Q2 <= aCnt(2);
36    Q1 <= aCnt(1);
37    Q0 <= aCnt(0);
38    --
39  END PROCESS;
```

A few preliminary observations:

(a) inputs *Ck*, *nCL*, *ENP*, *ENT* and *UD* (the ones that appear in the *sensitivity list*, line #1) are tested by *if-then-else* constructs (the keyword *elsif* is short for *else if*).
(b) line #2 defines a local variable *aCnt*, of type *unsigned 4-bit integer*, representing the counter state.
(c) the *aCnt := (others =>'0')* construct (as at line #4) sets to zero all the variable's bits.
(d) in the same way, *aCnt := (others =>'X')* (for instance at line #18) declares that the state of the counter is *unknown* (notice that this fact is relevant only for simulation, not synthesis).
(e) line #6 calls the library function *Ck'event* that detects a level transition of *Ck*. The argument of the *if* will be true in the case of a *rising edge*.
(f) two consecutive dashes '- -' signal the beginning of a comment that ends with the line.

Let's interpret now the code in natural language. If the input clear *nCL* is '0', the counter's state is cleared asynchronously (line #4). If *nCL* is at '1', the system waits for a rising edge of clock (line #5,6) and, therefore, the part of code following this control will be evaluated only when a rising edge arrives. Because of the rising edge check, the compiler must instance memory elements to store *aCnt* between two rising edges.

On the rising edge, the process controls the load *LD* input (line #7). If *LD* = '1', the value of the inputs *P3*, *P2*, *P1*, and *P0* is loaded in *aCnt*. Notice that the four inputs are combined together by the operator '&' to form a *4-bit variable*. If, instead, *LD* = '0', the process controls inputs *ENP* and *ENT* (line #9), which must be at '1' at the same time to enable counting.

If this condition is verified, next line controls input *UD* that sets the counting direction (if *UD* = '1', the statement *aCnt := aCnt + 1* is executed, else the other *aCnt := aCnt - 1*. Since the counting must be cyclical, *aCnt* is cleared (line #12) if *UD* = '1' and count has reached '1111'; else if *UD* = '0' and counter state *aCnt* is '0000', next state is assigned to '1111' (line #16).

Notice that the controls for possible unknown input signals make a bit more complex the interpretation of the code.

Following next's state logic, we find (line #29) the Boolean expression that assigns the output *Tc* (terminal count). Such expression is outside the construct that depends on *nCl* and the *Ck*. The compiler will translate it in a combinational network, function of the counter's state and *ENT* and *UD* inputs.

Last, we find the assignments of the outputs *Q3..Q0* that copy the bits of the state variable (lines #34..37). The special operator '<=' defines the assignment of a value to an output (inputs and outputs of a process are "SIGNALs" in VHDL jargon). Instead, in the previous lines of code the assignments to the variable *aCnt* have been defined using the operator ':='.

In the following, the difference between operators '<=' and ':=' will be considered. For now, it is enough to say that in VHDL the "SIGNALs" correspond to physical connections, while "VARIABLEs" concur to define the logic behavior of the network.

9.2.3 Finite State Machine

"Pulser_EN.vhd" describes the FSM with the same name but extension ".fsm" that has been designed as ASM chart. FSMs are exported in separated VHDL files (they are not included in the "Components.vhd"). Below is the "Pulser_EN" *entity declaration unit* with its input and output connections.

```
 1  ENTITY Pulser_EN IS
 2      PORT ( -------------------------------------->Clock & Reset:
 3          Ck:     IN std_logic;
 4          Reset:  IN std_logic;
 5          -------------------------------------->Inputs:
 6          i_TRG:  IN std_logic;
 7          i_TC:   IN std_logic;
 8          -------------------------------------->Outputs:
 9          o_OUT:  OUT std_logic;
10          o_LD:   OUT std_logic;
11          o_EN:   OUT std_logic
12          -------------------------------------------------
13          );
14  END Pulser_EN;
```

Next figure describes the FSM in general terms, by the three blocks with their connections. To facilitate the identification of inputs and outputs, *Deeds* added to their names a prefix '*i_*' or '*o_*'.

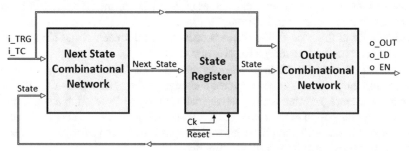

The code of the *architecture body unit*, below, recalls the three blocks.

```
 1  ARCHITECTURE behavioral OF Pulser_EN IS
 2    TYPE states is ( state_rs,
 3                     state_wt,
 4                     state_ps,
 5                     dummy_11 );
 6    SIGNAL State,
 7           Next_State: states;
 8  BEGIN
 9    -- Next State Combinational Logic --------------
10    FSM: process( State, i_TRG, i_TC )
11    begin
12                          -- omissis --
13    end process;
14
15    -- State Register ------------------------------
16    REG: process( Ck, Reset )
17    begin
18                          -- omissis --
19    end process;
20
21    -- Outputs Combinational Logic ----------------
22    OUTPUTS: process( State, i_TRG, i_TC )
23    begin
24                          -- omissis --
25    end process;
26  END behavioral;
```

Note at line #6,7 the definition of *State* and *Next_State*, both declared as "SIGNAL". The compiler will create physical connections for them.

State and *Next_State* are defined as "*states*", an ordinal type declared in the previous line #2..5 that defines the state names (*Deeds* has renamed the states instanced in the ASM chart with a prefix "*state_*").

The body of the code defines three processes, one for each of the blocks of the FSM general model (see lines #10, #16, and #21).

Let us first consider the *State Register* process (see below). The *sensitivity list* shows that the process is executed when a change of the input signals *Ck* and *Reset* occurs. The *state_rs* value is assigned asynchronously to *State* if the *Reset* input is '0', on the rising edge of the *Ck* input. If *Reset* = '1', *Next_State* is copied in *State*. This piece of code will compile as a parallel register that memorizes the FSM state. Note that "*rising_edge()*" is a library function.

```
 1  -- State Register ------------------------------
 2  REG: process( Ck, Reset )
 3  begin
 4    if (Reset = '0') then
 5             State <= state_rs;
 6    elsif rising_edge(Ck) then
 7             State <= Next_State;
 8        end if;
 9  end process;
```

The other two processes are compiled as combinational networks, because in their code all the outputs are completely specified as function of the inputs (see the next listing).

The *next state combinational logic* process declares *State*, *i_TRG*, and *i_TC* in the *sensitivity list*. The construct *case-when* is very similar to the C/C++ *switch* construct, or the Pascal *case*. In this code, the selector is *State*.

Trying to translate the code in natural language, we see that, if *State* is equal to *state_wt* (line #5), the value of the input *i_TRG* decides if *next_state* will be equal to *state_ps* or will remain equal to the current *state_wt*.

The other lines are very similar, except at line #23, that simply defines the default condition. If the current state could be different from the ones defined by design, the FSM will be forced into the reset state *state_rs*.

Note that if this last statement were missing, the compiler will understand that we would maintain memorized the outputs when none of the stated combination is there.

It will then generate a sequential circuit, instead of a combinational one. To be sure to avoid this, designers play it safe by inserting the *"when OTHERS"* clause even when not strictly necessary.

```
1  -- Next State Combinational Logic -------------
2  FSM: process ( State , i_TRG , i_TC )
3  begin
4    CASE State IS
5      when state_wt =>
6                  if (i_TRG = '1') then
7                      Next_State <= state_ps;
8                  else
9                      Next_State <= state_wt;
10                 end if;
11     when state_ps =>
12                 if (i_TC = '1') then
13                     Next_State <= state_rs;
14                 else
15                     Next_State <= state_ps;
16                 end if;
17     when state_rs =>
18                 if (i_TRG = '1') then
19                     Next_State <= state_rs;
20                 else
21                     Next_State <= state_wt;
22                 end if;
23     when OTHERS =>
24                 Next_State <= state_rs;
25   END case;
26 end process;
```

The last process describes the *Output Combinational Network*, where the outputs *o_OUT*, *o_LD*, and *o_EN* are defined as function of *State*, with a *CASE* statement, in a way similar to the previous process. Note that in this example the clause "*when OTHERS*" is used too.

```
1   -- Outputs Combinational Logic ----------------
2   OUTPUTS: process( State, i_TRG, i_TC )
3   begin
4     -- Set output defaults:
5     o_OUT <= '0';
6     o_LD  <= '0';
7     o_EN  <= '0';
8
9     -- Set output as function of current state and input:
10    CASE State IS
11      when state_wt =>
12                o_LD <= '1';
13      when state_ps =>
14                o_OUT <= '1';
15                if (i_TC = '0') then
16                    o_EN <= '1';
17                end if;
18      when OTHERS =>
19                o_OUT <= '0';
20                o_LD <= '0';
21                o_EN <= '0';
22    END case;
23  end process;
```

This piece of code allows us to get familiar with the usage of the operator '$<=$' in processes. For instance, pay attention to line #6, where we assign '0' to the signal *o_LD*.

Then, at line #12 the same output *o_LD* is set to '1', if *State = state_wt*. If this code were written in C/C++ or another procedural language, this would describe a sequence where "*before*" *o_LD* is reset, and "*after*", set. But this is VHDL, and a process defines a parallel behavior in which the assignments to *signals* should be understood in a different way.

In this code, when the process *executes*, it produces a logic value on the outputs, and no delay is implicit there. The assignment *o_LD* $<=$ '0' should be intended as a default output value for *o_LD*.

If no other logic condition applies, *o_LD* will be '0' on the process execution end. Otherwise, if a different condition will define *o_LD* at '1', this will be the value on the process execution end, without regard to the other assignment.

9.2.4 Top-Level Entity

In a VHDL project, the *top-level entity* represents the entire system and instances all the VHDL project entities defined in the project itself, in a *hierarchical* way. In the example that we have considered, the *top-level entity* is defined in the file "Pulser_DE0CV.vhd".

Deeds has generated it starting from the schematics, including in it all the needed references to the component and FSM entities, adding the description of all the connections between the blocks. Let's extract and comment a few excerpts from the code.

In the *entity declaration unit*, we find all the external connections of our circuit. *Deeds* added automatically some useful comment, just to remind the connections to the board devices.

In the example below, at line #7 the input *iTRG* has been connected, according to the user definitions, to the button 'Key[03]', through the 'PIN_M6' of the FPGA chip. Obviously, these comments are not relevant for the compiler, which will find these definitions embedded into the *Quartus® II* project file.

```
 1  ENTITY Pulser_DE0CV IS
 2    PORT (
 3      iCLOCK_50MHz:  IN std_logic;  --> PIN_V15
 4                                    --  "iCLK"  Clock: 2 Hz
 5                                    --  Sw[09],LEDR[09],Key[01]
 6                                    --  for Slow Clock Mode
 7      iTRG:          IN std_logic;  --> PIN_M6,
 8                                    --> Push-Button: Key[03]
 9      inReset:       IN std_logic;  --> PIN_P22,
10                                    --> Push-Button: Reset
11      iTIME_03:      IN std_logic;  --> PIN_T12,
12                                    --> Switch: Sw[03]
13      iTIME_02:      IN std_logic;  --> PIN_T13,
14                                    --> Switch: Sw[02]
15              -- omissis
16    );
17  END Pulser_DE0CV;
```

At line #3, we see the connection to the 50-MHz native board clock. Internally, this is connected to the clock scaler circuit (note that the generated comment reminds the user setting of the clock frequency and the *Slow Clock Mode*).

All the other definitions have been omitted here to shorten the code listing and make it more readable. Indeed, the number of entries is considerable and depends on the board's resources in use.

Deeds instances all the relevant output connections, even if not directly used by our project, to switch off all the displays, LEDs, and other unused output devices during the circuit test.

The code of the *architecture body unit*, shown in the next listing, declares the components used by the entity, for instance, the counter "Counter4b" (line #3), the FSM "Pulser_EN" (line #8), and other ones (here omitted).

The architecture description is *"structural"*, so all the SIGNALs used are declared together with the mapping of all the connections among the components. For instance, at line #13, a SIGNAL of name 'S001' is declared (the code generation is automated, so *Deeds* assign a different name to each net using its internal netlist identifier).

```
 1  ARCHITECTURE structural OF Pulser_DE0CV IS
 2                                      -- omissis
 3     COMPONENT Counter4b IS
 4        PORT(                         -- omissis
 5               );
 6     END COMPONENT;
 7
 8     COMPONENT Pulser_EN IS
 9        PORT(                         -- omissis
10               );
11     END COMPONENT;
12
13     SIGNAL S001: std_logic;
14                                      -- omissis
15     SIGNAL S017: std_logic;
16
17  BEGIN
18                                      -- omissis
19  END structural;
```

As an example, you see below the declaration of the *interface* of FSM component "Pulser_EN", which was shortened in the previous listing.

```
 1  COMPONENT Pulser_EN IS
 2     PORT( ------------------------->Clock & Reset:
 3            Ck:      IN std_logic;
 4            Reset:   IN std_logic;
 5            --------------------------->Inputs:
 6            i_TRG:   IN std_logic;
 7            i_TC:    IN std_logic;
 8            --------------------------->Outputs:
 9            o_OUT:   OUT std_logic;
10            o_LD:    OUT std_logic;
11            o_EN:    OUT std_logic
12            --------------------------
13            );
14  END COMPONENT;
```

Let's examine now the body of the architecture, shortened to focus the attention on the relevant elements. At line #3 and following, a few examples of usage of the '<=' operator , used outside processes, represent a simple *wired* connection.

```
 1   BEGIN
 2                         -- omissis --
 3       S005  <=  iCLK;
 4       S007  <=  inReset;
 5                         -- omissis --
 6       oQ3_Q0_00  <=  S010;
 7       oQ3_Q0_01  <=  S011;
 8                         -- omissis --
 9       S004  <=  '0';
10       S001  <=  '1';
11                         -- omissis --
12       C680:  Counter4b  PORT  MAP( S005,  S007,  S003,  S006,
13                                    S001,  S004,  S017,  S016,
14                                    S015,  S014,  S013,  S012,
15                                    S011,  S010,  S009 );
16       C704:  Pulser_EN  PORT  MAP( S005,  S007,  S002,  S009,
17                                    S008,  S003,  S006 );
18   END  structural;
```

Line #3 means that the input *iCLK* is connected to the internal SIGNAL named 'S005'. In a similar way, at line #6 the internal net 'S010' is connected to the output *oQ3_Q0_00*. As an example of constant setting, the internal SIGNAL 'S004' is set to '0' (line #9).

The architecture body ends with the effective connection of the components to the internal nets. In the VHDL jargon, this operation is named *"mapping"* and is obtained here with the statement "PORT MAP".

In this example, on line #17 an instance named 'C704' of the "Pulser_EN" component is connected to the internal SIGNALs declared in the arguments. Their order corresponds to that of the component declaration, so, for instance, its *Reset* terminal is connected to the SIGNAL 'S007'.

9.2.5   Other VHDL Examples

In the following, a few examples of VHDL code describe frequently used networks.

Decoder

A decoder activates the output corresponding to the input binary code, assuming the Enable input is active (as seen in Sect. 2.6.1). Below is the symbol of a $2 \rightarrow 4$ decoder and its *entity declaration unit*, as defined in *Deeds*.

```
 1   ENTITY  Decoder_2_4  IS
 2     PORT(  A1:  IN   std_logic;
 3            A0:  IN   std_logic;
 4            EN:  IN   std_logic;
 5            Y0:  OUT  std_logic;
 6            Y1:  OUT  std_logic;
 7            Y2:  OUT  std_logic;
 8            Y3:  OUT  std_logic  );
 9   END  Decoder_2_4;
```

The PORT declaration lists inputs *A1*, *A0*, *EN*, and outputs *Y0..Y3*. Below is a possible component's description using the construct *with-select-when*.

In this case, we did not use a process to describe the behavior. The language allows a considerable freedom of choice; in this case, it is convenient to adopt a descriptive approach, basically similar to a truth table.

In the VHDL code generated by *Deeds*, below, line #2 defines *aNumber* as a *vector* of *three wires* (indexed as 2,1,0), which on line #3 gathers together the three inputs *EN*, *A1*, and *A0*.

The binary value of *aNumber*, i.e., the group of the three input wires, selects the active output.

```
 1   ARCHITECTURE behavioral OF Decoder_2_4 IS
 2     SIGNAL aNumber: std_logic_vector( 2 downto 0 );
 3   BEGIN
 4     aNumber  <= EN & A1 & A0;
 5     with aNumber select
 6       Y0 <= '0' when "000", '0' when "001",
 7              '0' when "010", '0' when "011",
 8              '1' when "100", '0' when "101",
 9              '0' when "110", '0' when "111", 'X' when others;
10     with aNumber select
11       Y1 <= '0' when "000", '0' when "001",
12              '0' when "010", '0' when "011",
13              '0' when "100", '1' when "101",
14              '0' when "110", '0' when "111", 'X' when others;
15     with aNumber select
16       Y2 <= '0' when "000", '0' when "001",
17              '0' when "010", '0' when "011",
18              '0' when "100", '0' when "101",
19              '1' when "110", '0' when "111", 'X' when others;
20     with aNumber select
21       Y3 <= '0' when "000", '0' when "001",
22              '0' when "010", '0' when "011",
23              '0' when "100", '0' when "101",
24              '0' when "110", '1' when "111", 'X' when others;
25   END behavioral;
```

The rest of the code assigns a value to each output, on the bases of the selection dictated by *aNumber*. For instance, at lines #6..9, the output *Y0* is assigned to '0' for all the combinations of *aNumber*, except *when* it is equal to '100' (i.e., when *EN* = '1', *A1* = '0' and *A0* = '0').

Remember that the assignment operator '<=', when used outside the processes, represents a connection. The clause *'X' when others* is introduced only for simulation purposes, and it is not necessary for synthesis.

Multiplexer

As described in Sect. 2.6.2, a multiplexer selects which one of the inputs will be copied in the output, according to the binary value of the selection inputs. The $4 \rightarrow 1$ multiplexer symbol and the corresponding *entity declaration unit* follow.

```
1  ENTITY Multiplexer_4_1 IS
2    PORT ( I0: IN   std_logic;
3           I1: IN   std_logic;
4           I2: IN   std_logic;
5           I3: IN   std_logic;
6           S1: IN   std_logic;
7           S0: IN   std_logic;
8           Q:  OUT  std_logic );
9  END Multiplexer_4_1;
```

In this case, it is convenient to use the construct *when-else*. The behavior of the component is easily readable in the following code.

```
1  ARCHITECTURE behavioral OF Multiplexer_4_1 IS
2  BEGIN
3    Q <= I0 when ((S1 = '0') and (S0 = '0')) else
4         I1 when ((S1 = '0') and (S0 = '1')) else
5         I2 when ((S1 = '1') and (S0 = '0')) else
6         I3 when ((S1 = '1') and (S0 = '1')) else 'X';
7  END behavioral;
```

The output Q takes the value of one of the inputs $I0..I3$, according to the combinations of the input $S1$ and $S0$, this time represented as logic expressions. For instance, Q copies $I2$ if $(S1 = '1')$ and $(S0 = '0')$.

Demultiplexer

Input IN is copied on the output selected by $S1$ and $S0$ (see Sect. 2.6.3). An example of a $1 \rightarrow 4$ demultiplexer follows.

```
1  ENTITY Demultiplexer_1_4 IS
2    PORT ( I:  IN   std_logic;
3           S1: IN   std_logic;
4           S0: IN   std_logic;
5           Q0: OUT  std_logic;
6           Q1: OUT  std_logic;
7           Q2: OUT  std_logic;
8           Q3: OUT  std_logic );
9  END Demultiplexer_1_4;
```

A demultiplexer is the same as a decoder with enable input; therefore, the implementation of both is identical.

```
1  ARCHITECTURE behavioral OF Demultiplexer_1_4 IS
2    SIGNAL aNumber: std_logic_vector ( 2 downto 0 );
3  BEGIN
4    aNumber <= I & S1 & S0;
5    with aNumber select
6      Q0 <= '0' when "000", '0' when "001",
7            '0' when "010", '0' when "011",
8            '1' when "100", '0' when "101",
```

```
 9              '0' when "110", '0' when "111", 'X' when others;
10                           -- omissis ...
11   END behavioral;
```

Full Adder

The component, described in Sect. 3.9.2, adds two bits and the input carry and generates the sum and the output carry. On the right is the VHDL declaration of its terminals.

```
 1   ENTITY Adder_Full IS
 2      PORT ( CIN:  IN   std_logic;
 3             COUT: OUT  std_logic;
 4             A:    IN   std_logic;
 5             B:    IN   std_logic;
 6             S:    OUT  std_logic );
 7   END Adder_Full;
```

The construct *with-select-when* produces a list that reflects the component's truth table. *ABC*, defined as *vector* of three *signals*, groups together inputs *A*, *B*, and *CIN* (line #4).

```
 1   ARCHITECTURE behavioral OF Adder_Full IS
 2      SIGNAL ABC: std_logic_vector( 2 downto 0 );
 3   BEGIN
 4      ABC <= A & B & CIN;
 5      --
 6      with ABC select
 7      S <= '0' when "000",
 8           '1' when "001",
 9           '1' when "010",
10           '0' when "011",
11           '1' when "100",
12           '0' when "101",
13           '0' when "110",
14           '1' when "111",
15           'X' when others;
16      --
17      with ABC select
18      COUT <= '0' when "000",
19              '0' when "001",
20              '0' when "010",
21              '1' when "011",
22              '0' when "100",
23              '1' when "101",
24              '1' when "110",
25              '1' when "111",
26              'X' when others;
27   END behavioral;
```

Magnitude Comparator

It is a component used to compare the magnitude of two unsigned integer numbers (encountered in Sect. 8.3.6). The figure shows the 4-bit version. On the right is the corresponding entity in VHDL, as generated by *Deeds*.

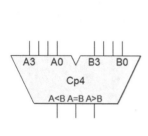

```
 1  ENTITY  Compar_4  IS
 2     PORT ( A3 :    IN    std_logic;
 3            A2 :    IN    std_logic;
 4            A1 :    IN    std_logic;
 5            A0 :    IN    std_logic;
 6            B3 :    IN    std_logic;
 7            B2 :    IN    std_logic;
 8            B1 :    IN    std_logic;
 9            B0 :    IN    std_logic;
10            MIN :   OUT   std_logic;
11            EQU :   OUT   std_logic;
12            MAJ :   OUT   std_logic );
13  END  Compar_4;
```

MIN is asserted when operand *A* is lower than operand *B*.

MAJ is set to one in the opposite case, and *EQU* when they are equal.

This description is easily readable also in the architecture description, shown below, where a process construct is used.

```
 1  ARCHITECTURE behavioral OF Compar_4 IS
 2  BEGIN
 3     Cmp4: PROCESS( A3, A2, A1, A0,
 4                    B3, B2, B1, B0 )
 5     variable A: unsigned( 3 downto 0 );
 6     variable B: unsigned( 3 downto 0 );
 7     BEGIN
 8        A := (A3 & A2 & A1 & A0);
 9        B := (B3 & B2 & B1 & B0);
10        --
11        if    (A > B) then MIN <= '0'; EQU <= '0'; MAJ <= '1';
12        elsif (A < B) then MIN <= '1'; EQU <= '0'; MAJ <= '0';
13        elsif (A = B) then MIN <= '0'; EQU <= '1'; MAJ <= '0';
14        else              MIN <= 'X'; EQU <= 'X'; MAJ <= 'X';
15        END IF;
16     END PROCESS;
17  END behavioral;
```

In the process body, two variables (*A* and *B*) group together the corresponding input wires (lines #8..9). In this way, we can compare in algebraic mode the variables and assert, or not, the output coherently (lines #11..13).

Flip-flop D-PET

Below, the VHDL description of a D-PET-type flip-flop, with $\overline{Clear}$ and $\overline{Preset}$.

```
 1  ENTITY DpetFF IS
 2     PORT ( D, Ck    : IN std_logic;
 3            nCL, nPR : IN std_logic;
 4            Q, nQ    : OUT std_logic );
 5  END DpetFF;
```

Note that the *D* input is not in the sensitivity list (line #3), because a change of that input only will not modify the flip-flop state. Lines #5..7 evaluate the asynchronous inputs $\overline{Clear}$ and $\overline{Preset}$ (*nCL* and *nPR*), changing the flip-flop outputs independently of the clock. If they are not active (line #8), on the positive clock edge the output *Q* will copy the value of *D*.

```
1   ARCHITECTURE behavioral OF DpetFF IS
2   BEGIN
3     Dff: PROCESS( Ck, nCL, nPR )
4     BEGIN
5       if    (nCL='0') and (nPR='0') then Q <= 'X'; nQ <= 'X';
6       elsif (nCL='0') and (nPR='1') then Q <= '0'; nQ <= '1';
7       elsif (nCL='1') and (nPR='0') then Q <= '1'; nQ <= '0';
8       elsif (nCL='1') and (nPR='1') then
9         if (Ck'event) AND (Ck='1') THEN -- Positive Edge
10            Q <= D;  nQ <= not D;
11        END IF;
12      else Q <= 'X'; nQ <= 'X';
13      END IF;
14    END PROCESS;
15  END behavioral;
```

Flip-Flop E-PET

The VHDL description of an E-PET-type flip-flop is obviously very similar to the one of the D-PET, with the addition of the input *E* condition.

```
1   ENTITY EpetFF IS
2     PORT (  D, E, Ck: IN std_logic;
3             nCL, nPR: IN std_logic;
4             Q, nQ    : OUT std_logic );
5   END EpetFF;
```

At line #12, the flip-flop outputs are updated on the positive clock edge only if *E* is active. Note that the line immediately below is there only for simulation purposes and states that if the input *E* is unknown, the outputs will be too.

```
1   ARCHITECTURE behavioral OF EpetFF IS
2   BEGIN
3     Eff: PROCESS( Ck, nCL, nPR )
4     BEGIN
5       if    (nCL='0') and (nPR='0') then Q <= 'X'; nQ <= 'X';
6       elsif (nCL='0') and (nPR='1') then Q <= '0'; nQ <= '1';
7       elsif (nCL='1') and (nPR='0') then Q <= '1'; nQ <= '0';
8       elsif (nCL='1') and (nPR='1') then
9         if (Ck'event) AND (Ck='1') THEN -- Positive Edge
10            if       (E = '1') then Q <= D;  nQ <= not D;
11            elsif not(E = '0') then Q <= 'X'; nQ <= 'X';
12            END IF;
13        END IF;
14      else Q <= 'X'; nQ <= 'X';
15      END IF;
16    END PROCESS;
17  END behavioral;
```

Flip-Flop JK-PET

The VHDL of JK-PET flip-flop is similar to the previous two.

```
1  ENTITY  JKpetFF  IS
2     PORT(   J,  K,  Ck:  IN std_logic;
3               nCL,  nPR:  IN std_logic;
4                Q,  nQ    :  OUT std_logic );
5  END  JKpetFF;
```

The presence of the JK *Toggle* modality needs a variable representing the state of the flip-flop (line #2).

```
1   ARCHITECTURE behavioral OF JKpetFF IS
2   BEGIN
3     JKff: PROCESS( Ck, nCL, nPR )
4       variable  OutQ: STD_LOGIC;
5       BEGIN
6       if     (nCL='0') and (nPR='1') then OutQ := '0';
7       elsif (nCL='1') and (nPR='0') then OutQ := '1';
8       elsif (nCL='1') and (nPR='1') then
9         if (Ck'event) AND (Ck='1') THEN
10          -- Positive Edge
11          if     (J = '0') AND (K = '1') THEN OutQ := '0';
12          elsif (J = '1') AND (K = '0') THEN OutQ := '1';
13          elsif (J = '1') AND (K = '1') THEN OutQ := not OutQ;
14          elsif not((J='0')AND(K='0')) THEN OutQ := 'X';
15          END IF;
16        END IF;
17      else                                OutQ := 'X';
18      END IF;
19      --
20      Q  <= (    OutQ);
21      nQ <= (not OutQ);
22      --
23    END PROCESS;
24  END behavioral;
```

On the positive clock edge, the *J* and *K* inputs are evaluated (lines #11..15) and the flip-flop state is updated. The output *Q* and $\overline{Q}$ are then updated on the end of the process code (lines #20..21).

Parallel Register

We have already encountered the VHDL of a parallel register, when we considered the *State Register* process of the FSM. Here is presented an example of a more general coding, defining a register with *Enable* and $\overline{Clear}$ inputs.

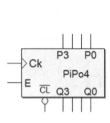

```
 1   ENTITY PiPoE4 IS
 2     PORT ( Ck  : IN  std_logic;
 3            nCL : IN  std_logic;
 4            E   : IN  std_logic;
 5            P3  : IN  std_logic;
 6            P2  : IN  std_logic;
 7            P1  : IN  std_logic;
 8            P0  : IN  std_logic;
 9            Q3  : OUT std_logic;
10            Q2  : OUT std_logic;
11            Q1  : OUT std_logic;
12            Q0  : OUT std_logic );
13   END PiPoE4;
```

To group together the register bits, we introduce the variable *aReg* (line #3). If $\overline{Clear}$ input is active (line #6), the variable is cleared. Else, on the positive clock edge, if the enable input *EN* is active (line #8 and 9), the input values *P3..P0* are assigned to the variable. The statements at lines #18..21 assign to the register outputs the updated variable bits.

```
 1   ARCHITECTURE behavioral OF PiPoE4 IS
 2   BEGIN
 3     RegPiPoE4: PROCESS ( Ck, nCL )
 4       variable aReg: std_logic_vector ( 3 downto 0 );
 5     BEGIN
 6       if    (nCL = '0') then     aReg := (others =>'0');
 7       elsif (nCL = '1') then
 8         if (Ck'event) AND (Ck='1') THEN -- Positive Edge
 9           if (E = '1') then
10               aReg := (P3 & P2 & P1 & P0);
11           elsif not (E = '0') then
12               aReg := (others =>'X');
13           END IF;
14         END IF;
15       else            aReg := (others =>'X');
16       END IF;
17
18       Q3 <= aReg(3);
19       Q2 <= aReg(2);
20       Q1 <= aReg(1);
21       Q0 <= aReg(0);
22
23     END PROCESS;
24   END behavioral;
```

Shift Register (S.I.P.O.)

An example of *Serial-Input Parallel-Output* shift register is shown on the figure. This 4-bit register has an *EN* input that enables the shifting. Data enters through the input *In* and shifts from *Q3* to *Q0*.

```
 1   ENTITY  SiPoE4  IS
 2      PORT (  I    :  IN  std_logic;
 3              Ck   :  IN  std_logic;
 4              nCL  :  IN  std_logic;
 5              E    :  IN  std_logic;
 6              Q3   :  OUT std_logic;
 7              Q2   :  OUT std_logic;
 8              Q1   :  OUT std_logic;
 9              Q0   :  OUT std_logic );
10   END  SiPoE4;
```

The VHDL description style resembles that of the parallel register see before, with the difference that, on the positive clock edge (and if the enable *E* is active), the variable bits are assigned coherently with the required function.

```
 1   ARCHITECTURE behavioral OF SiPoE4 IS
 2   BEGIN
 3     RegSiPoE4: PROCESS( Ck, nCL )
 4     variable aReg: std_logic_vector( 3 downto 0 );
 5     BEGIN
 6       if     (nCL = '0') then aReg := (others =>'0');
 7       elsif (nCL = '1') then
 8         if (Ck'event) AND (Ck='1') THEN -- Positive Edge
 9           if (E = '1') then
10             aReg := (I & aReg(3) & aReg(2) & aReg(1));
11           elsif not(E = '0') then
12             aReg := (others =>'X');
13           END IF;
14         END IF;
15       else aReg := (others =>'X');
16       END IF;
17       --
18       Q3 <= aReg(3);
19       Q2 <= aReg(2);
20       Q1 <= aReg(1);
21       Q0 <= aReg(0);
22       --
23     END PROCESS;
24   END behavioral;
```

On line #10, the new variable value is constructed joining the input *In* (*I* in the code) and the current bits in position 3, 2, and 1.

After the clock edge, we obtain the copy on the input in the higher bit, and the other ones shifted to the right.

9.3 FPGA Prototyping Exercises

This section offers a few exercises of system project and prototyping on FPGA. Others are included in the *Deeds* Web site's *Learning Materials*.

9.3.1 Synchronous Serial Communication System (8-bit)

Complete the design of a *synchronous serial communication system*, and then implement it on a FPGA board.

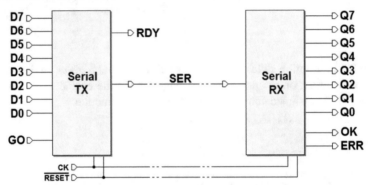

The TX unit reads an eight-bit parallel data (*D0..D7*); when the input *GO* goes from 0 to 1, data is serialized and transmitted on the line *SER*, in the format shown in the following figure:

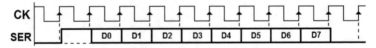

- The bit sequence is synchronous with the clock CK.
- The bit time duration is equal to one clock cycle.
- Each sequence begins with a start bit (high).
- Eight data bits follow (D0..D7).
- The bit packet ends with a Stop Bit (low).

The output *RDY* is activated when the transmitter is waiting for a *low to high transition* on the input *GO*.

The receiver RX waits for packets on the line *SER*. When a start bit is detected, the receiver processes the serial sequence and copies the *D0..D7* values on the outputs *Q0..Q7*.

The stop bit is evaluated. If correctly received, the system activates the *OK* output for the duration of a clock cycle, otherwise the receiver sets the *ERR* output, maintaining it active until *SER* returns to 0.

The proposed architecture is shown below (a template of the schematic and the controllers' FSM are available for downloading from *Deeds* Web site).

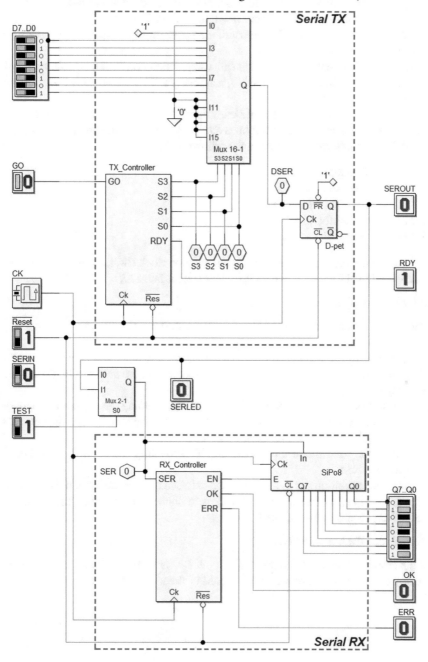

TX Design Guidelines

The TX circuit is divided into the usual structure *controller–datapath*, where the latter is composed of a multiplexer $16 \rightarrow 1$ and a D-PET flip-flop.

While waiting for the *GO* command, the transmitter generates a low level *(the idle state of the line)*.

On the positive edge of *GO*, it transmits sequentially on *SER* a '1' *(the start bit)*, the eight bits *D0..D7*, starting from *D0*, and finally a '0' *(the stop bit)*. Then, it waits for the next *GO* command.

A "Mux16-1" multiplexer (from *Deeds* library) provides the data, under the control of the FSM using the selection lines *S3..S0*. As shown in the previous figure, the multiplexer inputs are connected in the following order:

- $I0$ = low (the idle line);
- $I1$ = high (the start bit);
- $I2..I9 = D0..D7$ (the data bits);
- $I10$ = low (the stop bit);
- $I11..I15$ = low (not used).

To grant the synchronicity of the output with the clock *CK*, the multiplexer output is fed to a D-PET flip-flop, which drives the output line *SER*.

RX Design Guidelines

The RX datapath is represented by an eight-bit shift register ("SiPo8"), used to deserialize and store the data received on *SER*.

On the rising edge of the clock, if the register enable input *E* is active, data shifts by a position $(In \rightarrow Q7 \ ... \rightarrow Q1 \rightarrow Q0)$. Otherwise, the outputs *Q0..Q7* remain unchanged.

The controller synchronizes the shift operations, enabling the register when needed by activating the *EN* line. The controller waits for the start bit and then enables data shifting in the register, for each data bit.

Finally, the controller evaluates the stop bit, activating *OK* or *ERR*, according to the specifications.

Note that a multiplexer $2 \rightarrow 1$ was added between the transmitter and the receiver. The *TEST* input, when activated, allows us to connect the receiver directly to the transmitter to allow stand-alone testing.

Testing the Design on FPGA

If the circuit schematic template available on *Deeds* Web site is used, the I/O association for an *Terasic/Altera* DE0-CV board is already configured.

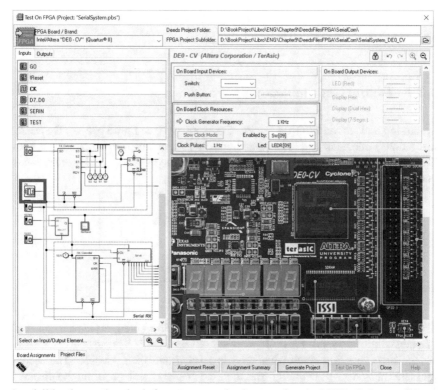

As visible above, the clock frequency set in the template is 1 KHz. It is possible to slow down the clock thanks to the *Slow Clock Mode* feature.

We set the switch "Sw[09]" to activate the mode, and the LED "LEDR[09]" visualizes the clock pulses. During the test, if the switch is at '0', the clock works at the regular frequency. If at '1', the clock slows down to 1 Hz, as defined in the setup.

To interact with the board resources, it could be convenient to refer to the figure below, which summarizes the associations.

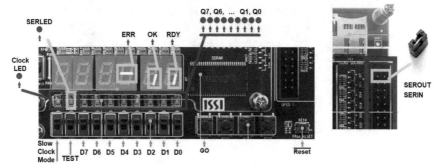

On right side of the figure are represented also the *SEROUT* and *SERIN* terminals, made available on the expansion header. They are useful to connect the transmitter to the receiver using a jumper or a cable (in this way, for example, we can test what happens if the line is disconnected during the test operations, or if the line is very long and exposed to external noise).

9.3.2 Digital Chronometer

We design and test on FPGA a *Digital Chronometer* (see figure below). Requested resolution is one hundredth of a second, maximum time measurable about one hour. Time is displayed with six decimal figures, two for the minutes, two for the seconds, and two for the hundredths of second.

The system has a push-button (*PLS*) and a lamp bulb (*LIT*). At system reset all displays' digits are set to zero. *LIT* is on for all the time *PLS* is pushed.

The time counting starts when *PLS* is pushed and then released. The second pressure on *PLS* stops counting, and the time elapsed can be read on the displays. The third pressure resets the display, and the timer waits for *PLS* to start a new counting sequence.

We assume to have available a clock *CK* with 100 Hz frequency and a FPGA board, such as the *Terasic/Altera* DE0-CV with six *seven-segment displays*.

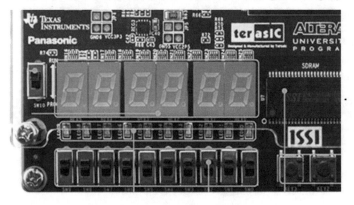

Design Guidelines

We suggest to use for the chronometer the structure *controller–datapath*. The latter is composed by the *time counter* and the associated display. The former handles the push-button functionality, the lamp, and the counter controls.

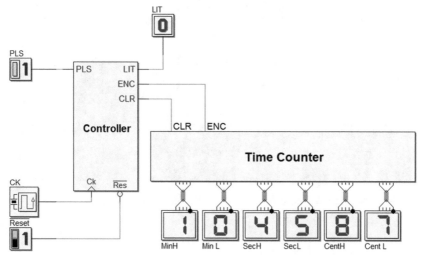

Observe that the *time counter* needs only two control inputs, *CLR* (to clear its contents) and *ENC* (to enable/stop counting). The controller can be implemented as a FSM that reads the push-button through the *PLS* input, activates the lamp with the output *LIT*, and generates the control signals *CLR* and *ENC* for the time counter.

We suggest to design a controller FSM that tracks the *PLS* input value, acting on its level changes as required by the specifications.

Time Counter

The time counter can be designed in many different ways. We suggest to separate the counter in six *Binary Coded Decimal* (BCD) elements, each for every digit, defining in *Deeds* a *Circuit Block Element* (CBE) block.

Following this approach, each digit will be driven by its own counter, receiving a count enable input *EN* from the component on the right side, and generating a *terminal count TC* to increment the digit placed on its left. Take also in account that the *seconds* and *minutes* should be counted *module 60*, so two of the blocks must generate *TC* on the '5' digit.

In the next figure, the CBE component to be completed, as available in the *digital contents* of this book, opened in the *Deeds-DcS* circuit editor.

The proposed CBE shows an added input *MOD* allowing to set the counting module (i.e., equal to 10 if *MOD* = '1', or 6 if *MOD* = '0'). The CBE receives also, as usual, the inputs *CK* and $\overline{RES}$.

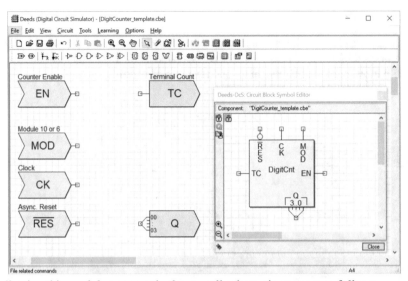

Following this *modular* approach, the overall schematic appears as follows.

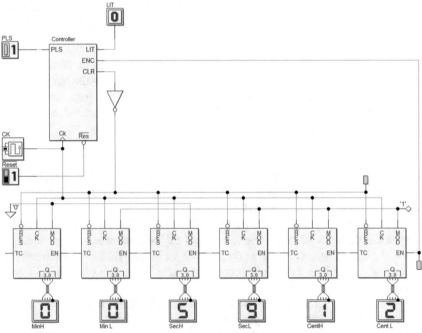

The CBE design, in turn, can be defined in different ways. A first, intuitive method can be to connect a 4-bit binary counter with a few gates around, to implement the module selection logic and the generation of *TC*. In this case, the approach must stand on the experience of the designers and their own creativity.

It is preferable to use a more systematic approach, designing the module in terms of FSM, therefore defining its behavior in an algorithmic way (as seen, for instance, in Sect. 7.2.2).

Testing the Chronometer on FPGA

A circuit to be completed is available on the *digital contents* of the book. If this file is used, the *I/O associations* are already configured (for an *Altera/Terasic* DE0-CV board). Following the specifications, the clock frequency is set to 100 Hz. The *"Slow Clock Mode"* setting allows us to feed the clock manually, for testing, pressing the button "Key[00]", if the switch "Sw[09]" is set to '1'. Normal operations will take place if the switch is left at '0'.

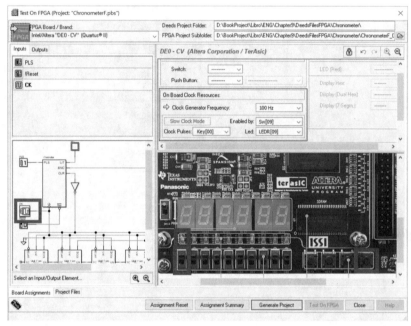

The following figure can be useful to interact with the board controls resources during the circuit test.

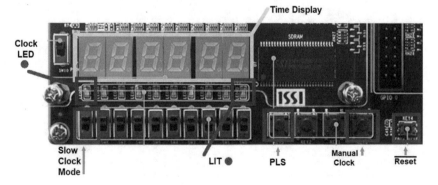

9.4   Solutions

9.4.1   *Synchronous Serial Communication System (8-bit)*

ASM Diagram of the Transmitter's Controller

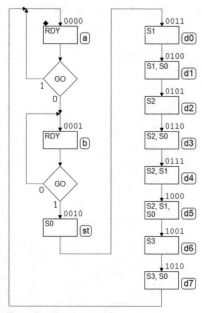

ASM Diagram of the Receiver's Controller

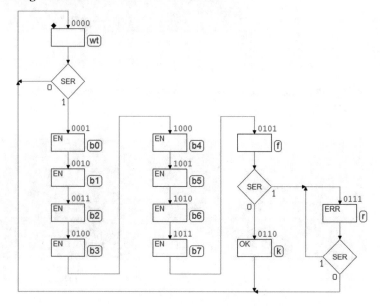

9.4.2 Digital Chronometer

Schematic

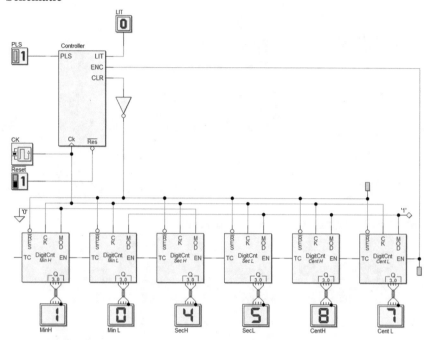

Controller ASM Diagram

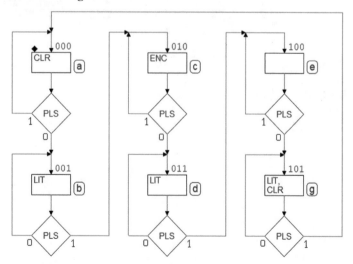

CBE Counter Module (Circuital Approach)

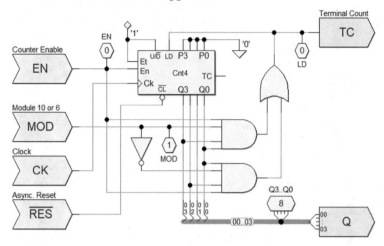

CBE Counter Module (Algorithmic Approach)

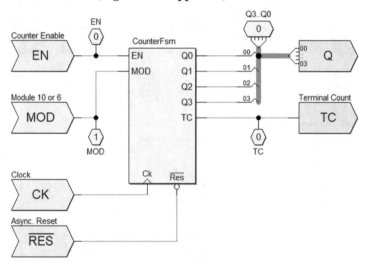

ASM Diagram (CBE Counter Module)

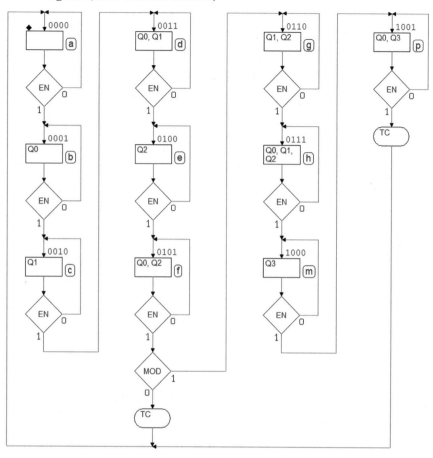

Appendix A
The Powers of 2

As we are dealing with binary values, it is useful to remember the values of the most significant powers of 2. Below are the smallest:

2^0	2^1	2^2	2^3	2^4	2^5	2^6	2^7	2^8	2^9	2^{10}
1	2	4	8	16	32	64	128	256	512	1024

We see that $2^{10} = 1024$ is equal to approximately $10^3 = 1000$. Let the prefix K (meaning Kilo) indicate the value 1024 even though the *International Electrotechnical Commission* (IEC) standard establishes the prefix Kibi (from Kilo binary):

2^{10}	2^{11}	2^{12}	2^{13}	2^{14}	2^{15}	2^{16}	2^{17}	2^{18}	2^{19}	2^{20}
$1K$	$2K$	$4K$	$8K$	$16K$	$32K$	$64K$	$128K$	$256K$	$512K$	$1024K$

Similarly 2^{20} let M (Mega) indicate (Mebi), equal to about 10^6:

2^{20}	2^{21}	2^{22}	2^{23}	2^{24}	2^{25}	2^{26}	2^{27}	2^{28}	2^{29}	2^{30}
$1M$	$2M$	$4M$	$8M$	$16M$	$32M$	$64M$	$128M$	$256M$	$512M$	$1024M$

Let G (Giga) (10^9) indicate (Gibi), which is equal to 2^{30}:

2^{30}	2^{31}	2^{32}	2^{33}	2^{34}	2^{35}	2^{36}	2^{37}	2^{38}	2^{39}	2^{40}
$1G$	$2G$	$4G$	$8G$	$16G$	$32G$	$64G$	$128G$	$256G$	$512G$	$1024G$

Let T (Tera) indicate 2^{40} (Tebi), about 10^{12}.

© Springer International Publishing AG, part of Springer Nature 2019
G. Donzellini et al., *Introduction to Digital Systems Design*,
https://doi.org/10.1007/978-3-319-92804-3

To be thorough, a table including all the prefixes follows.

IEC prefix		Representations			Customary prefix	
Name	Symbol	Base 2	Base 1024	Base 10	Name	Symbol
Kibi	Ki	2^{10}	1024^1	$= 1.024 \times 10^3$	Kilo	k or K
Mebi	Mi	2^{20}	1024^2	$\approx 1.049 \times 10^6$	Mega	M
Gibi	Gi	2^{30}	1024^3	$\approx 1.074 \times 10^9$	Giga	G
Tebi	Ti	2^{40}	1024^4	$\approx 1.100 \times 10^{12}$	Tera	T
Pebi	Pi	2^{50}	1024^5	$\approx 1.126 \times 10^{15}$	Peta	P
Exbi	Ei	2^{60}	1024^6	$\approx 1.153 \times 10^{18}$	Exa	E
Zebi	Zi	2^{70}	1024^7	$\approx 1.181 \times 10^{21}$	Zetta	Z
Yobi	Yi	2^{80}	1024^8	$\approx 1.209 \times 10^{24}$	Yotta	Y

Appendix B
State Diagrams

The *"State Diagram"* is one of several methods for describing the behavior and performing the synthesis of a FSM, alternative to the ASM method developed in the book. It is quite similar to the former and still used in textbooks and documentation materials.

As the ASM, it describes the machine state-to-state transitions and outputs. In the following, we show how to convert *ASM diagrams* into *State Diagrams*, using as examples state machines presented in the book as ASM.

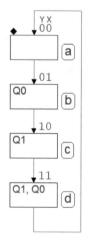

A first example of a *State Diagram* (right) is the module-4 counter described in Sect. 7.2.1 (left).

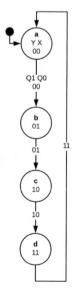

The circles, or *"bubbles"* represent the states, which are connected by *lines (arches)* with arrows that point to the direction of the transition.

The states are identified by a letter (a, b, c, d); each circle contains also the corresponding binary code. The lines are labelled with the $Q1\ Q0$ output values.

In other representations, in the case of a *Moore* machine, the outputs could also be written inside the state circles.

If there are inputs conditioning the state transitions, they are written by the lines, using digits separated by a slash. The first set of digits (before the slash) indicates the value of the inputs: with only one input there will be two lines exiting each state.

© Springer International Publishing AG, part of Springer Nature 2019
G. Donzellini et al., *Introduction to Digital Systems Design*,
https://doi.org/10.1007/978-3-319-92804-3

The figure below shows the bidirectional counter introduced in Sect. 7.2.2. There is now an input *DIR* that sets the count direction. If *DIR* = 1 the count is up, if *DIR* = 0 the count is down. The couple of digits after the slash shows the outputs, as in the previous case.

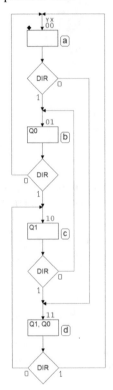
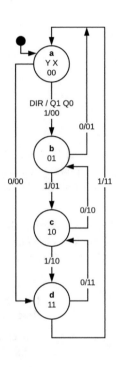

The edge detector (Sect. 7.2.2) is another example of a *Moore* machine. A line that starts and ends in the same state indicates a waiting loop.

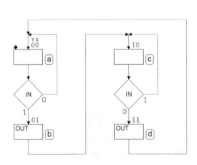
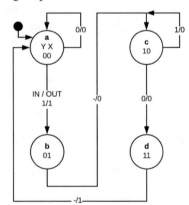

In a *Mealy* machine, such as the edge detector with conditioned outputs (Sect. 7.2.3), the outputs are function of both state and inputs. The *State Diagram* indicates, as before, the outputs in the lines after the slash, but in this case, outputs in the same state are different when conditioned by an input.

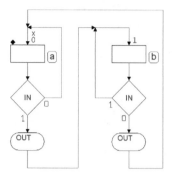

The figure below is another example of a *Mealy* machine. It represents the algorithm of the circuit seen in Sect. 7.3.9.

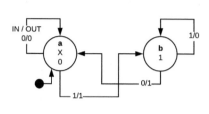

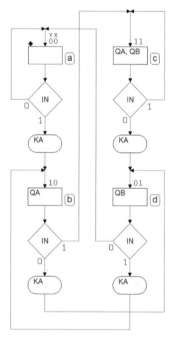

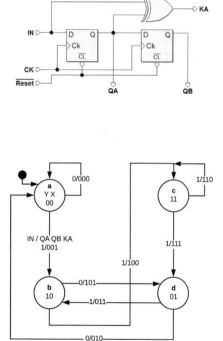

Appendix C
VHDL Code

In this pages, we list the VHDL codes described in Chap. 9. Listings are almost complete. The omitted parts can be obtained by exporting with *Deeds* the projects available in the digital contents pages (on *Deeds* Web site).

C.1 Code from the Pulse Generator Example

C.1.1 Schematic

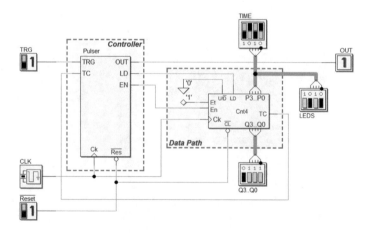

C.1.2 Top Entity

```
1   --------------------------------------------------------------------
2   -- Deeds (Digital Electronics Education and Design Suite)
3   -- VHDL Code generated on (10/03/2018, 14:50:44)
4   --       by the Deeds (Digital Circuit Simulator)(Deeds-DcS)
```

© Springer International Publishing AG, part of Springer Nature 2019
G. Donzellini et al., *Introduction to Digital Systems Design*,
https://doi.org/10.1007/978-3-319-92804-3

```vhdl
 5  --         Ver. 2.10.300 (Jan 19, 2018)
 6  -- Copyright(c)2002-2018 University of Genoa, Italy
 7  --       Web Site: https://www.digitalelectronicsdeeds.com
 8  ------------------------------------------------------------------------
 9  -- Code compiled for: "DE0-CV Board"
10  -- Chip FPGA: Intel/Altera Cyclone(r) V (5CEBA4F23C7)
11  -- Proprietary EDA Tool: Quartus(r) II (Ver = 12.1sp1)
12  ------------------------------------------------------------------------
13
14  LIBRARY ieee;
15  USE ieee.std_logic_1164.ALL;
16  USE ieee.numeric_std.all;
17
18  ENTITY Pulser_DE0CV IS
19    PORT(
20      ------------------------------------------------------------> Clocks:
21      iCLOCK_50MHz: IN  std_logic;    --> PIN_V15 --> "iCLK": 2 Hz (Sw[09], LEDR[09]
22                                      --> Key[01] for Slow Clock Mode)
23      ------------------------------------------------------------> Inputs:
24      iTRG:         IN  std_logic;    --> PIN_M6,  PButton: Key[03] H (if pressed)
25      inReset:      IN  std_logic;    --> PIN_P22, PButton: Reset L (if pressed)
26      iTIME_03:     IN  std_logic;    --> PIN_T12, Switch: Sw[03]
27      iTIME_02:     IN  std_logic;    --> PIN_T13, Switch: Sw[02]
28      iTIME_01:     IN  std_logic;    --> PIN_V13, Switch: Sw[01]
29      iTIME_00:     IN  std_logic;    --> PIN_U13, Switch: Sw[00]
30      ------------------------------------------------------------> Outputs:
31      oOUT:         OUT std_logic;    --> PIN_L2,  Red Led: LEDR[08]
32      oLEDS_03:     OUT std_logic;    --> PIN_Y3,  Red Led: LEDR[03]
33      oLEDS_02:     OUT std_logic;    --> PIN_W2,  Red Led: LEDR[02]
34      oLEDS_01:     OUT std_logic;    --> PIN_AA1, Red Led: LEDR[01]
35      oLEDS_00:     OUT std_logic;    --> PIN_AA2, Red Led: LEDR[00]
36      oQ3_Q0_03:    OUT std_logic;    --> PIN_U1,  Red Led: LEDR[07]
37      oQ3_Q0_02:    OUT std_logic;    --> PIN_U2,  Red Led: LEDR[06]
38      oQ3_Q0_01:    OUT std_logic;    --> PIN_N1,  Red Led: LEDR[05]
39      oQ3_Q0_00:    OUT std_logic;    --> PIN_N2,  Red Led: LEDR[04]
40      ------------------------------------------------------------> Added Inputs:
41      iSLOW_Sw09:   IN  std_logic;    --> PIN_AB12 --> "iCLK" Sw[09]
42      iPULSE_Key01: IN  std_logic;    --> PIN_W9  --> "iCLK" Key[01]
43      ------------------------------------------------------------> Added Outputs:
44      oCLOCK_LedR09: OUT std_logic;   --> PIN_L1,  Red Led: LEDR[09]
45      ------------------------------------------------------------> Default Outputs:
46      oPIN_U21:     OUT std_logic;    --> PIN_U21, Seven Segm. Display: HEX 0 0 (a)
47      oPIN_V21:     OUT std_logic;    --> PIN_V21, Seven Segm. Display: HEX 0 1 (b)
48      oPIN_W22:     OUT std_logic;    --> PIN_W22, Seven Segm. Display: HEX 0 2 (c)
49      oPIN_W21:     OUT std_logic;    --> PIN_W21, Seven Segm. Display: HEX 0 3 (d)
50      oPIN_Y22:     OUT std_logic;    --> PIN_Y22, Seven Segm. Display: HEX 0 4 (e)
51      oPIN_Y21:     OUT std_logic;    --> PIN_Y21, Seven Segm. Display: HEX 0 5 (f)
52      oPIN_AA22:    OUT std_logic;    --> PIN_AA22, Seven Segm. Display: HEX 0 6 (g)
53      oPIN_AA20:    OUT std_logic;    --> PIN_AA20, Seven Segm. Display: HEX 1 0 (a)
54      oPIN_AB20:    OUT std_logic;    --> PIN_AB20, Seven Segm. Display: HEX 1 1 (b)
55      oPIN_AA19:    OUT std_logic;    --> PIN_AA19, Seven Segm. Display: HEX 1 2 (c)
56      oPIN_AA18:    OUT std_logic;    --> PIN_AA18, Seven Segm. Display: HEX 1 3 (d)
57      oPIN_AB18:    OUT std_logic;    --> PIN_AB18, Seven Segm. Display: HEX 1 4 (e)
58      oPIN_AA17:    OUT std_logic;    --> PIN_AA17, Seven Segm. Display: HEX 1 5 (f)
59      oPIN_U22:     OUT std_logic;    --> PIN_U22, Seven Segm. Display: HEX 1 6 (g)
60      oPIN_Y19:     OUT std_logic;    --> PIN_Y19, Seven Segm. Display: HEX 2 0 (a)
61      oPIN_AB17:    OUT std_logic;    --> PIN_AB17, Seven Segm. Display: HEX 2 1 (b)
62      oPIN_AA10:    OUT std_logic;    --> PIN_AA10, Seven Segm. Display: HEX 2 2 (c)
63      oPIN_Y14:     OUT std_logic;    --> PIN_Y14, Seven Segm. Display: HEX 2 3 (d)
64      oPIN_V14:     OUT std_logic;    --> PIN_V14, Seven Segm. Display: HEX 2 4 (e)
65      oPIN_AB22:    OUT std_logic;    --> PIN_AB22, Seven Segm. Display: HEX 2 5 (f)
66      oPIN_AB21:    OUT std_logic;    --> PIN_AB21, Seven Segm. Display: HEX 2 6 (g)
67      oPIN_Y16:     OUT std_logic;    --> PIN_Y16, Seven Segm. Display: HEX 3 0 (a)
68      oPIN_W16:     OUT std_logic;    --> PIN_W16, Seven Segm. Display: HEX 3 1 (b)
69      oPIN_Y17:     OUT std_logic;    --> PIN_Y17, Seven Segm. Display: HEX 3 2 (c)
70      oPIN_V16:     OUT std_logic;    --> PIN_V16, Seven Segm. Display: HEX 3 3 (d)
71      oPIN_U17:     OUT std_logic;    --> PIN_U17, Seven Segm. Display: HEX 3 4 (e)
72      oPIN_V18:     OUT std_logic;    --> PIN_V18, Seven Segm. Display: HEX 3 5 (f)
73      oPIN_V19:     OUT std_logic;    --> PIN_V19, Seven Segm. Display: HEX 3 6 (g)
74      oPIN_U20:     OUT std_logic;    --> PIN_U20, Seven Segm. Display: HEX 4 0 (a)
75      oPIN_Y20:     OUT std_logic;    --> PIN_Y20, Seven Segm. Display: HEX 4 1 (b)
76      oPIN_V20:     OUT std_logic;    --> PIN_V20, Seven Segm. Display: HEX 4 2 (c)
77      oPIN_U16:     OUT std_logic;    --> PIN_U16, Seven Segm. Display: HEX 4 3 (d)
78      oPIN_U15:     OUT std_logic;    --> PIN_U15, Seven Segm. Display: HEX 4 4 (e)
79      oPIN_Y15:     OUT std_logic;    --> PIN_Y15, Seven Segm. Display: HEX 4 5 (f)
80      oPIN_P9:      OUT std_logic;    --> PIN_P9,  Seven Segm. Display: HEX 4 6 (g)
81      oPIN_N9:      OUT std_logic;    --> PIN_N9,  Seven Segm. Display: HEX 5 0 (a)
82      oPIN_M8:      OUT std_logic;    --> PIN_M8,  Seven Segm. Display: HEX 5 1 (b)
83      oPIN_T14:     OUT std_logic;    --> PIN_T14, Seven Segm. Display: HEX 5 2 (c)
84      oPIN_P14:     OUT std_logic;    --> PIN_P14, Seven Segm. Display: HEX 5 3 (d)
85      oPIN_C1:      OUT std_logic;    --> PIN_C1,  Seven Segm. Display: HEX 5 4 (e)
86      oPIN_C2:      OUT std_logic;    --> PIN_C2,  Seven Segm. Display: HEX 5 5 (f)
87      oPIN_W19:     OUT std_logic     --> PIN_W19, Seven Segm. Display: HEX 5 6 (g)
88    );
```

```vhdl
 89  END Pulser_DE0CV;
 90  ARCHITECTURE structural OF Pulser_DE0CV IS
 91    ---------------------------------------------------------------> Components:
 92    COMPONENT ClockScaler IS
 93      PORT( iMClk: IN  std_logic;   -- Master Clock
 94            iH4:  IN  std_logic;   -- iH4..iH0 = "high" frequency selection
 95            iH3:  IN  std_logic;
 96            iH2:  IN  std_logic;
 97            iH1:  IN  std_logic;
 98            iH0:  IN  std_logic;
 99            iL3:  IN  std_logic;   -- iL3..iL0 = "low" frequency selection
100            iL2:  IN  std_logic;   --                 and Button Modes
101            iL1:  IN  std_logic;
102            iL0:  IN  std_logic;
103            iSwch: IN  std_logic;  -- Switch
104            iBut:  IN  std_logic;  -- Button for manual pulsed Clock
105            oSClk: OUT std_logic;  -- Output Clock
106            oLed:  OUT std_logic   -- Slow "Clock Pulse" Led
107            );
108    END COMPONENT;
109    --
110    COMPONENT Counter4b IS
111      PORT( Ck : IN std_logic;
112            nCL: IN std_logic;
113            LD : IN std_logic;
114            ENP: IN std_logic;
115            ENT: IN std_logic;
116            UD : IN std_logic;
117            P3 : IN std_logic;
118            P2 : IN std_logic;
119            P1 : IN std_logic;
120            P0 : IN std_logic;
121            Q3 : OUT std_logic;
122            Q2 : OUT std_logic;
123            Q1 : OUT std_logic;
124            Q0 : OUT std_logic;
125            Tc : OUT std_logic );
126    END COMPONENT;
127    ----------------------------------------------------> Finite State Machine(s):
128    COMPONENT Pulser_EN IS
129      PORT( -------------------------------->Clock & Reset:
130            Ck:    IN std_logic;
131            Reset: IN std_logic;
132            -------------------------------->Inputs:
133            i_TRG: IN std_logic;
134            i_TC:  IN std_logic;
135            -------------------------------->Outputs:
136            o_OUT: OUT std_logic;
137            o_LD:  OUT std_logic;
138            o_EN:  OUT std_logic
139            ---------------------------------------
140            );
141    END COMPONENT;
142    ----------------------------------------------------------------------> Signals:
143    SIGNAL S001: std_logic;
144    SIGNAL S002: std_logic;
145    SIGNAL S003: std_logic;
146    SIGNAL S004: std_logic;
147    SIGNAL S005: std_logic;
148    SIGNAL S006: std_logic;
149    SIGNAL S007: std_logic;
150    SIGNAL S008: std_logic;
151    SIGNAL S009: std_logic;
152    SIGNAL S010: std_logic;
153    SIGNAL S011: std_logic;
154    SIGNAL S012: std_logic;
155    SIGNAL S013: std_logic;
156    SIGNAL S014: std_logic;
157    SIGNAL S015: std_logic;
158    SIGNAL S016: std_logic;
159    SIGNAL S017: std_logic;
160    ------------------------------------------------------------> Added Signals:
161    SIGNAL SSLOW_Sw09: std_logic;
162    SIGNAL SPULSE_Key01: std_logic;
163    SIGNAL iCLK: std_logic;
164    SIGNAL SCLOCK_LedR09: std_logic;
165
166  BEGIN -- structural
167    -----------------------------------------------------------------> Input:
168    S002 <= NOT iTRG;
169    S005 <= iCLK;
170    S007 <= inReset;
171    S014 <= iTIME_00;
172    S015 <= iTIME_01;
```

```
173    S016 <= iTIME_02;
174    S017 <= iTIME_03;
175    --------------------------------------------------------------------> Output:
176    oOUT <= S008;
177    oLEDS_00 <= S014;
178    oLEDS_01 <= S015;
179    oLEDS_02 <= S016;
180    oLEDS_03 <= S017;
181    oQ3_Q0_00 <= S010;
182    oQ3_Q0_01 <= S011;
183    oQ3_Q0_02 <= S012;
184    oQ3_Q0_03 <= S013;
185    --------------------------------------------------------------------> Constants:
186    S004 <= '0';
187    S001 <= '1';
188    oPIN_U21  <= '1';
189    oPIN_V21  <= '1';
190    oPIN_W22  <= '1';
191    oPIN_W21  <= '1';
192    oPIN_Y22  <= '1';
193    oPIN_Y21  <= '1';
194    oPIN_AA22 <= '1';
195    oPIN_AA20 <= '1';
196    oPIN_AB20 <= '1';
197    oPIN_AA19 <= '1';
198    oPIN_AA18 <= '1';
199    oPIN_AB18 <= '1';
200    oPIN_AA17 <= '1';
201    oPIN_U22  <= '1';
202    oPIN_Y19  <= '1';
203    oPIN_AB17 <= '1';
204    oPIN_AA10 <= '1';
205    oPIN_Y14  <= '1';
206    oPIN_V14  <= '1';
207    oPIN_AB22 <= '1';
208    oPIN_AB21 <= '1';
209    oPIN_Y16  <= '1';
210    oPIN_W16  <= '1';
211    oPIN_Y17  <= '1';
212    oPIN_V16  <= '1';
213    oPIN_U17  <= '1';
214    oPIN_V18  <= '1';
215    oPIN_V19  <= '1';
216    oPIN_U20  <= '1';
217    oPIN_Y20  <= '1';
218    oPIN_V20  <= '1';
219    oPIN_U16  <= '1';
220    oPIN_U15  <= '1';
221    oPIN_Y15  <= '1';
222    oPIN_P9   <= '1';
223    oPIN_N9   <= '1';
224    oPIN_M8   <= '1';
225    oPIN_T14  <= '1';
226    oPIN_P14  <= '1';
227    oPIN_C1   <= '1';
228    oPIN_C2   <= '1';
229    oPIN_W19  <= '1';
230    --------------------------------------------------------------------> Component Mapping:
231    SSLOW_Sw09 <= iSLOW_Sw09;
232    SPULSE_Key01 <= iPULSE_Key01;
233    oCLOCK_LedR09 <= SCLOCK_LedR09;
234    ClockScaler_iCLK: ClockScaler PORT MAP (
235        iCLOCK_50MHz, '1', '0', '1', '0', '0', '1', '1', '1', '1',
236        SSLOW_Sw09, SPULSE_Key01, iCLK, SCLOCK_LedR09 );
237
238    C680: Counter4b PORT MAP ( S005, S007, S003, S006, S001, S004, S017, S016,
239                               S015, S014, S013, S012, S011, S010, S009 );
240    C704: Pulser_EN PORT MAP ( S005, S007, S002, S009, S008, S003, S006 );
241  END structural;
```

C.1.3   Components

```
1    ---------------------------------------------------------------------
2    -- Deeds (Digital Electronics Education and Design Suite)
3    -- VHDL Code generated on (10/03/2018, 14:50:44)
4    --      by the Deeds (Digital Circuit Simulator)(Deeds-DcS)
5    --      Ver. 2.10.300 (Jan 19, 2018)
6    -- Copyright(c)2002-2018 University of Genoa, Italy
7    --      Web Site: https://www.digitalelectronicsdeeds.com
8    ---------------------------------------------------------------------
9    -- Code compiled for: "DE0-CV Board"
```

```
10    -- Chip FPGA: Intel/Altera Cyclone(r) V (5CEBA4F23C7)
11    -- Proprietary EDA Tool: Quartus(r) II (Ver = 12.1sp1)
12    ----------------------------------------------------------------------------
13
14    library ieee;
15    use ieee.std_logic_1164.all;
16    use ieee.numeric_std.all;
17
18    ENTITY Counter4b IS
19      PORT( Ck : IN std_logic;
20            nCL: IN std_logic;
21            LD : IN std_logic;
22            ENP: IN std_logic;
23            ENT: IN std_logic;
24            UD : IN std_logic;
25            P3 : IN std_logic;
26            P2 : IN std_logic;
27            P1 : IN std_logic;
28            P0 : IN std_logic;
29            Q3 : OUT std_logic;
30            Q2 : OUT std_logic;
31            Q1 : OUT std_logic;
32            Q0 : OUT std_logic;
33            Tc : OUT std_logic );
34    END Counter4b;
35
36    ----------------------------------------------------------------------------
37    ARCHITECTURE behavioral OF Counter4b IS
38    BEGIN
39      Count4b: PROCESS( Ck, nCL, ENP, ENT, UD )
40      variable aCnt: unsigned( 3 downto 0 );
41      BEGIN
42        if      (nCL = '0') then          aCnt := (others =>'0');
43        elsif (nCL = '1') then
44          if (Ck'event) AND (Ck='1') then
45            if     (LD = '1') then        aCnt := (P3 & P2 & P1 & P0); -- Load
46            elsif (LD = '0') then
47              if  (ENP = '1') and (ENT = '1')then
48                if    (UD = '1') then
49                  if (aCnt < "1111") then aCnt := aCnt + 1;
50                  else                    aCnt := (others =>'0');
51                  end if;
52                elsif (UD = '0') then
53                  if (aCnt > "0000") then aCnt := aCnt - 1;
54                  else                    aCnt := (others =>'1');
55                  end if;
56                else                      aCnt := (others =>'X'); -- (UD: Unknown)
57                END IF;
58              elsif not((ENP ='0')or
59                        (ENT ='0') ) then aCnt := (others =>'X'); -- (ENP: Unknown)
60              END IF;
61            else                          aCnt := (others =>'X'); -- (LD: Unknown)
62            END IF;
63          END IF;
64        else                              aCnt := (others =>'X'); -- (nCL: Unknown)
65        END IF;
66        --
67        Tc <= ENT and (   (aCnt(3) and aCnt(2) and aCnt(1) and aCnt(0) and UD) or
68                       (not(aCnt(3) or  aCnt(2) or  aCnt(1) or  aCnt(0) or  UD)) );
69        --
70        Q3 <= aCnt(3);
71        Q2 <= aCnt(2);
72        Q1 <= aCnt(1);
73        Q0 <= aCnt(0);
74      END PROCESS;
75    END behavioral;
76    ----------------------------------------------------------------------------
77    library ieee;
78    use ieee.std_logic_1164.all;
79    use ieee.numeric_std.all;
80
81    -- Clock Scaler (Altera DE1, DE2 and DE2-115 version, master clock = 50 MHz)
82    ENTITY ClockScaler IS
83      PORT( iMClk: IN  std_logic;   -- Master Clock
84            iH4:   IN  std_logic;   -- iH4..iH0 = "high" fr. sel.
85            iH3:   IN  std_logic;
86            iH2:   IN  std_logic;
87            iH1:   IN  std_logic;
88            iH0:   IN  std_logic;
89            iL3:   IN  std_logic;   -- iL3..iL0 = "low" freq. sel.
90            iL2:   IN  std_logic;   --           and Button Modes
91            iL1:   IN  std_logic;
92            iL0:   IN  std_logic;
93            iSwch: IN  std_logic;   -- Switch
```

```
94              iBut:  IN  std_logic;    -- Button for manual pulsed Clock
95              oSClk: OUT std_logic;    -- Output Clock
96              oLed:  OUT std_logic     -- Slow "Clock Pulse" Led
97              );
98      END ClockScaler;
99
100     -----------------------------------------------------------------------
101     ARCHITECTURE behavioral OF ClockScaler IS
102     BEGIN
103        --    ...omissis... (see the complete code generated by Deeds)
104     END behavioral;
```

C.1.4   Finite State Machine

```
1     -----------------------------------------------------------------------
2     -- Deeds (Digital Electronics Education and Design Suite)
3     -- VHDL Code generated on (10/03/2018, 14:50:44)
4     --     by the Deeds (Finite State Machine Simulator)(Deeds-FsM)
5     --        Ver. 2.10.300 (Jan 19, 2018)
6     -- Copyright(c)2002-2018 University of Genoa, Italy
7     --       Web Site: https://www.digitalelectronicsdeeds.com
8     -----------------------------------------------------------------------
9
10    LIBRARY ieee;
11    USE ieee.std_logic_1164.ALL;
12
13    ENTITY Pulser_EN IS
14       PORT( -----------------------------------> Clock & Reset:
15             Ck:    IN std_logic;
16             Reset: IN std_logic;
17             ---------------------------------->Inputs:
18             i_TRG: IN std_logic;
19             i_TC:  IN std_logic;
20             ---------------------------------->Outputs:
21             o_OUT: OUT std_logic;
22             o_LD:  OUT std_logic;
23             o_EN:  OUT std_logic );
24    END Pulser_EN;
25
26    ARCHITECTURE behavioral OF Pulser_EN IS        -- (Behavioral Description)
27       TYPE states is ( state_rs,
28                        state_wt,
29                        state_ps,
30                        dummy_11 );
31       SIGNAL State,
32              Next_State: states;
33    BEGIN
34       -- Next State Combinational Logic ------------------------------------
35       FSM: process( State, i_TRG, i_TC )
36       begin
37         CASE State IS
38           when state_wt =>
39                      if (i_TRG = '1') then
40                         Next_State <= state_ps;
41                      else
42                         Next_State <= state_wt;
43                      end if;
44           when state_ps =>
45                      if (i_TC = '1') then
46                         Next_State <= state_rs;
47                      else
48                         Next_State <= state_ps;
49                      end if;
50           when state_rs =>
51                      if (i_TRG = '1') then
52                         Next_State <= state_rs;
53                      else
54                         Next_State <= state_wt;
55                      end if;
56           when OTHERS =>
57                      Next_State <= state_rs;
58         END case;
59       end process;
60
61       -- State Register ----------------------------------------------------
62       REG: process( Ck, Reset )
63       begin
64         if (Reset = '0') then
65                   State <= state_rs;
66         elsif rising_edge(Ck) then
67                   State <= Next_State;
```

```
68            end if;
69      end process;
70
71      -- Outputs Combinational Logic ----------------------------------
72      OUTPUTS: process( State, i_TRG, i_TC )
73      begin
74          -- Set output defaults:
75          o_OUT <= '0';
76          o_LD  <= '0';
77          o_EN  <= '0';
78
79          -- Set output as function of current state and input:
80          CASE State IS
81            when state_wt =>
82                       o_LD <= '1';
83            when state_ps =>
84                       o_OUT <= '1';
85                       if (i_TC = '0') then
86                          o_EN <= '1';
87                       end if;
88            when OTHERS =>
89                       o_OUT <= '0';
90                       o_LD  <= '0';
91                       o_EN  <= '0';
92          END case;
93      end process;
94   END behavioral;
```

C.2   Other VHDL Examples

C.2.1   Decoder

```
1    --------------------------------------------------------------------------
2    library ieee;
3    use ieee.std_logic_1164.all;
4
5    ENTITY Decoder_2_4 IS
6      PORT( A1: IN   std_logic;
7            A0:  IN   std_logic;
8            EN:  IN   std_logic;
9            Y0:  OUT  std_logic;
10           Y1:  OUT  std_logic;
11           Y2:  OUT  std_logic;
12           Y3:  OUT  std_logic );
13   END Decoder_2_4;
14
15   ARCHITECTURE behavioral OF Decoder_2_4 IS
16     SIGNAL aNumber: std_logic_vector( 2 downto 0 );
17   BEGIN
18     aNumber <= EN & A1 & A0;
19     with aNumber select
20       Y0 <= '0' when "000", '0' when "001",
21             '0' when "010", '0' when "011",
22             '1' when "100", '0' when "101",
23             '0' when "110", '0' when "111", 'X' when others;
24     with aNumber select
25       Y1 <= '0' when "000", '0' when "001",
26             '0' when "010", '0' when "011",
27             '0' when "100", '1' when "101",
28             '0' when "110", '0' when "111", 'X' when others;
29     with aNumber select
30       Y2 <= '0' when "000", '0' when "001",
31             '0' when "010", '0' when "011",
32             '0' when "100", '0' when "101",
33             '1' when "110", '0' when "111", 'X' when others;
34     with aNumber select
35       Y3 <= '0' when "000", '0' when "001",
36             '0' when "010", '0' when "011",
37             '0' when "100", '0' when "101",
38             '0' when "110", '1' when "111", 'X' when others;
39   END behavioral;
```

C.2.2 Multiplexer

```
1    --------------------------------------------------------------------------
2    library ieee;
3    use ieee.std_logic_1164.all;
4
5    ENTITY Multiplexer_4_1 IS
6      PORT ( I0: IN   std_logic;
7             I1: IN   std_logic;
8             I2: IN   std_logic;
9             I3: IN   std_logic;
10            S1: IN   std_logic;
11            S0: IN   std_logic;
12             Q: OUT  std_logic );
13   END Multiplexer_4_1;
14
15   ARCHITECTURE behavioral OF Multiplexer_4_1 IS
16   BEGIN
17     Q <= I0 when ((S1 = '0') and (S0 = '0')) else
18          I1 when ((S1 = '0') and (S0 = '1')) else
19          I2 when ((S1 = '1') and (S0 = '0')) else
20          I3 when ((S1 = '1') and (S0 = '1')) else 'X';
21   END behavioral;
```

C.2.3 Demultiplexer

```
1    --------------------------------------------------------------------------
2    library ieee;
3    use ieee.std_logic_1164.all;
4    ENTITY Demultiplexer_1_4 IS
5      PORT (  I: IN   std_logic;
6             S1: IN   std_logic;
7             S0: IN   std_logic;
8             Q0: OUT  std_logic;
9             Q1: OUT  std_logic;
10            Q2: OUT  std_logic;
11            Q3: OUT  std_logic );
12   END Demultiplexer_1_4;
13
14   ARCHITECTURE behavioral OF Demultiplexer_1_4 IS
15     SIGNAL aNumber: std_logic_vector( 2 downto 0 );
16   BEGIN
17     aNumber <= I & S1 & S0;
18     with aNumber select
19       Q0 <= '0' when "000", '0' when "001",
20             '0' when "010", '0' when "011",
21             '1' when "100", '0' when "101",
22             '0' when "110", '0' when "111", 'X' when others;
23     with aNumber select
24       Q1 <= '0' when "000", '0' when "001",
25             '0' when "010", '0' when "011",
26             '0' when "100", '1' when "101",
27             '0' when "110", '0' when "111", 'X' when others;
28     with aNumber select
29       Q2 <= '0' when "000", '0' when "001",
30             '0' when "010", '0' when "011",
31             '0' when "100", '0' when "101",
32             '1' when "110", '0' when "111", 'X' when others;
33     with aNumber select
34       Q3 <= '0' when "000", '0' when "001",
35             '0' when "010", '0' when "011",
36             '0' when "100", '0' when "101",
37             '0' when "110", '1' when "111", 'X' when others;
38   END behavioral;
```

C.2.4 Full Adder

```
 1  ---------------------------------------------------------------
 2  library ieee;
 3  use ieee.std_logic_1164.all;
 4
 5  ENTITY Adder_Full IS
 6    PORT( CIN:  IN   std_logic;
 7          COUT:OUT  std_logic;
 8          A:    IN   std_logic;
 9          B:    IN   std_logic;
10          S:    OUT  std_logic );
11  END Adder_Full;
12
13  ARCHITECTURE behavioral OF Adder_Full IS
14    SIGNAL ABC: std_logic_vector( 2 downto 0 );
15  BEGIN
16    ABC <= A & B & CIN;
17    with ABC select
18    S  <= '0' when "000",
19          '1' when "001",
20          '1' when "010",
21          '0' when "011",
22          '1' when "100",
23          '0' when "101",
24          '0' when "110",
25          '1' when "111",
26          'X' when others;
27    with ABC select
28    COUT  <= '0' when "000",
29             '0' when "001",
30             '0' when "010",
31             '1' when "011",
32             '0' when "100",
33             '1' when "101",
34             '1' when "110",
35             '1' when "111",
36             'X' when others;
37  END behavioral;
```

C.2.5 Comparator

```
 1  library ieee;
 2  use ieee.std_logic_1164.all;
 3  use ieee.numeric_std.all;
 4
 5  ENTITY Compar_4 IS
 6    PORT( A3:   IN   std_logic;
 7          A2:   IN   std_logic;
 8          A1:   IN   std_logic;
 9          A0:   IN   std_logic;
10          B3:   IN   std_logic;
11          B2:   IN   std_logic;
12          B1:   IN   std_logic;
13          B0:   IN   std_logic;
14          MIN:  OUT  std_logic;
15          EQU:  OUT  std_logic;
16          MAJ:  OUT  std_logic );
17  END Compar_4;
18
19  ARCHITECTURE behavioral OF Compar_4 IS
20  BEGIN
21    Cmp4: PROCESS( A3, A2, A1, A0, B3, B2, B1, B0 )
22    variable A: unsigned( 3 downto 0 );
23    variable B: unsigned( 3 downto 0 );
24    BEGIN
25      A := (A3 & A2 & A1 & A0);
26      B := (B3 & B2 & B1 & B0);
27      --
28      if    (A > B) then MIN <= '0'; EQU <= '0'; MAJ <= '1';
29      elsif (A < B) then MIN <= '1'; EQU <= '0'; MAJ <= '0';
```

```
30        elsif (A = B) then MIN <= '0'; EQU <= '1'; MAJ <= '0';
31        else              MIN <= 'X'; EQU <= 'X'; MAJ <= 'X';
32      END IF;
33    END PROCESS;
34 END behavioral;
```

C.2.6   Flip-Flop D-PET

```
1  library ieee;
2  use ieee.std_logic_1164.all;
3
4  ENTITY DpetFF IS
5    PORT(  D, Ck   : IN std_logic;
6           nCL, nPR: IN std_logic;
7           Q, nQ    : OUT std_logic );
8  END DpetFF;
9
10 ARCHITECTURE behavioral OF DpetFF IS
11 BEGIN
12   Dff: PROCESS( Ck, nCL, nPR )
13   BEGIN
14     if    (nCL='0') and (nPR='0') then Q <= 'X'; nQ <= 'X';
15     elsif (nCL='0') and (nPR='1') then Q <= '0'; nQ <= '1';
16     elsif (nCL='1') and (nPR='0') then Q <= '1'; nQ <= '0';
17     elsif (nCL='1') and (nPR='1') then
18       if (Ck'event) AND (Ck='1') THEN -- Positive Edge
19         Q <=  D;  nQ <= not D;
20       END IF;
21     else Q <= 'X'; nQ <= 'X';
22     END IF;
23   END PROCESS;
24 END behavioral;
```

C.2.7   Flip-Flop E-PET

```
1  library ieee;
2  use ieee.std_logic_1164.all;
3
4  ENTITY EpetFF IS
5    PORT(  D, E, Ck: IN std_logic;
6           nCL, nPR: IN std_logic;
7           Q, nQ    : OUT std_logic );
8  END EpetFF;
9  ARCHITECTURE behavioral OF EpetFF IS
10 BEGIN
11   Eff: PROCESS( Ck, nCL, nPR )
12   BEGIN
13     if    (nCL='0') and (nPR='0') then Q <= 'X'; nQ <= 'X';
14     elsif (nCL='0') and (nPR='1') then Q <= '0'; nQ <= '1';
15     elsif (nCL='1') and (nPR='0') then Q <= '1'; nQ <= '0';
16     elsif (nCL='1') and (nPR='1') then
17       if (Ck'event) AND (Ck='1') THEN -- Positive Edge
18         if        (E = '1') then Q <=  D;  nQ <= not D;
19         elsif not(E = '0') then Q <= 'X'; nQ <= 'X';
20         END IF;
21       END IF;
22     else Q <= 'X'; nQ <= 'X';
23     END IF;
24   END PROCESS;
25 END behavioral;
```

C.2.8   Flip-Flop JK-PET

```
1  library ieee;
2  use ieee.std_logic_1164.all;
3
4  ENTITY JKpetFF IS
```

```
 5     PORT(   J,  K,  Ck:  IN  std_logic;
 6             nCL,  nPR:  IN  std_logic;
 7             Q,  nQ    :  OUT  std_logic );
 8   END JKpetFF;
 9
10   ARCHITECTURE behavioral OF JKpetFF IS
11   BEGIN
12     JKff: PROCESS( Ck, nCL, nPR )
13       variable  OutQ: STD_LOGIC;
14     BEGIN
15       if    (nCL='0') and (nPR='1') then OutQ := '0';
16       elsif (nCL='1') and (nPR='0') then OutQ := '1';
17       elsif (nCL='1') and (nPR='1') then
18         if (Ck'event) AND (Ck='1') THEN
19           -- Positive Edge
20           if    (J = '0') AND (K = '1') THEN OutQ := '0';
21           elsif (J = '1') AND (K = '0') THEN OutQ := '1';
22           elsif (J = '1') AND (K = '1') THEN OutQ := not OutQ;
23           elsif not((J='0')AND(K='0'))  THEN OutQ := 'X';
24           END IF;
25         END IF;
26       else                                      OutQ := 'X';
27       END IF;
28       --
29       Q  <= (     OutQ);
30       nQ <= (not OutQ);
31       --
32     END PROCESS;
33   END behavioral;
```

C.2.9 · *Parallel Register*

```
 1   library ieee;
 2   use ieee.std_logic_1164.all;
 3
 4   ENTITY PiPoE4 IS
 5     PORT(  Ck : IN std_logic;
 6            nCL: IN std_logic;
 7            E  : IN std_logic;
 8            P3 : IN std_logic;
 9            P2 : IN std_logic;
10            P1 : IN std_logic;
11            P0 : IN std_logic;
12            Q3 : OUT std_logic;
13            Q2 : OUT std_logic;
14            Q1 : OUT std_logic;
15            Q0 : OUT std_logic );
16   END PiPoE4;
17   ARCHITECTURE behavioral OF PiPoE4 IS
18   BEGIN
19     RegPiPoE4: PROCESS( Ck, nCL )
20       variable aReg: std_logic_vector( 3 downto 0 );
21     BEGIN
22       if    (nCL = '0') then    aReg := (others =>'0');
23       elsif (nCL = '1') then
24         if (Ck'event) AND (Ck='1') THEN -- Positive Edge
25           if (E = '1') then
26                 aReg := (P3 & P2 & P1 & P0);
27           elsif not(E = '0') then
28                 aReg := (others =>'X');
29           END IF;
30         END IF;
31       else          aReg := (others =>'X');
32       END IF;
33
34       Q3 <= aReg(3);
35       Q2 <= aReg(2);
36       Q1 <= aReg(1);
37       Q0 <= aReg(0);
38
39     END PROCESS;
40   END behavioral;
```

C.2.10 Shift Register (S.I.P.O.)

```
 1  library ieee;
 2  use ieee.std_logic_1164.all;
 3
 4  ENTITY SiPoE4 IS
 5    PORT( I   : IN std_logic;
 6          Ck  : IN std_logic;
 7          nCL: IN std_logic;
 8          E   : IN std_logic;
 9          Q3  : OUT std_logic;
10          Q2  : OUT std_logic;
11          Q1  : OUT std_logic;
12          Q0  : OUT std_logic );
13  END SiPoE4;
14
15  ARCHITECTURE behavioral OF SiPoE4 IS
16  BEGIN
17    RegSiPoE4: PROCESS( Ck, nCL )
18    variable aReg: std_logic_vector( 3 downto 0 );
19    BEGIN
20      if    (nCL = '0') then aReg := (others =>'0');
21      elsif (nCL = '1') then
22        if (Ck'event) AND (Ck='1') THEN -- Positive Edge
23          if (E = '1') then
24            aReg := (I & aReg(3) & aReg(2) & aReg(1));
25          elsif not(E = '0') then
26            aReg := (others =>'X');
27          END IF;
28        END IF;
29      else aReg := (others =>'X');
30      END IF;
31      --
32      Q3  <= aReg(3);
33      Q2  <= aReg(2);
34      Q1  <= aReg(1);
35      Q0  <= aReg(0);
36      --
37    END PROCESS;
38  END behavioral;
```